Radioactivity in the Environment

Volume 19

Social and Ethical Aspects of Radiation Risk Management

A companion series to the Journal of Environmental Radioactivity

Series Editor
M.S. Baxter
Ampfield House
Clachan Seil
Argyll, Scotland, UK

Volume 1:	Plutonium in the Environment (A. Kudo, Editor)
Volume 2:	Interactions of Microorganisms with Radionuclides (F.R. Livens and M. Keith-Roach, Editors)
Volume 3:	Radioactive Fallout after Nuclear Explosions and Accidents (Yu.A. Izrael, Author)
Volume 4:	Modelling Radioactivity in the Environment (E.M. Scott, Editor)
Volume 5:	Sedimentary Processes: Quantification Using Radionuclides (J. Carroll and I. Lerche, Authors)
Volume 6:	Marine Radioactivity (H.D. Livingston, Editor)
Volume 7:	The Natural Radiation Environment VII (J.P. Mclaughlin, S.E. Simopoulos and F. Steinhäusler, Editors)
Volume 8:	International Conference on Isotopes and Environmental Studies (P.P. Povinec and J.A. Sanchez-Cabeza, Editors)
Volume 9:	Deep Geological Disposal of Radioactive Waste (W.R. Alexander and L.E. Mckinley, Editors)
Volume 10:	Radioactivity in the Terrestrial Environment (G. Shaw, Editor)
Volume 11:	Analysis of Environment Radionuclides (P.P. Povinec, Editor)
Volume 12:	Radioactive Aerosols (C. Papastefanou, Author)
Volume 13:	U-Th Series Nuclides in Aquatic Systems (S. Krishnaswami and J. Kirk Cochran, Editors)
Volume 14:	Remediation of Contaminated Environments (G. Voigt and S. Fesenko, Editors)
Volume 15:	Airborne Radioactive Contamination in Inhabited Areas (K.G. Andersson, Editor)
Volume 16:	Environmental Radionuclides: Tracers and Timers of Terrestrial Processes (K. Froehlich, Editor)
Volume 17:	Technologically Enhanced Natural Radiation (A.S. Paschoa and F. Steinhäusler, Authors)
Volume 18:	Tropical Radioecology (John R. Twining, Editor)

Radioactivity in the Environment

Volume 19

Social and Ethical Aspects of Radiation Risk Management

Edited by

Deborah Oughton
Centre for Environmental Radioactivity,
Norwegian University of Life Sciences, Ås, Norway

Sven Ove Hansson
Royal Institute of Technology (KTH),
Division of Philosophy, Stockholm, Sweden

ELSEVIER

AMSTERDAM • BOSTON • HEIDELBERG • LONDON
NEW YORK • OXFORD • PARIS • SAN DIEGO
SAN FRANCISCO • SINGAPORE • SYDNEY • TOKYO

Elsevier
The Boulevard, Langford Lane, Kidlington, Oxford OX51 GB, UK
Radarweg 29, PO Box 211, 1000 AE Amsterdam, The Netherlands
225 Wyman Street, Waltham, MA 02451, USA

British Library Cataloguing in Publication Data
A catalogue record for this book is available from the British Library

Library of Congress Cataloging-in-Publication Data
A catalog record for this book is available from the Library of Congress

ISBN: 978-0-08-045015-5
ISSN: 1569-4860

This book has been manufactured using Print On Demand technology. Each copy is produced to order and is limited to black ink. The online version of this book will show colour figures where appropriate.

Contents

Preface xiii
Contributors xv

Introduction

1. Introduction
Sven Ove Hansson and Deborah Oughton

1.1 Part I. Ethical Principles for Radiation Protection 4
1.2 Part II. Putting Protection to Practice 6
1.3 Part III. Nuclear Accidents 8
1.4 Part IV. Proliferation and the Nuclear Fuel Cycle 10
1.5 Part V. Public Participation 11

Part I
Ethical Principles for Radiation Protection

2. Radiation Risks and the ICRP
Jack Valentin

2.1 What is ICRP? 17
2.2 The Aims and Scope of ICRP Recommendations 18
2.3 The Early History and Development of ICRP Recommendations 21
2.4 The Development of the System of Radiological Protection and Current ICRP Recommendations 23
2.5 Ethical Underpinning of the Evolution of ICRP Recommendations 26
2.6 Some Moot Points 29
References 30

3. Moral Thinking and Radiation Protection
Sven Ove Hansson

3.1 Introduction 33
3.2 Individual vs Collective Approaches 35
3.3 Weighing vs Limit-Setting 39
3.4 The de minimis Issue 43

3.5 Valuing Future Effects 46
3.6 Protecting the Most Sensitive People 47
3.7 Conclusion 49
References 49

4. A Cross-Cultural Approach to Radiation Ethics
Friedo Zölzer

4.1 Background 53
4.2 Ethics of Radiation Protection in a Globalizing World 55
4.3 "Principles of Biomedical Ethics" as an Example 56
4.4 Possible Sources of Cross-Cultural Ethics 57
4.5 The Need for Cross-Cultural Discourse 58
4.6 The Relevance of the "Four Principles" for Cross-Cultural Radiation Protection Ethics 59
4.7 Further Cross-Culturally Accepted Principles with Relevance for Radiation Protection 62
4.8 Conclusion 66
References 67

5. Ethical Aspects of Ecological Risks from Radiation
Deborah Oughton

5.1 Introduction 71
5.2 Valuing the Environment: Philosophical Theories 73
5.3 Common Ethical Principles 78
5.4 Harms and Values in Practical Radiation Protection 78
5.5 Assigning Monetary Value to the Environment 80
5.6 Conclusion: Relevance of the Value Debate to Ecological Radiological Protection 81
References 82

6. Why Chemical Risk Assessment can Learn from Radiation Risk Assessment
Carl F. Cranor

6.1 Introduction 87
6.2 Some Principles and Presumptions of Radiation Protection 88
6.3 Contamination 90
6.4 The Developmental Basis of Disease 92
6.5 Contamination of Developing Children 92
6.6 Adverse Health Effects 93
6.7 Particular Substances have No Obvious Thresholds 98
6.8 A Unified Approach to Dose-response Assessment 99
6.9 Conclusion 100
References 101

Part II
Putting Protection to Practice

7. Ethical Issues in Clinical Radiology
Jim Malone

7.1 Introduction 107
7.2 Ethics in Medical Radiology 109
7.3 Medical, Social, and Legal Context for Radiology 114
7.4 Risk, Uncertainty, Communication, and Skeptical Doctors 116
7.5 Justification Issues 118
7.6 Some Special Concerns 121
7.7 Conclusions 124
Acknowledgments 125
References 125

8. Ethics in Practice—Protecting Workers
Christopher J. Kalman

8.1 Introduction 131
8.2 Radiation Protection in Workers 132
8.3 Ethical Requirements for Occupational Radiation Health Research 137
8.4 Radiation Science and Workers 139
8.5 Summary 140
References 141

9. ALARA: What is Reasonably Achievable?
Sven Ove Hansson

9.1 Introduction 143
9.2 ALARA and Optimization in the Radiation Protection System 145
9.3 A Three-Levelled Model 146
9.4 What is "Reasonably Achievable?" 148
9.5 The Scope of Economic Compromises 149
9.6 ALARA and Cost-Benefit Analysis 151
9.7 Related Principles of Protection 152
9.8 Conclusion 154
References 154

Part III
Lessons Learned from Nuclear Accidents

10. Lessons Learned from the Chernobyl Accident in Norway
Astrid Liland and Lavrans Skuterud

10.1 Introduction 159
10.2 Radioactive Contamination—A Societal Challenge 161

10.3 Early Lessons from the Chernobyl Accident 162
10.4 Long-Term Challenges and Lessons Learned 167
10.5 Health Concerns 172
References 174

11. Lessons from the Fukushima Daiichi Nuclear Disaster
Michio Miyasaka

11.1 What Happened at Fukushima 177
11.2 "Safety Culture" as a Monoculture 184
11.3 Applied Ethics for Nuclear Science 188
Acknowledgments 194
References 194

12. Environmental Injustice Inherent in Radiation Dose Standards
Kristin Shrader-Frechette

12.1 Introduction 197
12.2 The Fukushima Accident 198
12.3 Fukushima Environmental-Injustice Victims 199
12.4 Poor People: EIJ Victims 200
12.5 Poor People and Nuclear/Radiation Regulations 201
12.6 Children: FD EIJ Victims 203
12.7 Children and Nuclear-Radiation Regulations 204
12.8 Nuclear Workers: FD EIJ Victims 206
12.9 Workers and Nuclear-Radiation Regulations 207
12.10 Future Generations, FD, and Nuclear-Radiation Regulations 208
12.11 Conclusions 210
References 211

13. Safety Culture and Safety Quality
Carl Rollenhagen

13.1 Introduction 215
13.2 Personal and Organizational Values 219
13.3 Safety Culture/Climate 223
13.4 Safety Quality and Human Reliability Analysis 226
13.5 Implications for Safety Management 227
13.6 General Discussion 230
References 233

Part IV
Proliferation and the Nuclear Fuel Cycle

14. The Legacies of Soviet Nuclear Testing in Kazakhstan

Fallout, Public Health and Societal Issues

Susanne Bauer, Boris Gusev, Tatyana Belikhina, Timur Moldagaliev and Kazbek Apsalikov

14.1 The Semipalatinsk Nuclear Test Site	244
14.2 Research into Radiation Effects of Nuclear Testing Near Semipalatinsk during Soviet Time	245
14.3 Radiation Risk Research in Kazakhstan during the Early Post-Soviet Years	248
14.4 Addressing Nuclear Legacies in the New Economy of Kazakhstan	253
14.5 Conclusions	255
References	256

15. Moral Dilemmas of Uranium and Thorium Fuel Cycles

Behnam Taebi

15.1 Introduction	260
15.2 Existing Nuclear Fuel Cycle: Uranium	261
15.3 The Closed Fuel Cycle and Intergenerational Justice Dilemmas	267
15.4 Is Thorium a Viable Substitute or Supplement for Nuclear Fuel?	273
15.5 Conclusions	277
Acknowledgments	278
References	279

Part V
Public Participation

16. Social Identities and Public Uptake of Science

Chernobyl, Sellafield, and Environmental Radioactivity Sciences

Brian Wynne

16.1 Introduction	283
16.2 Sheep Farmers, Scientists, and Radiation Hazards: The Background	285
16.3 Scientific Knowledge and Social Identities	288
16.4 Public Belief and Private Dissent	291
16.5 Credibility: The Social Dimension	299
16.6 Conclusions: Lay Reflexivity and Social Identities	303
Acknowledgments	307
References	307

17. Stakeholder Engagement in Regaining Decent Living Conditions after Chernobyl
Jacques Lochard

17.1 Introduction 311
17.2 A Key Feature of the Situation of the Population in the Contaminated Territories: The Loss of Control on Daily Living Conditions 313
17.3 The ETHOS Project 316
17.4 Key Lessons 322
17.5 New Approach to Long-Term Rehabilitation 325
17.6 Conclusion 326
Acknowledgments 327
References 328
Conclusions and Recommendations of the International Conference Held In Stolyn – Republic of Belarus, 15–16 November 2001 (Translation from The Russian Original Text) 330

18. Public Participation—Potential and Pitfalls
Sven Ove Hansson and Deborah Oughton

18.1 Introduction 333
18.2 What is Participation? 335
18.3 The Role of Participation in Democracy 335
18.4 Two Ways to Justify Participation 338
18.5 Quality Criteria for Participative Procedures 339
18.6 Conclusion 343
References 343

19. Compensation or Bribery? Ethical Issues in Relation to Radwaste Host Communities
Mike Hannis and Kate Rawles

19.1 Introduction 347
19.2 Host Communities 350
19.3 Ethical Acceptability and Public Acceptability 350
19.4 Site Selection 351
19.5 Effects on the Host Community—Negative Impacts 352
19.6 Positive Impacts 354
19.7 Appropriate Compensation: Process 355
19.8 Appropriate Compensation: Outcome 356
19.9 The Inadequacy of Existing Guidelines 357
19.10 Compensation, Bribery, and Incentives 358
19.11 Is Bribery Unavoidable in This Context? 359
19.12 Mitigation and New Build 361
19.13 Bribery, Benefits, and Independence 363

19.14 Financial Compensation 364
19.15 Ethical Decision Making and Commensurability: Against Algorithms 366
19.16 Responses 368
19.17 Problems with Cost-Benefit Analysis 369
19.18 Conclusions 371
19.19 Afterword 372
Acknowledgments 373
References 373

Index 375

Preface

This cooperation began in an elevator in Prague in late August, 2011. As participants in Friedo Zölzer's symposium Ethics of Environmental Health, we were both impressed by the quality of many of the contributions. In the time required for the elevator to take us from the first to the fourth floor, we decided to jointly edit a volume on social and ethical aspects of radiation protection. Everything was in place to give the book project a flying start. In addition to the seminar presentations, we both had ideas on other potential contributors. Deborah already had contacts with a publisher who was actively looking for such a book.

We would like to thank all the authors for their devotion to the project, their hard work, and their forbearance with our editorial deadlines and directions. Thanks are also due to Murdoch S. Baxter for his patience and support and for including the book in his series, and to Jill Cetel and her colleagues at Elsevier for all their help. It has been a pleasurable experience to cooperate with you all in making this book. Let us hope that our joint efforts will contribute to more ethical and socially responsible practices in the radiation protection in the years to come.

Stockholm and Oslo, March 30, 2013
Sven Ove Hansson, Deborah Oughton

Contributors

Kazbek Apsalikov, M.D., Ph.D. (Radiation Hygiene), is Director of the Research Institute for Radiation Medicine and Ecology in Semei, Kazakhstan. He has conducted large-scale clinical and epidemiological research projects among the population in the Semipalatinsk region, including INCO-COPERNICUS project "Semipalatinsk Follow-up", a UNFPA-supported project on reproductive health in relation to radiation exposure due to nuclear testing near Semipalatinsk, and collaborative projects with Japanese scientists.

Susanne Bauer, MSc (Environmental Sciences), Ph.D. (Public Health), is associate professor at the Department of Social Sciences, Goethe University Frankfurt, Germany. She has gained extensive research experience as an environmental epidemiologist, before she moved on to the social studies of science.

Tatyana Belikhina holds an M.D. and a Ph.D. in Hygiene. She has been head of the department of research management at the Research Institute for Radiation Medicine and Ecology Semei, Kazakhstan, for five years. Her research experience includes radiation epidemiology, clinical investigations and coordination of different research directions in the field of radiation medicine.

Carl F. Cranor is Distinguished Professor of Philosophy and member of the Environmental Toxicology Graduate Program at the University of California, Riverside. His research focuses on philosophic issues concerning risks, science and the law. He has written *Regulating Toxic Substances* (Oxford, 1993), *Toxic Torts: Science, Law and the Possibility of Justice* (Cambridge, 2006), and *Legally Poisoned: How the Law Puts Us at Risk from Toxicant* (Harvard, 2011) and served on numerous science advisory panels.

Boris Gusev, M.D., Ph.D. (Radiation Biology), has long-standing experience in researching the radiation-hygienic, medical, and demographic consequences of nuclear bomb testing in Kazakhstan and participated in joint investigations with scientists from Hiroshima University (Japan), NCI (USA), and the Federal Office for Radiation Protection (Germany). He was Chief physician of the Dispensary № 4 from 1962 to 1990 and director of the Research Institute for Radiation Medicine and Ecology from 1991 to 1997.

Mike Hannis holds an honorary research fellowship in environmental politics and philosophy at Keele University (UK), and an associate lectureship at the Open University. He is also an editor of *The Land* magazine.

Sven Ove Hansson is professor in philosophy at the Royal Institute of Technology (KTH) in Stockholm. He is editor-in-chief of *Theoria* and was president of the Society for Philosophy and Technology in 2011-2013. He has written extensively on the philosophy of risk, and his book *The Ethics of Risk*, is scheduled for publication in late 2013.

Christopher J. Kalman is a career occupational physician, currently employed as Director of Occupational Health and Safety for NHS Lothian, a large integrated health-care provider within the state owned National Health Service in Scotland. He has a specialist interest in ionising radiation, with postgraduate qualifications in radiobiology. He filled senior medical roles in the support of the UK's Royal Navy nuclear propulsion programme, and also as a Chief Medical Officer for a nuclear power utility in Scotland. He is an Honorary Clinical Senior Lecturer in Public Health at both the Universities of Edinburgh and Glasgow, and sat on the UK Government's Expert Radioactive Waste Advisory Committee.

Astrid Liland is a nuclear chemist who has thirteen years' experience in research and research management in radiation protection, radioecology and emergency preparedness and recovery. She has worked on the wider consequences of nuclear accidents in Norway, Belarus and Japan. She is currently leading the section for radiation research at the Norwegian Radiation Protection Authority.

Jacques Lochard is director of the Nuclear Protection Evaluation Centre (CEPN) in Fontenay-aux-Roses, France. He was educated as economist and has a long experience in practical radiation protection particularly in relation with the protection of people in the areas contaminated by the Chernobyl accident in Belarus. He is currently member of the Main Commission of the International Commission of Radiological Protection and Chair of the Committee 4 in charge of the application of the Commission's Recommendations.

Jim Malone is Professor (Emeritus) of Medical Physics, Trinity College, Dublin, Consultant with the International Atomic Energy Agency (IAEA), Vienna, regular contributor at the European School of Medical Physics in Geneva and Director of the Robert Boyle Foundation. He was Dean of Medical School/Faculty of Health Sciences and Director of Graduate Studies at Trinity College and a board member of about ten corporate bodies. He has wide interests in the humanities.

Michio Miyasaka is professor of health care ethics at School of Health Sciences, Niigata University. He received a Ph.D. from the University of Tokyo in 1998. He has published three books in Japanese, which include *Methods in Health Care Ethics: Principles, Procedures, and Narratives* (in Japanese, Tokyo, 2005) and *The Records of Hansen's Disease Prison* (in Japanese, Tokyo, 2006). He has published more than 60 journal articles and book chapters in Japanese and English.

Timur Moldagaliev is a Ph.D.candidate in public health at the Medical University of Semei, Kazakhstan. His research interests include the psychosocial aspects of living with nuclear legacies in areas adjacent to the Semipalatinsk nuclear test site.

Deborah Oughton is professor in nuclear and environmental chemistry, and research director of the Centre of Environmental Radioactivity, at the Norwegian University of Life Sciences (UMB). She currently holds a professorship in Research Ethics at the University of Oslo and is deputy head of the National Committee on Research Ethics in Science and Technology. She has published widely in both science and ethics, and has been a strong advocate of putting ethics into practice in radiation risk management.

Kate Rawles has a background in environmental philosophy and ethics. She has worked as an independent consultant for Nirex UK on ethical questions in relation to radioactive waste management. She currently works half-time as a senior lecturer in Outdoor Studies at the University of Cumbria and half-time as a writer, campaigner and outdoor philosopher. Her book, *The Carbon Cycle; crossing the great divide* was published in the UK by Two Ravens Press in 2012 and is forthcoming in North America with Rocky Mountain Books in 2013.

Carl Rollenhagen is adjunct professor at the Royal Institute of Technology (KTH). He has worked extensively with safety management and human factors in the nuclear industry and published books and articles about event investigation and safety culture.

Kristin Shrader-Frechette is O'Neill Professor, Department of Biological Sciences and Department of Philosophy, University of Notre Dame. Her research, translated into thirteen languages, focuses on methodological/ethical problems in radiobiology, quantitative risk assessment, and public health. The US National Science Foundation has funded her research for 28 years. Her 16 books include *Taking Action, Saving Lives* (Oxford UP, 2007), and *What Will Work: Fighting Climate Change with Renewable Energy, Not Nuclear Power* (Oxford UP, 2011). *Practical Philosophy of Science* will appear in 2014.

Lavrans Skuterud is biophysicist and environmental chemist at the Norwegian Radiation Protection Authority. He has twenty years' experience in studies of radioactivity in the environment, in population groups in Norway, Russia, Ukraine and Belarus, and in management of the Chernobyl consequences in Norway. He is also a consultant for the Norwegian Reindeer Husbandry Administration, and served as a consultant for UN/FAO during the Fukushima crisis.

Behnam Taebi is assistant professor in ethics of technology at Delft University of Technology. With a Master's degree in material science and a Ph.D. in philosophy of technology, he has worked and published on the social and ethical aspects of nuclear power production. He is currently working on a monograph on nuclear justice. He is also the coordinating editor of a special issue of *Journal of Risk Research* on "Socio-technical challenges of nuclear power production and waste management after Fukushima" (co-editor Prof. Ibo van de Poel).

Jack Valentin, an Honorary Fellow of the Society for Radiological Protection and a Fellow of the Institute of Physics, has a Ph.D. degree in genetics. He used to be a senior regulator at the then Swedish Radiation Protection Authority, then became the Scientific Secretary of the International Commission on Radiological Protection (ICRP). Being retired, he now enjoys life as an aviation and classic car buff, but continues to do radiological protection work as a private consultant.

Brian Wynne is professor of science studies at Lancaster University, where he is a founder-member and ex-Director of the Centre for Science Studies, and the Centre for the Study of Environmental Change. With a Masters degree and Ph.D. in materials science, he retrained in history, philosophy and sociology of science, and has worked and published on the sociology and culture of scientific knowledge in public arenas, such as risk assessment and environmental policy, as well as on public experiences of scientific expertise.

Friedo Zölzer is professor of radiation biology at the University of South Bohemia in Ceske Budejovice, Czech Republic. He is head of the Department of Radiology, Toxicology and Civil Protection. While his main research interest is in the effects of low doses of radiation, he has become involved with related ethical issues more recently, organizing an International Symposium on Ethics of Environmental Health in Prague 2011 and in Ceske Budejovice in 2014.

Introduction

Chapter 1

Introduction

Sven Ove Hansson,[1,*] and Deborah Oughton,[2,*]
[1]*Royal Institute of Technology (KTH), Division of Philosophy, Stockholm, Sweden*
[2]*Centre for Environmental Radioactivity, Norwegian University of Life Sciences, ÅS, Norway*
**Corresponding authors: E-mails: soh@kth.se, deborah.oughton@umb.no*

Chapter Outline

1.1 Part I. Ethical Principles for Radiation Protection 4
1.2 Part II. Putting Protection to Practice 6
1.3 Part III. Nuclear Accidents 8
1.4 Part IV. Proliferation and the Nuclear Fuel Cycle 10
1.5 Part V. Public Participation 11

There was a time when radiation protection was believed to be an issue for natural science alone. Scientific experts could determine radiation levels below which there would be no adverse health effects. These limits could be used in combination with dosimetry to make sure that no one was exposed to dangerous levels of radiation.

But three parallel developments have made this approach hopelessly inadequate. First, developments in science have given us strong reasons to assume that exposure limits alone cannot provide a guarantee against radiation health effects. The probability of getting cancer decreases with reduced exposure, but we have no way to identify a limit below which they disappear completely. The consensus view is that we had better assume that no excess exposure is free of risk.

Secondly, a series of technological failures in nuclear facilities have made radiological safety a public issue. Although the radioactive releases from the Three Mile Island accident in 1979 were limited, it had a worldwide impact on the public opinion. The accidents in Chernobyl in 1986 and in the Fukushima Daiichi Nuclear Power Plant in 2011 demonstrated the potential damaging effects of radiation accidents and showed that these could go far beyond the direct radiological impacts.

Radioactivity in the Environment, Volume 19
ISSN 1569-4860, http://dx.doi.org/10.1016/B978-0-08-045015-5.00001-0

Thirdly, at least since the "rebellious generation" of the 1960s there has been a general change in public attitudes toward authorities. This has affected all social areas, not just radiation and the nuclear sector. Throughout the world, the public are less willing to accept any authority and more prone to demand full information and the opportunity to participate in decisions that affect them.

In combination, these three developments have radically changed the scene for radiation protection. They have made radiation protection more difficult and conflict-ridden but also more intellectually challenging and socially important. As a consequence of this, it is now widely recognized in the radiation protection community that social and ethical aspects have a central role in their professional undertakings. This book is a response to the often expressed need for in-depth discussions of these new perspectives on radiation protection.

Making decisions more ethical and making them more participatory are two distinct demands. But there are also many connections between the two. Therefore, we have found it appropriate to combine these two aspects in one and the same book. In the selection of themes, contributors, and viewpoints we have strived for plurality rather than uniformity. This is a subject-area that has largely arisen out of discontent with previous approaches that were conceived as too one-sided and technocratic. That is a good reason to strive for pluralism and for an open and inclusive discussion.

1.1 PART I. ETHICAL PRINCIPLES FOR RADIATION PROTECTION

The book begins with a series of chapters devoted to the underlying ethical principles of radiation protection. We begin by having a close look at the recommendations of the International Commission on Radiological Protection (ICRP) that have a dominant role in radiation protection throughout the world. In Chapter 2 *Jack Valentin* describes the development of these recommendations from 1928 onward. The considerable changes in the recommendations that have taken place in this period depend to a large part on increased knowledge of the biological effects of radiation, but he shows with concrete examples that reorientations in social and ethical outlooks have also had a major role. As Valentin emphasizes, decisions on exposure standards and protective measures cannot be based on science alone. Values also have a role, whether explicit or implicit. Although ethical discussions are rare in the ICRP publications, Valentin (who is a former Scientific Secretary of the organization) ensures us that ethical issues have often been discussed before new documents were released, and he also reminds us that several leading representatives of the organization have written papers on the basic philosophical issues in radiation protection.

In conclusion he brings up four controversial issues that have been discussed both within and outside of the ICRP. First, can the difference in dose limits between occupational and nonoccupational exposures be justified? In his view, it can. Secondly, should dose limits be calculated for average individuals or for the most sensitive group of individuals? Interestingly, Valentin expresses support for basing standards on the best estimate of risks for women (who are more sensitive to radiation than men), rather than as presently on the best estimate of risks to a person of "average sex". Thirdly, can the use of cost-benefit analysis be defended against the criticism that it puts an economic value on human life? He says that cost-benefit analysis is acceptable since the monetary values that it operates with represent the cost of saving a life, not the value of a life. Finally, is it appropriate for the ICRP to base its recommendations on "Western" ethical theories? Valentin answers this question by raising doubts about its premiss. The formal recommendations of the ICRP do not mention any specific ethical theory, he says, and legislators in Africa and Asia have made thorough use of the ICRP recommendations.

The practical applicability of ethical theories has often been put to question. In Chapter 3 *Sven Ove Hansson* concedes that their applicability is often more limited than what one might have wished, but he claims that radiation protection is an exception. This is because radiation protection operates with doses that are measured in numerical units. These numbers can be added up just like the utilities of utilitarianism, and they can be compared to exact limits in ways that conform with deontological ideas. The chapter identifies five contact points, or parallel issues, between moral philosophy and radiation protection. First, should we consider risks and benefits separately for each individual, or should we make one great sum for all effects, irrespective of who is affected by them? Secondly, should the acceptability of risks be determined by weighing against benefits or by the application of general limits for the size of the risks themselves? Thirdly, should all effects be considered, or should sufficiently small effects such as small risks and small radiation doses be left out of our moral deliberations? Fourthly, how should we value future effects? Fifthly and finally, how should individual differences be treated in an ethical discourse aiming at justice?

Most of the discussion on radiation ethics refers to moral theories such as utilitarianism and deontology. In Chapter 4 *Friedo Zölzer* questions whether it is adequate to base the system of radiation protection exclusively on theories of ethics such as utilitarianism and deontology that have been developed within a secular Western context. Is there, he asks, a "common morality" for humanity as a whole? He finds the four so-called middle-level principles of bioethics promoted by Beauchamp and Childress (autonomy, nonmaleficence, beneficence, and justice) to have a strong support in the major religions of the world and also in many nonreligious oral and written traditions. He sees this as a good sign of their cross-cultural validity. In addition, he finds widespread support in religious sources for three more middle principles that are important for radiation protection, namely concern for the underprivileged, intergenerational equity, and precaution. In summary, he proposes that attention to the common basis of religious traditions in ethics can provide radiation protection with underpinnings that are more generally acceptable to people of different cultural backgrounds than those that can be derived from Western philosophical traditions.

Historically, the main focus of radiation protection has been on humans, but recent developments have called for an assessment of the potential environmental impacts of ionizing radiation. In Chapter 5 *Deborah Oughton* shows that these changes raise ethical and well as scientific challenges. The paper looks at three perspectives on environmental ethics: anthropocentrism, biocentrism and ecocentrism. Rather than trying to ground the ethics of ecological radiation protection in one specific environmental philosophy or worldview, Oughton argues that all three views offer strong reasons to include impacts on plants and animals in radiation risk assessment. Moreover, it is important to respect the diversity in cultural views on environmental value that the different philosophical views represent. However, the different views can result in challenges in practice. One such challenge is the assessment of harm to nonhuman biota and in particular the ways in which different value systems impact on environmental assessment. Another challenge is what role monetary valuation of environmental goods should have. It is suggested that we need a holistic evaluation of the environmental impacts of ionizing radiation. It should consider not only the direct consequences on the health of humans and nonhuman species, but also the more complex social, ethical, and economic consequences of both human and nonhuman exposures.

The regulations for chemicals and radiation differ in many respects. In Chapter 6 *Carl Cranor* proposes that these differences ought to be much smaller and that most of the adjustment should go in one direction: Toxicological risk assessment should make much more use of the no-threshold assumption that is promoted by the ICRP and by national radiation protection agencies throughout the world. According to this assumption, the risk of cancer decreases with decreasing exposure, but it does not go down to zero until the exposure does so. As Cranor points out, for many adverse effects of chemicals the no-threshold assumption is biologically highly plausible. For others, there may be a threshold but it is not relevant in practice because it is very low or because it has already been exceeded by current exposures. In particular, this applies to a wide range of chemical substances that cause diseases in fetuses and children at much lower concentrations than in adults. The situation is very similar for radiation; fetuses and children are much more sensitive than adults to radiation-induced malignancies. Cranor's conclusion is that a no-threshold presumption should be applied to chemical risks. It can be replaced by a threshold approach only if the biology and the exposure conditions give us good reasons to do so.

1.2 PART II. PUTTING PROTECTION TO PRACTICE

In the second part, we investigate how ethical principles for radiation protection can be put to practical use. We begin with the dominant form of anthropogenic radiation exposure, namely medical radiology that gives rise to around

98 per cent of manmade human radiation exposure. We tend to take for granted that the benefits of such exposures outweigh the risks. There are good reasons to believe this to be true in most cases, but unfortunately not in all cases.

Currently, doses from diagnostic radiology are increasing. The major reason for this is that computerized tomography (CT), while significantly improving the diagnosis and therefore also the treatment of many conditions, also gives rise to much higher doses than traditional plain radiographs. In Chapter 7 *Jim Malone* assesses the current status of radiation protection in clinical radiology and proposes improvements that are important from an ethical point of view. Specifically, he directs our attention at four ethically problematic areas that are in need of special attention. One of these is diagnostic interventions that take place in spite of being medically unjustified, due for instance to insufficient training of referring physicians or to the practice of so-called defensive medicine. The second is diagnostic radiology undertaken for nonmedical purposes such as crime prevention, detection of swallowed contraband, or litigation. The third is radiology on pregnant or possibly pregnant patients that may be harmful to the fetus. The fourth is the need for accurate information to patients and proper informed consent for radiological procedures, something that is very often neglected in practice.

Sources of ionizing radiation are ubiquitous in modern life, and most public and occupational exposures do not result from the nuclear industry. In Chapter 8 *Chris Kalman* discusses the protection of workers from radiation exposure, emphasizing the need for consistent standards of protection from ionizing radiation, irrespective of source, if worker confidence and equitable protection is to be maintained. He recounts the early experience of high exposures leading to deterministic damage for instance among radiologists and radium dial painters. Major improvements have taken place since then, but nevertheless the present situation is not satisfactory. To the contrary, "shockingly, we have seen even the recurrence of deterministic (cell killing) effects or suggestions of dose sharing as a radiation protection tool." He identifies physicians performing intervention radiology as a group where statutory over exposure occurs, and sewage workers exposed to the discharges from medical facilities as a group receiving doses higher than the public receive from discharges from nuclear power facilities. The acute health effects in the Chernobyl liquidators are another reminder that radiation protection yet has important problems to solve. Kalman puts emphasis on the employer's duty of care toward the workers, which includes protection against radioactive hazards. Based on his experience as an occupational physician, he strongly emphasizes the need to provide workers with adequate information. When adequately informed, workers are well able to discuss and suggest principles for radiation protection.

Kalman also discusses the ethical requirements on research with exposed workers as participants. He puts focus on an often neglected aspect of research ethics, namely the participants' right to timely information. Instead of receiving information through mass media, he says, the participants should be offered prepublication briefings about findings concerning the risks they are exposed to. As he shows with an example, this is fully possible to implement and to combine with standard norms of scientific publication.

One of the basic principles for radiation protection promoted by the ICRP and others is the requirement to keep exposures as low as reasonably achievable (ALARA). But what does that mean in practice? In Chapter 9 *Sven Ove Hansson* shows that the principle has been interpreted differently in different practical applications, and discusses four ethically important issues of interpretation. First, does ALARA require exposures to be "as low as possible" or does is require some sort of optimal balance that it is undesirable to deviate from also in the direction of lower doses? Secondly, should ALARA be applied even to very low doses or is there a dose level below which it is not applicable? Thirdly, should judgments on what is reasonably achievable be made separately for each company, or in a unified manner for a whole branch of industry or perhaps even for society as a whole? Fourthly, can ALARA be operationalized with cost-benefit calculations, and in that case how?

1.3 PART III. NUCLEAR ACCIDENTS

This section begins with two chapters analyzing what we can learn from the two most serious nuclear accidents that we have experienced. In Chapter 10 *Astrid Liland* and *Lavrans Skuterud* summarize the experiences from Norway of the Chernobyl accident in 1986. Due not least to meteorological conditions, Norway is the country outside the former Soviet Union that was most severely affected by radioactive fallout. When the local food became too contaminated to eat, the very way of life of farmers and in particular Sámi reindeer herders was threatened. Being a country without a nuclear industry of its own, Norway was not well prepared to deal with an accident in that industry. In the early period after the accident the authorities had difficulties in dealing with the situation, and there were instances of conflicting messages and withheld information. However, the authorities learned from their shortcomings, and in the years that followed they developed several measures that served the double purposes of protecting human health and preserving traditional lifestyles and methods of food production. In order for such measures to be successful, it was necessary for the authorities to co-operate closely with the local population. Still today, almost three decades after the accident, reindeer herders have to take special measures to limit the dietary intake of radionuclides.

It is still difficult to know what happened at the Fukushima Daiichi power station on March 11, 2011. In Chapter 11 *Michio Miyasaka* explains these difficulties. Not surprisingly, high radiation levels still preclude detailed investigations of some aspects of the accident. In addition, the evaluation is made complicated by the vested interests of the major actors. The Government had issued fairly detailed regulations on protection against seismic motion, but the regulations on protection against tsunamis were much less clear. Therefore, if it is concluded that the accident was caused by seismic motion, then the focus of the discussion will be whether the company complied fully with the regulations. On the other hand, if the accident is deemed to be caused primarily by the tsunami, then the company can be seen as not really responsible for the accident since it has followed the government's (insufficient) requirements in this respect.

Miyasaka points out, however, that this analysis is based on a limited view of the company's responsibility. As he notes, "from the point of view of TEPCO, the company had built its plant in full compliance with government safety directives and, if some situation were to occur beyond what is envisioned by such directives, then it should be the responsibility of the government to take care of it." He contrasts this view with the safety culture approach according to which a company is responsible not only for complying with the detailed government regulations but also for doing whatever it takes to ensure that its operations are safe. According to Miyasaka, it would be useful to make sure that applied ethics is practiced in the nuclear sector as a means to make operations more safe. He bases this proposal in part on the introduction of bioethics into medical practice in Japan, that has led to a much stronger emphasis on the patient's right to information and on informed consent. Similarly, increased focus on ethics in the nuclear sector could arguably lead to a stronger emphasis on residents' and other affected persons' right to adequate information and to have a say in decision making that concerns them.

In Chapter 12 *Kristin Shrader-Frechette* argues that nuclear energy gives rise to massive environmental injustice, both after accidents as exemplified by Fukushima and more generally also under normal operating conditions. She points out four groups of people that are the major victims of environmental injustice. Poor people, in particular the sick and elderly poor, are victims since they do not have the economic means to move out of contaminated areas. Children are victims to environmental injustice since their bodies are considerably more sensitive to radiation than those of adults. Nuclear workers are victims due to their higher exposures. This applies in particular to the postaccident workers at Fukushima who are largely temporary contract workers receiving much higher doses than the regular workers in nuclear facilities. Many of these workers, she says, are poor and desperate for work and did not genuinely consent to the health risks they are exposed to. The fourth and final group of victims are future generations that are affected by the delayed, transgenerational effects of ionizing radiation. Shrader-Frechette concludes that these environmental injustices are largely inherent in regulations and that society should be able to overcome such regulations-driven injustices.

Experience from the nuclear industry and other safety-critical industries has shown that neither technical factors nor mistakes by single individuals can fully explain major accidents. The concept of safety culture has been developed to capture the influence of the larger organizational context on safety outcomes. In Chapter 13 *Carl Rollenhagen* reviews the safety culture concept and how it is related to other concepts used in safety research. Safety culture refers to how safety is valued in an organization. Rollenhagen warns that we should not focus on "the value of safety perceived as an absolute attribute." Instead, the crucial issue is how safety is valued in relation to other values and how the value of safety is implemented in actions actually taken. Strong statements about the priority of safety are not enough if the day-to-day management focuses on competing values such as production and cost containment at the expense of safety. In addition to a safety culture, a successful organization must have what he calls safety quality, which includes concrete arrangements to support safety such as training arrangements, technological resources, expertise and time resources for safety, etc. In summary, the chapter shows how difficult it is to implement the simple slogan "safety first" in organizations with strong competing values, and gives some indication of what is needed to overcome these difficulties.

1.4 PART IV. PROLIFERATION AND THE NUCLEAR FUEL CYCLE

We then turn to the risks associated with the military uses of nuclear fission. In Chapter 14 *Susanne Bauer*, *Boris Gusev*, *Tatyana Belikhina*, *Timur Moldagaliev*, and *Kazbek Apsalikov* discuss the effects on the local population of the Soviet nuclear tests at the Semipalatinsk test site in today's Kazakhstan. Between 1949 and 1989, more than 110 atmospheric and 340 underground tests were performed in the area. Secret Soviet assessments of the health impacts have become available, showing increased rates of cancer, including childhood cancer. Retrospective studies by both Kazakhstani and Western researchers have confirmed these findings and shown the expected correlations between cancer frequency and radioactive exposure. The paper also describes how the legacy of nuclear weapons testing has influenced the whole region, giving rise to difficult psychosocial issues.

The risk of nuclear weapons proliferation depends to a large part on the choice of a reactor technology and a nuclear fuel cycle. That choice also has important implications for the radiation risks in nuclear energy production. In Chapter 15 *Behnam Taebi* performs an ethical analysis of these technological choices. He employs four dimensions of analysis, namely safety (largely radiation protection), security (largely nonproliferation), sustainability (environmental friendliness and resource durability), and economic viability. Using these dimensions, he investigates four major fuel cycles. The open uranium cycle is the one currently used in most commercial reactors. Its major characteristic is that the spent fuel is treated as waste that ultimately has to be permanently stored. In the closed uranium cycle, spent fuel is sent to reprocessing, a chemical process that extracts components including plutonium that can be reused. Taebi also discusses two thorium-based technologies, both of which are yet technologically undeveloped but based on solid physical principles. One of these employs fuel rods that contain both thorium and uranium. The other uses a liquid fuel consisting of thorium fluoride and uranium fluoride. The comparison shows that none of these technologies is better than all the others in all respects.

The closed uranium cycle is better than the open uranium cycle for future generations since its waste problems are smaller, but it increases safety and security concerns and economic burdens on the current generation. Thorium cycles appear to be better than uranium cycles from the viewpoint of security (nonproliferation) but worse in terms of safety and economic viability. Taebi concludes that the choice of a fuel cycle is a clear case of a moral dilemma.

1.5 PART V. PUBLIC PARTICIPATION

We begin our section on public participation with a study of severe failures in the communication between on the one hand administrators and scientists, on the other hand affected members of the public. In Chapter 16 *Brian Wynne* summarizes fieldwork among hill sheep farmers in the Lake District of Northern England. Their farming practices were drastically affected by government restrictions in response to hazards from radioactive cesium isotopes originating in the Chernobyl accident. But the farmers were frustrated by the top-down attitudes of scientific experts and government officials. Attempts to achieve a dialogue failed, and the farmers found that their own knowledge of the local environments and the natural conditions of sheep farming were neglected. In addition, scientists' advice and predictions repeatedly turned out to be inaccurate, and as a consequence of this some of the farmers lost confidence in both scientists and government. When scientists had been wrong several times before, why should they be right the next time? One of the many interesting conclusions drawn in this study is that some of the sheep farmers exhibited a better understanding of scientific uncertainty than what the scientists assigned to inform them did.

But the worst environmental effects of the Chernobyl catastrophe in 1986 were of course those in the immediate vicinity of the site of accident. In the beginning of the 1990s, it became more and more clear that the countermeasures taken after the Chernobyl accident failed to adequately protect the population that had chosen to continue living in (moderately) contaminated areas. In spite of efforts by authorities to reduce exposures, there was a significant increase in radiation doses, in particular among children. Residents resented the top-down approach taken by authorities. They felt increasingly powerless, and many of them adopted attitudes that led them to neglect basic protection measures. This of course resulted in further increases in radiation exposures.

The ETHOS project that is described by *Jacques Lochard* in Chapter 17 was constructed to deal with this situation. The project was run in the years 1996–2001 in five villages in southern Belarus. Its aim was to empower the local inhabitants. They were given means so that they could themselves measure and monitor radiation levels in foodstuffs and the environment. They were also encouraged to form working groups focusing on practical issues such as the production of clean milk, improving radiation protection at school, etc. The overall result of this was a significant improvement in the radiological situation: decreased internal contamination of the children, less contamination in locally produced foodstuff, etc. The detailed measurements performed by the villagers turned out to provide much better guidance for countermeasures than the average values used by the authorities. Well-informed individual protective actions in the day-to-day life were efficient complements to the large-scale actions taken by the authorities. However, the participants of the ETHOS project have remaining questions about the suitability of the contaminated areas as places for humans to continue living in for generations ahead. Lochard proposes that this be further considered in a systematic research effort conducted on the general health status of the population and its evolution over the years in co-operation with the local population.

The importance of stakeholder participation in activities that give rise to radiation exposure has repeatedly been emphasized. However, it is often unclear why participation is undertaken and what its purpose is. In Chapter 18 *Sven Ove Hansson* and *Deborah Oughton* discuss participation from the viewpoint of democratic theory. Democratic decision-making has to take place in stable units such as nations and regions. Often the group of persons affected by a particular decision does not coincide with any of these units. In such cases, other means are needed to ensure that those affected are involved in the decision-making process and have a real influence on its outcome. To achieve this, participatory procedures are needed. This means that such procedures are a necessary component in a well-functioning democracy, making up for the misfit between decision-making units and the groups affected by many decisions. Based on this analysis, five quality criteria for participatory procedures in a democratic society are proposed: representativeness, transparency, impact on the decision, early involvement, and full access to expert knowledge.

Current plans for the disposal of radioactive waste mean that the waste will be placed in some geological repository. Such siting is usually seen as a disadvantage to people living in the proximity of the disposal facility. A strong case can be made in favor of compensating them for this. But on the other hand, an equally strong case can be made in favor of not "bribing" them into consent. In Chapter 19 *Mike Hannis* and *Kate Rawles* discuss the fine line between compensation and bribery and in particular how bribery can be avoided. They propose three procedural criteria for compensation to steer clear of corruption. First, potential sites should be selected according to robust and transparent criteria. It would be inappropriate to even consider the question of fair compensation unless the community has been chosen in a fair process. Secondly, the interests of future generations and nonhuman species must be adequately represented in the deliberative processes. Thirdly, compensation should be negotiated and offered by a trusted body that is independent of the nuclear industry.

In addition to these procedural criteria, the authors also discuss what types of compensation can be fair, as judged by substantial criteria. A fair compensation has to be proportionate in quantity, neither too small nor too large. (A too small compensation is an offense, and a too large one is a bribe.) It also has to be of the right kind. Direct financial compensation is often problematic. The risk that a community will be stigmatized by the presence of radioactive waste can arguably not be compensated for by money, whereas improved infrastructure and the development of high status employment opportunities may be more appropriate. But according to these authors calculative approaches to compensation based on cost-benefit analysis will not in general produce satisfactory outcomes, for several reasons including value incommensurability and the "crowding out" of intrinsic motivations.

Part I

Ethical Principles for Radiation Protection

Chapter 2

Radiation Risks and the ICRP

Jack Valentin[a]
Jack Valentin Radiological Protection, Stockholm, Sweden
E-mail: jack.valentin@telia.com

Chapter Outline

2.1 What is ICRP? 17
2.2 The Aims and Scope of ICRP Recommendations 18
2.3 The Early History and Development of ICRP Recommendations 21
2.4 The Development of the System of Radiological Protection and Current ICRP Recommendations 23
2.5 Ethical Underpinning of the Evolution of ICRP Recommendations 26
2.6 Some Moot Points 29
2.6.1 Different Dose Limits for Occupational or Public Exposures 29
2.6.2 Protecting Average Individuals or the Most Sensitive Ones? 29
2.6.3 Pricing Life? 30
2.6.4 Cultural Differences in Ethical Terminology 30

2.1 WHAT IS ICRP?

The International Commission on Radiological Protection, ICRP for short, is an advisory nongovernmental organization, established to advance for the public benefit the science of radiological protection, in particular by providing recommendations and guidance on all aspects of protection against ionizing radiation. Its recommendations form the basis of the basic safety standards documents issued by the United Nations and the European Commission and are reflected worldwide in legislation and regulations concerning radiation.

ICRP was formed in 1928 by the International Congress of Radiology, with the name of the International X-ray and Radium Protection Committee (IXRPC), following a decision by the Second International Congress of Radiology.

a. Although the author was the Scientific Secretary of ICRP from 1997 through 2008, the views presented here are his own and do not necessarily represent those of ICRP.

Radioactivity in the Environment, Volume 19
ISSN 1569-4860, http://dx.doi.org/10.1016/B978-0-08-045015-5.00002-2

In 1950 it was restructured and renamed as now. It is an independent charity (i.e. a nonprofit-making organization), registered in the United Kingdom. It works closely with its sister body, the International Commission on Radiation Units and Measurements (ICRU), has official relationships with several United Nations bodies, the European Commission, and the Nuclear Energy Agency of the OECD, and works with organizations such as the major international standardization bodies (ISO and IEC). ICRP maintains contact with the professional radiological community through links with the International Radiation Protection Association and takes account of progress reported by national organizations (ICRP, 2007a).

ICRP comprises a Main Commission with a Chairperson, 12 other members, and a Scientific Secretary; five Committees (each with 10–20 members) dealing with various aspects of radiological protection; and a number of ad hoc Task Groups and Working Parties drafting new advisory documents. At any one time, some 200–250 people worldwide are engaged within ICRP. More information about ICRP and its membership is available at its web site, www.icrp.org.

Early on in the history of ICRP, it became apparent that its independence and scientific integrity could be jeopardized by demands from special interest groups and others with vested interests. Its second ever meeting, in 1934, involved pressures concerning membership that were the first, but certainly not the last, examples of demands or covert criticisms aimed at gaining outside control the membership and/or policies of ICRP. Wary of such attempts, ICRP maintains as its strict policy that members are elected by the Commission itself. Outside nominations are accepted as a means to achieve the widest possible range of expertise, but the actual elections are made by the Commission alone, and solely on the grounds of scientific merit, not as representatives of any country, organization, or other entity (Clarke & Valentin, 2009).

2.2 THE AIMS AND SCOPE OF ICRP RECOMMENDATIONS

According to its current general recommendations, the primary aim of the Commission's recommendations is to contribute to an appropriate level of protection for people and the environment against the detrimental effects of radiation exposure without unduly limiting the desirable human actions that may be associated with such exposure (ICRP, 2007a).

Obviously, as underlined in the 2007 Recommendations, this aim cannot be achieved solely on the basis of scientific knowledge on radiation exposure and its health effects. Scientific data are a necessary prerequisite, but societal and economic aspects of protection have also to be considered. All of those concerned with radiological protection have to make value judgments about the relative importance of different kinds of risk and about the balancing of risks and benefits. In this, radiological protection is not different from other fields concerned with the control of hazards. Thus, ICRP (2007a) states that the basis for, and distinction between, scientific estimations and value judgments should be made clear whenever possible, so as to increase the transparency, and thus the understanding, of how decisions have been reached.

Nevertheless, the formal ethical bases on which the Recommendations of ICRP rest are not explicitly mentioned in the formal Recommendations, and touched upon in just a few ICRP documents. General discussions are provided in ICRP (1999, Annex D) and Clarke and Valentin (2009); specific aspects of medical ethics are mentioned in ICRP (1996) and the particular ethical considerations in the context of volunteers in medical research are discussed in some detail in ICRP (1992).

In order to understand the ethical issues involved in radiological protection, one needs to know that in general, ionizing radiation causes two types of harmful effect. High doses will cause harmful tissue reactions, often called *deterministic effects*. These are often of an acute nature, and they will usually only appear if the dose exceeds a threshold value. Both high and low doses may cause *stochastic effects* (cancer or heritable effects). Spontaneous cancers and genetic damage occur frequently, and at the present state of scientific knowledge it is not possible to distinguish radiation-induced cases from spontaneous ones. However, while the specific cases cannot be attributed to radiation, given sufficiently large exposed populations and sufficiently high doses, statistically detectable increases in the population incidence of cancer occur long after the exposures.

For various statistical reasons no such significant increase in the incidence of heritable effects has yet been demonstrated, but unequivocal evidence from animal and plant experiments, population genetic considerations, and observations of damage to the genetic material in human somatic cells all prove beyond any shadow of a doubt that heritable effects are also produced—albeit at a frequency which is too low to demonstrate statistically in man against the background of our considerable genetic burden of natural variation.

The ICRP system of radiological protection aims primarily to protect human health, and historically, this was the only aspect considered in any way at all until the ICRP (1977) Recommendations. The health objectives are to manage and control exposures to ionizing radiation so that deterministic effects are prevented, and the risks of stochastic effects are reduced to the extent reasonably achievable (ICRP, 2007a).

The protection of other species and the environment is a more recent, and somewhat more complicated, issue. ICRP (1977) claimed that "the level of safety required for the protection of all human individuals is thought likely to be adequate to protect other species, although not necessarily individual members of those species. The Commission therefore believes that if man is adequately protected then other living things are also likely to be sufficiently protected." In a marginally more sophisticated phrase, ICRP (1991) assumed that "the standard of environmental control needed to protect man to the degree currently thought desirable will ensure that other species are not put at risk. Occasionally, individual members of nonhuman species might be harmed, but not to the extent of endangering whole species or creating imbalance between species."

However, ICRP (2007a) observed that "there is no simple or single universal definition of "environmental protection" and the concept differs from country to country and from one circumstance to another." Other ways of considering radiation effects are therefore likely to prove to be more useful for nonhuman species—such as those that cause early mortality, or morbidity, or reduced reproductive success. The Commission's aim is now that of preventing or reducing the frequency of deleterious radiation effects to a level where they would have a negligible impact on the maintenance of biological diversity, the conservation of species, or the health and status of natural habitats, communities, and ecosystems. In achieving this aim, however, the Commission recognizes that exposure to radiation is but one factor to consider, and is often likely to be a minor one.

The ICRP system of radiological protection applies to all ionizing radiation exposures from any natural or man-made source, regardless of its size and origin. This however does not mean that all exposures, sources, and human actions, can or need to be equally considered. Instead, the approach should be graded according to the amenability of a particular source or exposure situation to regulatory controls, and the level of exposure/risk associated with that source or situation (ICRP, 2007b).

Thus, exposures that are not amenable to control, regardless of their magnitude, are *excluded* from radiological protection legislation. For instance, exposure to the natural radionuclide ^{40}K incorporated into the human body cannot be restricted by any conceivable regulatory action, and control of exposure to cosmic rays at ground level is obviously impractical. In short, some exposures cannot be regulated.

Furthermore, exposures that are such that the effort to control them is judged to be excessive compared to the associated risk should be *exempted* from some or all radiological protection regulatory requirements. For instance, while it is important to control the manufacture and supply of smoke detectors containing radioactive material, it makes sense to exempt their use in homes from regulatory licensing requirements. In short, some exposures (or more often, some aspects of some exposures) need not be regulated.

This does not necessarily mean that exempted exposures equal small doses. Even a small dose is worth removing, if the effort to do so is small. Conversely, if no reasonable control procedure can achieve significant dose reductions, exemption is warranted even if the doses are not trivially small.

Given the multitude of different exposure situations, the enormous range of possible doses, and the extreme sensitivity of radiation measuring equipment (compared with instruments for most other noxious agents), the need for the concepts of exclusion and exemption is usually taken as an indisputable fact in the public debate. The choice of exposures and exposure situations that merit exclusion is usually also reasonably uncontroversial. In contrast, exemption decisions may be rather more difficult, depending for instance on different perceptions of benefits and risks of particular situations involving radiation.

As an added complication, exemption may sometimes be seen as a means to achieve conservation of resources, which is usually regarded as ethically commendable. "Clearance" is a special case of exemption where regulatory control is relinquished because it is no longer warranted. For instance, much material (tools, clothes...) is taken into controlled areas of nuclear installations. Such material will by default be regarded as contaminated (and often is). Much of it can be returned outside the installation and reused after appropriate decontamination and/or measurements. While the concept of such clearance is not highly controversial, the levels and conditions to be applied can be the subject of heated debate.

2.3 THE EARLY HISTORY AND DEVELOPMENT OF ICRP RECOMMENDATIONS

The discoveries of X-rays in 1895 and radioactivity in 1896 immediately spawned numerous practical applications of these phenomena, particularly in medicine. The capacity of these radiations to cause serious damage to human tissues (what we now usually call deterministic effects or tissue reactions) also became apparent within months (e.g. Drury, 1896), and by 1902, Frieben extended the observation to include the induction of cancer. However, ignorance about the risks was widespread, there were numerous injuries over the next two decades, and several hundred deaths of medical staff (Molineus, Holthusen, & Meyer, 1992).

This was the backdrop that caused the 2nd International Congress of Radiology, in Stockholm 1928, to establish the "International X-ray and Radium Protection Committee" (IXRPC), which later developed into ICRP. As indicated above, the carcinogenic effect of ionizing radiation was already known, and in the previous year, Muller (1927) had reported that X-rays induce mutations in the genetic material.

Nevertheless, at this initial stage the protection philosophy was focused entirely on deterministic effects (described in the first Recommendations, IXRPC, 1928; as "injuries to superficial tissues, derangements of internal organs and changes in the blood"). The main emphasis of the 1928 Recommendations was on practical physical protection, such as shielding. No form of dose limit was proposed, but a prolonged holiday and a limitation of the working hours of medical staff were recommended. The occupational annual effective dose to medical staff at the time may have been in the order of 1000 mSv (cf. Clarke & Valentin, 2009)—this corresponds to about 400 times the average dose due to natural sources and is 50 times higher than the current recommended limit on average annual effective dose for occupational exposures.

The first "dose limit" (actually, a recommended limit on exposure rate for X-rays) was promulgated with the IXRPC (1934) Recommendations. It was still based entirely on the desire to avoid deterministic effects, and the recommendations clearly implied the concept of a safe threshold below which no untoward effects were expected. In modern terms and units, the limit would have corresponded to an annual effective dose of approximately 500 mSv. This would likely have achieved what the limit set out to do, i.e. to prevent deterministic harm (at least to healthy adults), but of course was inadequate with respect to stochastic harm.

These IXRPC Recommendations led to a great improvement in the standard of occupational radiation safety. The "dose limits" also served as the basis for the safety measures applied when nuclear energy programs were first developed during and immediately after World War II (Sowby, 1981). Sowby stresses that thanks to this, there were very few radiation injuries among the many thousands of workers involved in the early days of nuclear energy, despite the large amounts of radioactive material they handled.

The next, 1950, set of Recommendations appeared under the Commission's new name, ICRP (1951). Again, the quantitative restriction on exposures became more stringent; the new recommended limits correspond in modern terminology to an annual limit on occupational effective dose of approximately 150 mSv (although the concept of a limit was somewhat different than the current limits). Health effects that "should be kept under review" now included not just deterministic effects, but also, e.g. leukemia, malignant tumors, and genetic effects.

However, it is not immediately apparent that the inclusion of stochastic effects among those health parameters that should be monitored actually influenced the recommendations as such or, in particular, the "dose limit". The reduction from ~500 mSv in a year to ~150 mSv in a year may have reflected that ICRP considered the possibility of individual variations in radiosensitivity. Genetic harm had been known for many years already to occur at quite low doses in experimental organisms, and the 1950 Recommendations "strongly recommended that every effort be made to reduce exposures to all types of ionizing radiation to the lowest possible level"—but, inconsistently, the text was also full of expressions like "permissible levels", "maximum permissible exposure", and "the probable threshold for adverse effects". All of these implied the existence of a safe threshold below which there would be no deleterious effects.

The situation started to change with the next, 1954, set of Recommendations (ICRP, 1955) which claimed that: "whilst the values proposed for maximum permissible doses are such as to involve a risk which is small compared to the other hazards of life … it is strongly recommended that every effort be made to reduce exposure to all types of ionizing radiation to the lowest possible level." ICRP now recognized (albeit somewhat vaguely) the need to protect not just radiation workers but also the general public, e.g. with nuclear energy expected to be an expanding industry. The major problem was believed to be hereditary harm, but the occurrence of leukemia among radiologists and among the survivors in Hiroshima and Nagasaki also contributed to the decision to recommend that "in the case of the prolonged exposure of a large population, the maximum permissible levels should be reduced by a factor of 10 below those accepted for occupational exposures."

In a short amendment ICRP (1957) made several points reflecting significant ethical decisions. Thus, it was recommended that the dose restrictions for members of the public should apply to staff working outside "controlled areas" within an enterprise involving radiation. In other words, only those employees who were actually working with radiation (and would usually benefit from pertinent training) were to be regarded as occupationally exposed. This clarification must have had a significant effect in terms of reduced doses to other staff.

Furthermore, for the first time specific advice concerning pregnant women was provided: "Since ... the embryo is very radiosensitive, special care should be exercised to make sure that pregnant women are not occupationally exposed ... through some accident or otherwise ... to large doses of penetrating radiation."

This was soon followed by a major revision in the shape of the 1958 Recommendations, also somewhat quaintly called "Publication 1" (ICRP, 1959). They proposed new limitations of dose for occupational exposure and, for the first time in formal terms, for members of the public. The occupational limit was expressed as a restriction on the dose accumulated at any particular age in years, and corresponded to an average annual effective dose of 50 mSv, while the public limit was expressed was simply set per year at what is now termed 5 mSv. The dose limit for the public reflected the understanding that stochastic effects had to be taken into account, and that for such effects no safe threshold dose could be taken for granted. There was not yet any clear dose–response model, malignant tumors were not really considered, and leukemia was regarded as possibly not stochastic in nature, so the concern regarding stochastic effects was focused on genetic harm. Yet the 1958 Recommendations constituted a paradigm shift that subsequently evolved into the current system of radiological protection.

2.4 THE DEVELOPMENT OF THE SYSTEM OF RADIOLOGICAL PROTECTION AND CURRENT ICRP RECOMMENDATIONS

The increasing understanding of stochastic effects soon necessitated further revisions. The ICRP (1966) Recommendations, Publication 9, ventured a speculation on the possible dose–response relationship for stochastic effects: "the Commission sees no practical alternative, for the purposes of radiological protection, to assuming a linear relationship between dose and effect, and that doses act cumulatively. The Commission is aware that the assumptions of no threshold and of complete additivity of all doses may be incorrect, but is satisfied that they are unlikely to lead to the underestimation of risks."

Thus the default assumption of a safe threshold was rejected. For stochastic effects, primarily the probability rather than the severity of the effect is proportional to the size of the dose. Therefore, the objective of radiological protection refocused onto reducing and limiting the probability of harm, rather than preventing harm. As a logical consequence, it was no longer sufficient to aim at keeping doses below a limit. The concept of optimization of protection was signaled in the statement in ICRP (1966) that "as any exposure may involve some degree of risk, the Commission recommends that any unnecessary exposure be avoided and that all doses be kept as low as is readily achievable, economic and social consequences being taken into account."

The 1966 Recommendations also introduced a distinction between "normal operations" and accidents where the exposure "can be limited in amount only, if at all, by remedial action." Again, this raised new ethical issues, not least concerning the protection aims for emergency staff that might have to deal with situations entailing high dose rates.

The requirement that doses be reduced even below the dose limits necessitated further guidance. In a report, ICRP (1973) tightened up the requirement by stating that doses should be kept as low as *reasonably* achievable, rather than readily achievable, and suggested that differential cost–benefit analysis (CBA) could be used to ensure that protection was indeed optimized.

The next set of Recommendations, ICRP (1977), developed this by stating that doses "as low as reasonably achievable" would correspond to a *collective dose* so low that "any further reduction in dose would not justify the incremental cost required to accomplish it." The Recommendations went on to recommend that this be analyzed with CBA with collective dose as the independent variable and with a monetary value assigned to a unit of collective dose. Thus, the question that was asked (and hopefully answered by the use of CBA) was "How much does it cost and how many lives are saved?" This also meant that optimization of protection was the main means of radiological protection. Dose limits were no longer a primary regulatory tool, although they had to be retained in order to protect the individual from the combined exposure from all controlled sources.

The 1977 Recommendations also established the formal *System of radiological protection*, which is essentially still in place today (with some amendments and shifting accents as described below). The system includes three basic principles, viz.,

justification—no practice shall be adopted unless it produces a positive net benefit;

optimization—doses shall be as low as reasonably achievable, and economic and social factors taken into account;

application of limits—doses to individuals shall not exceed recommended limits.

The levels of dose at the limits were assumed to represent exceptional cases. Thus, although it was claimed that an average annual dose of 1 mSv would entail a risk that members of the public were likely to regard as acceptable, the public annual dose limit of 5 mSv was retained. It was argued that a public annual dose limit of 5 mSv would achieve lifetime doses corresponding to an average of 1 mSv per year in "critical [=highly exposed] groups." Similarly, the dose limit for workers was argued on a comparison of average doses, and therefore risk, in the workforce, with average risks in industries that would be recognized as being "safe", and not on maximum risks to be accepted.

The following, 1990, set of Recommendation (ICRP, 1991) included both revised risk estimates and significant amendments to the system of protection. Both the epidemiology and the dosimetry concerning cancer among survivors from Hiroshima and Nagasaki showed that the risk of cancer per unit dose of radiation had to be adjusted upwards by a factor of about 3. As a consequence, the 1990 Recommendations re-emphasized the need to keep doses as low as reasonably achievable. ICRP also reduced the annual dose limits from 50 mSv to 20 mSv for occupational exposure (averaged over 5-year periods with a maximum of 50 mSv in any one year), and from 5 to 1 mSv for public exposure.

These reductions should not be construed as directly proportional to the increased risk estimates. The concept of an acceptable risk, as discussed in the 1977 Recommendations, was no longer regarded as satisfactory. People tend to accept or reject activities (with their attendant benefits and risks) rather than specific risk values. ICRP (1991) used a much more sophisticated multi-attribute study to illuminate different risk dimensions associated with exposures at the dose limits.

Furthermore, the three basic principles were rephrased, and the most significant amendment concerned the optimization principle: "In relation to any particular source within a practice, the magnitude of individual doses, the number of people exposed, and the likelihood of incurring exposures where these are not certain to be received should all be kept as low as reasonably achievable, economic and social factors being taken into account. This procedure should be constrained by restrictions on the doses to individuals (dose constraints), or on the risks to individuals in the case of potential exposures (risk constraints) so as to limit the inequity likely to result from the inherent economic and social judgments." Thus, while the primary aim of optimization is to reduce the collective dose, ICRP added a restriction on individual dose to the process.

The current ICRP (2007a) Recommendations reiterate the system of radiological protection. Thus, they represent continuity rather than change to the fundamental features, and they again emphasize the importance of optimization. However, they also extend the concept. Added guidance on how to use constraints in planned exposure situations (such as the normal operation of practices using radiation) is complemented by the extension of optimization, with constraints, to other situations (i.e. emergencies and existing exposure situations). Furthermore, the 2007 Recommendations include a commitment to environmental protection.

2.5 ETHICAL UNDERPINNING OF THE EVOLUTION OF ICRP RECOMMENDATIONS

While explicit discussions of formal ethical underpinnings are rare in ICRP documents, this certainly does not mean that ICRP is unaware of the importance of such discussions. For instance, the ICRP (1966) Recommendations state that "as any exposure may involve some degree of risk, the Commission recommends that any unnecessary exposure be avoided and that all doses be kept as low as is readily achievable, economic and social consequences being taken into account." Here, ICRP considered ethical considerations to be implied in the word "social" (or, in the corresponding version of the same statement in the current, 2007, Recommendations, "societal").

Sometimes, debaters suggest that ICRP (or other organizations) should refrain from proposing a particular ethical approach. However, one cannot lift oneself from the floor by the bootstraps, i.e. protection recommendations will inevitably represent an ethical position, irrespective of whether that position is explicit, tacitly implied, or unpremeditated. Better then to specify the position, thus allowing users of the recommendations to adapt them to their own ethical predilections! Ethical issues are often discussed within ICRP before new documents are released for publication, and several leading representatives of ICRP have produced papers reflecting the basic tenets as well as their own views and interpretations (e.g. Beninson, 1996; González, 2011; Lindell, 1988; Silini, 1992; Taylor, 1957).

However, initially the level of sophistication of the ethical considerations was not very advanced. Between 1896, when deleterious effects of ionizing radiation were first identified as such, and the mid-1950s, when public concern about radiation risks increased and the focus of protection shifted toward stochastic harm, the purpose of radiological protection was just to avoid deterministic harm. The principle that was applied in order to achieve this was simply to keep individual doses below pertinent threshold values. Low doses of radiation were not a concern; if anything, they were regarded as beneficial. There was a plethora of radioactive consumer products.

Little is known about any ethical discussions within ICRP during this period; the protection philosophy appears to have been based loosely on Aristotelian virtue ethics. In other words, protective actions should be "good" and follow from an inner sense of moral orientation.

The first documented instance of a discussion within ICRP of the philosophy and principles of radiological protection is provided in the "Prefatory Review" of the 1958 Recommendations (ICRP, 1959). Neither ethics nor morals are mentioned explicitly, but in a section called "Objectives of Radiation Protection" the Recommendations state specifically that these are "to prevent or minimize somatic injuries [i.e. *deterministic effects and leukemia*] and to minimize the deterioration of the genetic constitution of the population." This constitutes a statement of the ethical basis of the Recommendations, even though it is not phrased as such, and heralds the balancing of utilitarian and deontological ethics that (as we shall see) characterizes later and current ICRP Recommendations.

The 1977 Recommendations (ICRP, 1977) focused on deciding what is reasonably achievable in dose reduction. The principle of justification aims to do more good than harm, and that of optimization aims to maximize the margin of good over harm for society as a whole. As pointed out by Hansson (2007) and Clarke and Valentin (2009), they therefore satisfy the utilitarian principle of ethics (consequence ethics), emphasizing what is best for society. Actions were judged by their overall consequences, usually by comparing, in monetary terms, the relevant benefits (e.g. statistical estimates of lives saved) obtained by a particular protective measure with the net cost of introducing that measure.

On the other hand, the principle of applying dose limits aims to protect the rights of the individual not to be exposed to an excessive level of harm, even if this could cause great problems for society at large. This principle therefore satisfies the deontological principle of ethics (duty ethics), which emphasizes the strictness of moral limits.

The 1990 Recommendations updated the principle of optimization significantly by introducing the concept of a constraint, i.e. an individual-related dose (or risk) criterion constituting an added restriction on the optimization process. In ethical terms, the utilitarian ambition to reduce the collective dose was modulated by the addition of the deontological constraint, which prevented minimization of the collective dose from being achieved through a very uneven (unfair) distribution of individual doses.

Thus, concern for the protection of the individual was being strengthened. Publication 77, an ICRP (1998a) report, further reflected what ICRP perceived as changing societal values, with more concern about individual welfare. It stated that "the perception of optimization of protection has become too closely linked to differential cost–benefit analysis," and that "the unlimited aggregation of collective dose over time and space into a single value is unhelpful because it deprives the decision maker of much necessary information. The levels of individual dose and the time distribution of collective dose may be significant factors in making decisions."

The most recent Recommendations, ICRP (2007a), and supporting reports such as ICRP (2006) on optimization, further emphasize controls on the maximum dose or risk to the individual and put less emphasis on collective dose and cost-benefit analysis. Overall, this reflects a modified ethical position, paying less attention to utilitarian values and accentuating deontological duty ethics, focusing on what is best for the individual.

Lest this contrast be perceived as overly theoretical, it should be recalled that it is not uncommon in occupational radiological protection, particularly perhaps in the nuclear industry. Sometimes the lowest collective dose is achieved if a given task is performed by just a few workers, each of whom get a fairly high individual dose, rather than by a larger team where doses to all members are low. This causes a conflict of interest between the interest of individual workers and the interest of society (or at least all concerned workers). The most common purpose of using a dose constraint in occupational contexts is to add protection of the individual (as prescribed by deontological duty ethics) while strict minimization of collective dose protects society and emphasizes utilitarian consequence ethics. The choice of value of a dose constraint decides the balancing of duty versus consequence ethics. At the same time, since operators are supposed to make the choice, it encourages them to assume more responsibility.

In the context of public exposure, ICRP dose constraints on optimization more frequently serve the purpose of ensuring that the combined exposure from several sources remains acceptable (which usually requires that the dose constraint is set by the regulator). This again represents duty ethics, but in this context not as an alternative to a "competing" utilitarian approach.

ICRP (1991, 2007a) also recommends the use of risk constraints, i.e. limitations on the probability of untoward events occurring. As yet, few countries are using formal risk constraints so there is little practical experience. In ethical terms, risk constraints represent duty ethics just like dose constraints.

Inevitably, radiological protection (and indeed any form of regulation or protection against some noxious agent) will require a balancing between utilitarian and deontological approaches. Recommendations or regulations concerning other hazards than radiation sometimes purport to be based on one of the principles alone. However, in real life, no practical protection work can represent just one of the principles, but one can give more emphasis to one of the principles without discarding the other one (Hansson, 2007). The development from ICRP (1977) to ICRP (2007a,b) represents a transition in emphasis from one half of this spectrum to the other half, but both sets of Recommendations rest squarely on the combined application of both utilitarian and deontological ethics.

The ICRP (2007a) Recommendations also constitute a commitment to protection of the environment, essentially nonhuman species. This is not due to any serious concern about existing radiation hazards. It is rather a matter of filling a conceptual gap by providing scientific evidence, rather than just assumptions, to show that other species are adequately protected.

But what is "adequate" in this context? The formal ethical basis for environmental protection is less well developed than that concerning human health. IAEA (2002) provides an overview of the various ideas and principles that might apply. ICRP has not yet made any formal statement concerning the ethical foundation for its environmental protection program. However, its aims are to prevent or reduce the frequency of deleterious radiation effects in the environment to a level where they would have a negligible impact on the maintenance of biological diversity, the conservation of species, or the health and status of natural habitats, communities, and ecosystems (ICRP, 2007a, 2008). These aims are in line with the possible considerations mentioned by IAEA (2002).

2.6 SOME MOOT POINTS

ICRP stresses that its recommendations and advice are based on a combination of scientific estimations and value judgments. Therefore, it is not surprising that sometimes, alternative views and interpretations are forwarded. A few selected examples of such issues follow below.

2.6.1 Different Dose Limits for Occupational or Public Exposures

Sometimes, this difference is called into dispute (e.g. Persson & Shrader-Frechette, 2001). However, ICRP has never argued that there would be any reason (such as knowing the hazard, and/or receiving a high salary including some form of risk premium) to permit "more" radiation in occupational contexts. Instead, when protection aims were widened to include stochastic effects (which can be induced at low doses, necessitating dose limitation for members of the public), the Commission's view was and is that "less" radiation must be permitted for the general public. One reason is that an average lifetime is about twice as long as the maximum exposure period of a worker. Furthermore, while workers are usually healthy adults, the general public includes more sensitive persons such as children and those suffering from diseases.

2.6.2 Protecting Average Individuals or the Most Sensitive Ones?

Hansson (2009) notes that ICRP bases its recommendations on a population average and acknowledges but does not take into account differences in radiosensitivity (between men and women, between adults and children, or between different genetic constitutions). He argues that the ICRP position is difficult to defend from an ethical point of view and identifies two alternative approaches: special standards for the sensitive groups, or ensuring that standards that apply to everybody are sufficiently stringent to protect the sensitive groups.

He also notes that this does not necessarily mean that current standards need to be changed (i.e. existing dose limits may be stringent enough, but whether this is so is not clear from the reasoning provided by ICRP). Such a position seems to imply that general standards need not be changed to protect the few individuals with extreme genetically determined radiosensitivity. This would be in line with the ICRP (1998b) conclusion that because of the high risk of spontaneous cancer in familial disorders, low doses of radiation (say 100 mSv) are most unlikely to impact significantly on life-time cancer risk in an affected individual.

The higher radiosensitivity of women than men and of children than adults is a more complicated issue, which is often subject to discussion within ICRP.

As soon as one moves away from the "average", one faces the very difficult question of how far to move. The present author's personal view is that it would probably be better if ICRP used the best estimate of the risk to women, rather than the current estimate of risk to a person of "average sex", as its nominal risk estimate, and that this probably would only rarely require any practical changes to radiological protection measures. Children pose additional questions; it should be recalled that ICRP does take account their anatomy and physiology when computing radiation doses per unit release of radioactive material into the environment (but not of the difference in sensitivity per unit dose).

2.6.3 Pricing Life?

Differential cost–benefit analysis in optimization requires that a monetary value be assigned to a unit of collective dose. This sometimes raises the question whether ICRP tries to assign a monetary value to human life.

However, the amount in question does not represent the value of a human life, and it represents the cost of saving a life. Since societal resources are not endless, there will be occasions when the amount spent will be limited even though it might be possible technically to do more. Optimization of protection, and the cautious use of CBA as one of several tools in this optimization, facilitates a fair distribution of life-saving resources. This position is shared by the Pontifical Academy of Sciences (1985).

2.6.4 Cultural Differences in Ethical Terminology

Zölzer (2012) queries whether it is appropriate for advisory bodies such as ICRP to base their recommendations on "Western" ethical theories, pointing out that 70% of the world population live in Africa or Asia.

There is probably scope for increasing the transparency of ICRP recommendations worldwide by considering different traditions and terminologies. However, the formal Recommendations of ICRP (such as ICRP, 2007a) do not mention explicitly any particular set of ethical theories underpinning the advice, and radiological protection legislation in African and Asian countries appears to be thoroughly based on ICRP Recommendations.

REFERENCES

Beninson, D. J. (1996). Risk of radiation at low doses. The 1996 Sievert Lecture. *Health Physics*, *71*, 122–125.

Clarke, R. H., & Valentin, J. (2009). The history of ICRP and the evolution of its policies. *Annals of the ICRP*, *39*(1), 75–110.

Drury, H. C. (1896). Dermatitis caused by Roentgen X-rays. *British Medical Journal*, *2*, 1377–1378.

Frieben, A. (1902). Demonstration eines Cancroid des rechten Handrückes, das sich nach langdauernder Einwirkung von Röntgenstrahlen entwickelt hat. *Fortschritte auf dem Gebiet der Röntgenstrahlen und der bildgebenden Verfahren*, *6*, 106–111.

González, A. J. (2011). The Argentine approach to radiation safety: Its ethical basis. *Science and Technology of Nuclear Installations*, 2011, 1–15. http://dx.doi.org/10.1155/2011/910718.

Hansson, S. O. (2007). Ethics and radiation protection. *Journal of Radiological Protection*, *27*, 147–156.

Hansson, S. O. (2009). Should we protect the most sensitive people? *Journal of Radiological Protection*, *29*, 211–218.

IAEA. (2002). *Ethical considerations in protecting the environment from the effects of ionizing radiation*. TECDOC 1270, Vienna, Austria: International Atomic Energy Agency.

ICRP. (1951). International recommendations on radiological protection. Revised by the International Commission on Radiological Protection and the 6th International Congress of Radiology, London, 1950. *British Journal of Radiology*, *24*, 46–53.

ICRP. (1955). Recommendations of the International Commission on Radiological Protection. *British Journal of Radiology*, (Suppl. 6), 100.

ICRP. (1957). Reports on amendments during 1956 to the Recommendations of the International Commission on Radiological Protection (ICRP). *Acta Radiologica*, *48*, 493–495.

ICRP. (1959). *Recommendations of the International Commission on Radiological Protection. ICRP Publication 1*. Oxford, UK: Pergamon Press.

ICRP. (1966). *Recommendations of the International Commission on Radiological Protection. ICRP Publication 9*. Oxford, UK: Pergamon Press.

ICRP. (1973). *Implications of Commission Recommendations that doses be kept as low as readily achievable. ICRP Publication 22*. Oxford, UK: Pergamon Press.

ICRP. (1977). Recommendations of the International Commission on Radiological Protection. ICRP Publication 26. *Annals of the ICRP*, *1*(3).

ICRP. (1991). 1990 Recommendations of the International Commission on Radiological Protection. ICRP Publication 60. *Annals of the ICRP*, *21*(1–3).

ICRP. (1992). Radiological protection in biomedical research. ICRP Publication 62. *Annals of the ICRP*, *22*(3).

ICRP. (1996). Radiological protection and safety in medicine. ICRP Publication 73. *Annals of the ICRP*, *26*(2).

ICRP. (1998a). Radiological protection policy for the disposal of radioactive waste. ICRP Publication 77. *Annals of the ICRP*, *27*(Suppl.).

ICRP. (1998b). Genetic susceptibility to cancer. ICRP Publication 79. *Annals of the ICRP*, *28*(1–2).

ICRP. (1999). Protection of the public in situations of prolonged radiation exposure. ICRP Publication 82. *Annals of the ICRP*, *29*(1–2).

ICRP. (2006). The optimisation of radiological protection - broadening the process. ICRP Publication 101b. *Annals of the ICRP*, *36*(3).

ICRP. (2007a). The 2007 Recommendations of the International Commission on Radiological Protection. ICRP Publication 103. *Annals of the ICRP*, *37*(2–4).

ICRP. (2007b). Scope of radiological protection control measures. ICRP Publication 104. *Annals of the ICRP*, *37*(5).

ICRP. (2008). Environmental protection - the concept and use of Reference Animals and Plants. ICRP Publication 108. *Annals of the ICRP*, *38*(4–6).

IXRPC. (1928). X-ray and radium protection. Recommendations of the 2nd International Congress of Radiology, 1928. *British Journal of Radiology*, *12*, 359–363.

IXRPC. (1934). International Recommendations for X-ray and Radium Protection. Revised by the International X-ray and Radium Protection Commission and adopted by the 4th International Congress of Radiology, Zürich, July 1934. *British Journal of Radiology*, *7*, 1–5.

Lindell, B. (1988). *How safe is safe enough?* Lauriston S. Taylor Lectures No.12. Bethesda, Md, USA: National Council on Radiation Protection and Measurements.

Molineus, W., Holthusen, H., & Meyer, H. (1992). *Ehrenbuch der Radiologen aller Nationen* (3rd ed.). Berlin, Germany: Blackwell Wissenschaft.

Muller, H. J. (1927). Artificial transmutation of the gene. *Science*, *66*, 84–87.

Persson, L., & Shrader-Frechette, K. (2001). An evaluation of the ethical principles of the ICRP's radiation protection standards for workers. *Health Physics*, *80*, 225–234.

Pontifical Academy of Sciences. (1985). *Biological implications of optimization in radiation procedures.* Working Group, 2–5 May 1983. Publications, Documenta 014, Conclusions, Vatican: Pontifical Academy of Sciences.

Silini, G. (1992). Ethical issues in radiation protection. The 1992 Sievert Lecture. *Health Physics*, *63*, 139–148.

Sowby, F. D. (1981). Radiation protection and the International Commission on Radiological Protection (ICRP). *Radiation Protection Dosimetry*, *1*, 237–240.

Taylor, L. S. (1957). The philosophy underlying radiation protection. *American Journal of Roentgenology*, *77*, 914–919.

Zölzer, F. (2012). *A cross-cultural approach to questions of ethics in radiation protection. IRPA 13 Glasgow presentation*. Retrieved 31.01.13 from: http://s281354445.websitehome.co.uk/Glasgow/Projects/IRPA2012/NewWebsite/wp-content/uploads/2012/06/Presentations/Dochart/thu/0940%20thu%20dochart%20Zolzer.ppt.

Chapter 3

Moral Thinking and Radiation Protection

Sven Ove Hansson
Royal Institute of Technology (KTH), Division of Philosophy, Stockholm, Sweden
E-mail: soh@kth.se

Chapter Outline

3.1 Introduction 33
3.2 Individual vs Collective Approaches 35
3.3 Weighing vs Limit-Setting 39
3.4 The de minimis Issue 43
3.5 Valuing Future Effects 46
3.6 Protecting the Most Sensitive People 47
3.7 Conclusion 49

3.1 INTRODUCTION

Moral thinking is part of our everyday lives, but it has also been condensed into specialized discourses. There are two major types of such discourses. One of them can be called fundamental ethics. It is usually conducted by philosophers and has a strong emphasis on the search for comprehensive basic principles for morality. Several moral theories have been put forward that purport to contain all the moral information needed to answer any and all ethical questions. Two of the most important groups of such theories are the utilitarian and the deontological ones. In utilitarian theories, it is assumed that the goodness or badness of alternative courses of action can be measured with some number, and acting rightly consists in choosing an alternative with a maximal degree of goodness. According to deontological theories, morality is based on a set of duties or obligations, and acting rightly consists in satisfying the duties that one has. Both deontological and (in particular) utilitarian theories come in many variants, and there are also several additional classes of moral theories, such as those that are based on rights and on contractual relationships. To put it somewhat bluntly, moral philosophers tend to agree that one of the many available moral theories is the one and only correct theory. However, they do not agree on which that theory is.

Radioactivity in the Environment, Volume 19
ISSN 1569-4860, http://dx.doi.org/10.1016/B978-0-08-045015-5.00003-4

The other type of ethical discourse is usually called applied ethics. It is devoted to the practical ethical issues that arise in various specialized types of human activities. Most of the areas of applied ethics refer to the activities of the members of a particular profession, such as physicians, nurses, research scientists, engineers, business managers, etc. The ethical discourses in these areas have their origin in initiatives in professional organizations. Applied ethics is usually conducted both by members of the concerned profession and by moral philosophers. In addition to the profession-related areas, there are also a few areas of applied ethics that do not have their origin in discussions within a profession. Environmental ethics and food ethics are examples of this. However, most social activities that lack a strong and unified profession also lack a well-developed ethical discourse, even if they have urgent ethical issues that need to be investigated. Traffic safety, welfare provision, and foreign aid are examples of such areas (Hansson, 2009b). Radiation protection is (hopefully) currently in a transition phase, developing a specialized ethical tradition of its own.

One might expect that applied ethics should proceed largely by applying fundamental moral theories to practical problems, in much the same way that applied mathematicians and physicians apply mathematical and physical theory to practical problems. But in practice, applied ethics is seldom performed in that way. Instead of applying all-encompassing theories like utilitarianism or deontology to solve their problems, applied ethicists tend to appeal either directly to our moral intuitions or to principles developed specifically for the subject-area in question (Hansson, 2003b). The reason for this is that in spite of their pretensions of complete coverage, fundamental moral theories have surprisingly little to say on the practical problems to be dealt with in applied ethics. This has become particularly evident in biomedical ethics. Experience shows that the fundamental theory that a moral philosopher adheres to has little or no predictive power for her standpoints in concrete issues in biomedical ethics (Heyd, 1996; Kymlicka, 1993). You can for instance find a utilitarian and a deontologist who agree on most of the ethical issues in healthcare, although they have different underpinnings for their standpoint. Similarly, two adherents of the same moral theory can disagree vehemently in practical moral issues since they apply it in different ways. The reason for this is that moral theories operate on an abstract level, and most practical moral problems cannot be connected in an unequivocal way to principles or standpoints on that level (Hansson, 2013).

But there is one major exception to this: the ethics of radiation protection. Many of the most important issues in radiation protection turn out to correspond to well-known problems in fundamental moral philosophy. Indeed, some of these problems have been discussed in parallel in both radiology and moral philosophy for many years, with few if any contacts between the two discussions (Hansson, 2007). The major reason for this connection between the two disciplines is that radiation protection refers to doses that are measured in numerical terms and added just like utilities are added in utilitarianism. (Provided, of course, that the linear nonthreshold assumption is used.) From a mathematical viewpoint, the difference between minimizing doses and maximizing the good is trivial; it is just a matter of a minus sign. Therefore, radiation protection can be used as a test lab for moral theories. We can for instance try out different principles for the distribution of goods by applying them to the distribution of radiation doses.

The rest of this chapter is devoted to five particularly important parallel issues in moral philosophy and radiation protection.

3.2 INDIVIDUAL VS COLLECTIVE APPROACHES

The first of the five contact points between moral philosophy and radiation protection is the way in which we weigh risks and benefits against each other.

A useful method to prepare ourselves for a decision is to identify and weigh the advantages and disadvantages of each of the options that are open to us. A practicable way to do this was proposed by Benjamin Franklin in 1772 in a letter to the chemist Joseph Priestley:

"When these difficult Cases occur... my Way is, to divide half a Sheet of Paper by a Line into two Columns, writing over the one Pro, *and over the other* Con. *Then during three or four Days Consideration I put down under the different Heads short Hints of the different Motives that at different Times occur to me for or against the Measure. When I have thus got them all together in one View, I endeavour to estimate their respective Weights; and where I find two, one on each side, that seem equal, I strike them both out: If I find a Reason* pro *equal to some two Reasons* con, *I strike out the three... and if after a Day or two of farther Consideration nothing new that is of Importance occurs on either side, I come to a Determination accordingly."*

(Franklin, 1970; pp. 437–438)

Franklin struck out items or group of items with equal weight. From this the step is not big to assigning a number to each item, representing its weight, and adding up these numbers in each column. This is the moral decision procedure proposed by Jeremy Bentham (1748–1832):

"Sum up all the values of all the pleasures *on the one side, and those of all the pains on the other. The balance, if it be on the side of pleasure, will give the* good *tendency of the act upon the whole, with respect to the interests of that* individual *person; if on the side of pain, the* bad *tendency of it upon the whole.*

Take an account of the number *of persons whose interests appear to be concerned; and repeat the above process with respect to each.* Sum up *the numbers... Take the* balance *which if on the side of* pleasure, *will give the general* good tendency *of the act, with respect to the total number or community of individuals concerned; if on the side of pain, the general* evil tendency, *with respect to the same community."*

(Bentham, 1780, pp. 27–28)

Bentham used the word "utility" for "that property in any object, whereby it tends to produce benefit, advantage, pleasure, good, or happiness"

(Bentham, 1780, p. 2). Therefore, moral theories based on this type of calculus are called "utilitarian". But neither Bentham nor any of his successors have been able to come up with a method to actually measure the moral values of options. Therefore, the literature on utilitarianism does not contain actual calculations of utility in real life, only hypothetical calculations in the style of "Suppose person A receives 3 units of utility and person B loses 2 units..." In this respect, radiation protection is more concrete. Radiation doses are summed up for each individual person, and then these values are in their turn summed up for the total number of "persons whose interests appear to be concerned", just as Bentham prescribed. But there are at least two major differences between dosimetry and Benthamite utility calculus. First, the latter is devoted to both positive and negative values (both of Franklin's columns) whereas the radiation protector only has negative values to record. Secondly, whereas the moral calculus is only a hypothetical exercise, dosimetry is a well-established empirical practice based on reasonably reliable dosimeters.

In the passage quoted above, Bentham proposed that we perform two procedures in order to compile information for utilitarian calculations. First, we sum up the values that pertain to each concerned individual, collecting so to say the values in one basket for each individual (See Figure 3.1). In the second procedure, we pour together the contents of all the individual baskets into one big, collective basket. This second step is an essential part of the utilitarian idea. It has the effect that an advantage or a disadvantage (such as a radiation dose) will be counted the same irrespectively of whom it affects. This was probably a major reason why Bentham proposed that we blend the contents of all the baskets. He was a strong advocate of equality. In his view, every person—nobleman or commoner, rich or poor, man or woman—should count for one and no one should count for more than anyone else (Guidi, 2008; Williford, 1975).

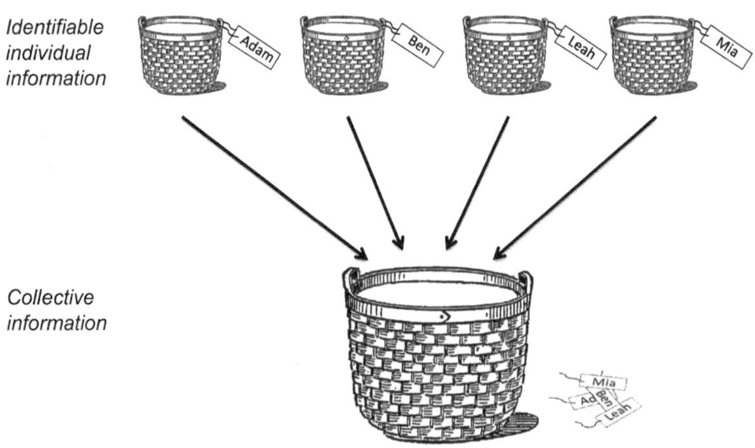

FIGURE 3.1 Bentham's method for compiling utility information.

But the pouring together of all the baskets also has another effect that is quite problematic from an egalitarian or otherwise justice-seeking point of view. In the one-basket approach, advantages and disadvantages will count the same irrespective of who receives them. Therefore, a disadvantage to one person will always be outweighed by a somewhat larger advantage to another person. This runs contrary to the idea of equality. From an egalitarian point of view, it is better to provide a disadvantaged person with a certain advantage than to grant an already advantaged person a somewhat larger advantage. The one-basket approach also effaces the distinctions that are necessary to make sense of moral categories such as compensation and desert. Inflicting an injury on you in order to gain an advantage for myself will count the same as inflicting that same injury on myself in order to gain the same advantage.

But we can avoid these drawbacks while still treating everyone equally. Instead of pouring all the individual baskets together we can keep them separated, but "anonymize" them. More precisely, the baskets should carry no information of the type that justice requires us to disregard such as whether they pertain to a man or a woman, a person from the upper or the lower classes, etc. To illustrate this, we can see Bentham's second step as actually consisting of two steps (Figure 3.2). First we remove the labels from the baskets. At that stage, we know how the contents are distributed but we do not know which

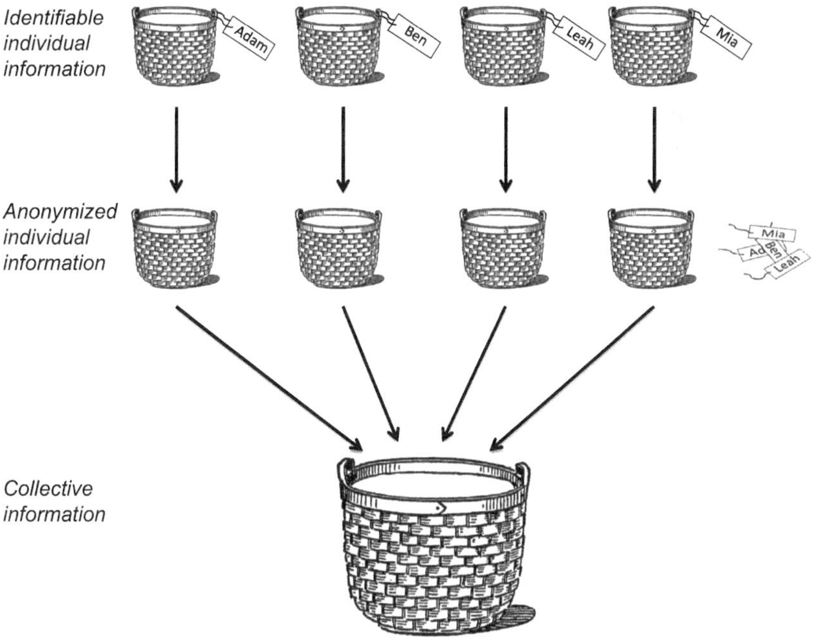

FIGURE 3.2 A more detailed account of Bentham's method, showing the possibility of an intermediate step (that may also be taken as the final step).

basket each person receives. In the second step, we pour all the baskets together, arriving at the same end result as with Bentham's method. The obvious advantage of this more detailed description is of course that it opens up the interesting option of only performing the first step, i.e. remove the labels but not pour together the baskets.

Having observed this, we can distinguish between two major ways to weigh pros and cons, or risks and benefits, against each other: individual and collective weighing. Individual weighing is concerned with the balance between advantages and disadvantages for each individual person, whereas collective weighing compares the total sum of all advantages to the total sum of all disadvantages (Hansson, 2004b). Individual weighing can be performed on either the upper or the middle level in Figure 3.2 (labeled or unlabeled individual baskets) whereas collective weighing takes place on the lower level (one collective basket).

Both of these approaches are commonly used in various social practices. Clinical medicine is perhaps the application area in which individual weighing of risks and benefits is most consistently used. In order to choose treatment recommendations for their patients, physicians weigh the expected positive treatment effects against the negative side effects. With few exceptions (infectious disease prevention being one of them) advantages pertaining to persons other than the patient do not enter the calculation. As one example of this, it is considered unethical to sedate a patient in order make him/her easier for the staff to handle; like other medical interventions sedation has to be justified with appeal to the patient's interest. Another example is that in medical research ethics as codified in the Helsinki declaration, a patient should not be offered to take part in a clinical trial if there is some treatment available that is known to be better than one of the treatments to which the patient can be randomized in the trial. This is often expressed as a requirement that there should be *clinical equipoise* between the different treatments, by which is meant the absence of any compelling reason from the viewpoint of the individual patient's interests to choose one treatment over the other. A patient should not participate in a clinical trial if that would be to her disadvantage, even if the total effect of the trial would be positive due to the expected benefits to future patients (Hansson, 2006).

However, outside of clinical medicine risk analysis is dominated by methods that employ collective risk-weighing. Disadvantages are measured in terms of the expectation value (probability-weighted value) of the number of fatalities. Values derived from different sources of risk are added to obtain a measure of the total "risk", i.e. sum of such expectation values. Suppose that a certain operation is associated with a 1% probability of an accident that will kill five persons, and also with a 2% probability of another type of accident that will kill one person. Then the total expectation value is $1\% \times 5 + 2\% \times 1 = 0.07$ deaths. In similar fashion, the expected number of deaths from a nuclear power plant is equal to the sum of the expectation values of each of the various types of accidents that can occur in the plant. One author has described this as "[t]he only meaningful way to evaluate the riskiness of a technology" (Cohen, 2003, p. 909).

One interesting example of the dominance of collective risk-weighing is the common criticism against the so-called NIMBY (not in my backyard) phenomenon. By this is meant that a person or group of persons protest against the siting in their neighborhood of a facility that will be disadvantageous to themselves but advantageous to society as a whole. Risk analysts who condemn NIMBY reactions seem to take it for granted that collective risk-weighing is justified in these cases. But the common assumption that NIMBY represents some type of irrational thinking only seems plausible if the discussion on siting of facilities refers to the big basket, not if it refers to the full information that is available if we retain information about the distribution of advantages and disadvantages (Hermansson, 2007; Luloff, Albrecht, & Bourke, 1998).

Radiation protection differs from most other areas in combining the individual and the collective methods of weighing. There is a long tradition of attending both to individual and collective doses. For individual doses, maximum allowable exposures have been specified. For collective doses, the major approach is expressed by the so-called ALARA principle for dose reduction ("as low as reasonably achievable", see Chapter 9). There is consensus in the radiation protection community that both these levels of analysis are needed, although their relative importance has been subject to debate (Wikman, 2004). This combination of two levels of analysis gives rise to a more nuanced—and consequently more complex—structure than if only one of the two levels of analysis is chosen. This may be one of the thought patterns in radiation protection that moral philosophers have something to learn from.

3.3 WEIGHING VS LIMIT-SETTING

In the case with only collective information (a single large basket), there is an obvious decision rule to apply:

The collective weighing principle (Hansson, 2003a):

An option is acceptable to the sum of all extent that the sum of all individual disadvantages that it gives rise to is outweighed by the sum of all individual advantages that it gives rise to.

If we have chosen to retain individual information, then the choice of a decision rule is less obvious. There is a simple case, namely that in which decisions can be made separately for each individual, one at a time. In such cases the following decision rule can be used:

The individualist weighing principle for a single concerned individual:

An option is acceptable to the extent that the sum of all disadvantages that it gives rise to for the concerned individual is outweighed by the sum of all advantages that it gives rise to for that same individual.

This is the rule commonly applied in clinical medicine in the choice between treatments that differ in their therapeutic and adverse effects.

These two decision rules apply to simple cases in which there is only one basket to consider, either because we only take one individual into account or because we have decided to pour together the contents of all baskets into one. The tricky problems arise when we have decided to take the separate interests of more than one individual into account, i.e. when we are on the top or middle level in Figure 3.2. We have to stay on one of these levels if we wish to account for moral considerations such as equality, justice, and individual rights. These levels also represent the types of situation that the radiation protector has to manage. Having received the dosimeter readings from all the employees of a plant, you should of course add up all these doses in order to see what the total (collective) dose was. Discussions on how to reduce that sum are self-evident parts of the established practice in radiation protection. But so is also a focused discussion on the highest individual doses and what can be done in particular to reduce them, even if such measures do not coincide with the most straight-forward and most economical ways to reduce the collective dose. The radiation protector is therefore in the same situation as the egalitarian who worries not only about the total welfare of a society (conventionally but very defectively measured as the gross national product), but also about the welfare of individual residents, in particular those who are worst-off.

The problem how best to take several individuals' interests into account has both a substantial and a procedural component. The substantial issue concerns how good or bad different outcomes are, if by an outcome we mean a state of affairs defined by what is in each individual's basket. The procedural issue concerns how the decision on such distributions should be made. Although the procedural issue is of paramount importance (see Chapters 16–19), here the discussion will be restricted to the substantial one. Both these aspects are so complex that it is often helpful to discuss them one at a time.

A common and conceptually quite simple solution is to set an individual limit and require (only) that each individual be on the right side of that limit:

The individual limit principle:

An option is acceptable if and only if each individual's situation is above a certain limit that is the same for all individuals.

In general social policies, this corresponds to the idea that each individual should be above a certain level, often called the "poverty line" or "poverty threshold". It is usually identified with the amount of resources necessary to obtain sufficient food, clothing, health care, and shelter. According to this view, once everyone is above the poverty line, the situation is acceptable, and there is no further need to worry about inequalities in income and resources. In radiation protection, this would correspond to a policy that only requires that all individual doses be below the dose limits and has no further requirements on the reduction of doses.

Both in social policies and radiation protection, such a policy can be criticized both for demanding too little and for being too uncompromising. It demands too little since it provides no stimulus to further improvements once the limit has been reached. In social policies, it makes no distinction between a society in which everyone is just above the poverty line and one in which everyone is far above that level. It would seem strange, to say the least, to be unbothered by such a difference. In radiation protection, we have a corresponding problem: The individual limit principle does not distinguish between a workplace in which every employee's exposure is just below the exposure limit and one in which everyone's exposure is a small fraction of the exposure limit. A radiation protector who does not worry about that difference could hardly be said to take her professional duties seriously.

It is the sharpness or absoluteness of the limit that makes this principle open to the criticism of being too uncompromising (and by its very nature, a limit has to be sharp in order to be unambiguously applicable). Consider a society in which everyone is well above the poverty line except very few who are just below it. Most of us would probably prefer such a society to one in which everyone is just above the line. Similarly, in radiation protection, consider a situation in which everyone's exposure is very small except a few persons whose exposure is just above the limit. We would probably prefer this to a situation in which everyone's exposure is barely below the limit—at any rate this is how radiation protectors would assess the two situations.

These examples show that the individual limit principle is too crude. It has the advantage over the collective weighing principle that it takes individual allotments seriously, but it has the serious disadvantage of not making any other distinctions than that between values above and below the limit. In radiation protection, we want to distinguish between different doses below the dose limit, and also between different doses above it. For obvious reasons, the corresponding nuances are also needed in moral philosophy and its application to social policies. In both cases, we need to combine concern for individual allotments with concern for gradations beyond that of being above or below a single limit.

Before attending to how that can be done it is worth noting that the individual limit principle is much akin to—and arguably expressible as a form of—one of the major alternatives to utilitarianism, namely deontological ethics, also called duty ethics. Deontologists such as Immanuel Kant (1724–1804) have proposed that an adequate moral theory should be based on strict moral limits that we are never allowed to transgress. A Kantian approach to radiation protection could be based on the precept that a duty-holder such as an employer is required to ensure that each individual's radiation exposure satisfies a precise criterion such as that of being below the dose limit. The same stipulation could also be expressed in the terms of a closely related type of moral theories namely *rights-based ethics*. The central postulate would then be that each individual has a right not to be exposed to doses above the limit.

Hence, in terms of moral theories, a radiation protector who only worried about collective doses would apply utilitarian thought patterns whereas one whose attention was limited to individual doses would follow deontological or rights-based thought patterns (Hansson, 2007). In moral philosophy, the general approach is to treat the different types of moral theories as mutually exclusive alternatives that one has to choose between. Moral philosophers typically identify themselves as adherents of one of these theories. On a conference in moral philosophy utilitarians will argue that deontology is a misconceived form of moral philosophy, deontologists will say the same about utilitarianism, and adherents of various other moral theories (such as virtue ethics) will claim that both utilitarianism and deontology are fundamentally flawed. In contrast, on a conference in radiation protection, we will usually not find proponents of collective dose minimization who consider individual dose limits to be useless or proponents of individual dose limits who see collective doses as irrelevant. Instead, we will find radiation protectors who try to combine the two lines of thought in various ways, although they may disagree on the relative priorities and on how the two principles are best combined. The common approach in radiation protection is to see to it that (1) the individual dose limits are upheld and (2) given that, the collective dose is minimized. This amounts to the following more general principle:

Combined individual limit and collective weighing:

An option is acceptable to the extent that (1) each individual's situation is above a certain limit that is the same for all individuals, and (2) the sum of all individual disadvantages that it gives rise to is outweighed by the sum of all individual advantages that it gives rise to.

Hence, whereas moral philosophers have discussed whether to choose a utilitarian or a deontological approach, radiation protectors have attempted to find ways to combine them. Since both thought-patterns have strong support in our moral intuitions, such a combinative line of thought may very well turn out to be the most useful and constructive one also for a wider field of applications than radiation protection. Moral philosophers may have something to learn from radiation protectors in this respect.

But we need not settle with the last-mentioned principle. There are other ways than this to combine the limitation of individual doses with that of collective doses. Another way is to modify collective dose minimization so that it gives higher weight to the reduction of high doses. For illustration, this can be done in a very simple way by tripling the part of a dose that exceeds, say, 10 mSv/year. We can call the resulting number the severity of the exposure. Hence, if the dose is 5 mSv then the severity is 5, but if the dose is 20 mSv then the severity is 40. Now consider the following two exposure patterns:

Dose pattern A: Eleven persons receive 10 mSv/y.
Dose pattern B: One person receives 50 mSv/y and ten persons receive 5 mSv/y.

According to collective dose minimization (and our collective weighing principle) dose pattern B is slightly better than dose pattern A since the collective dose is somewhat lower (100 respectively 110 mSv/y). However, according to the "tripling" criterion, B is by a wide margin worse than A since it scores higher on the severity measure (180 respectively 110). This, as far as I can see, corresponds better to how most radiation protection professionals would respond to the two scenarios.

The "tripling" function is just a very crude example of a way to combine the two criteria. Such methods need to be developed with more attention to practical implications (Wikman-Svahn, Peterson, & Hansson, 2006). This is somewhat related to the so-called extended cost-benefit analysis that has been discussed in radiation protection (IARC 1989; pp. 25–27). It is also akin to the so-called prioritarianist approach in moral philosophy, according to which the moral value of an outcome should be calculated by adding the values it has for all concerned individuals, but with extra weight given to the worse-off individuals (Parfit, 1997). But much work remains to be done in order to investigate how weighing and limit-setting principles can best be combined, in radiological protection as well as in a wider moral context.

3.4 THE DE MINIMIS ISSUE

Radiation protection standardly assumes the linear no-threshold assumption. This means that a smaller dose is assumed to give rise to a proportionately smaller risk: half the dose means half the risk, a hundredth of the dose means a hundredth of the risk, etc. It follows from this assumption that as the dose becomes diminutive, so does the risk, but it never disappears until the dose is zero. In spite of this one might very well ask whether very small radiation doses should at all be taken into account. Is there a level below which they can just be neglected?

Proponents of such a limit have often used the term "de minimis" to denote doses that are allegedly too small for serious consideration. The discussion on such doses has often been connected with ideas about a general limit below which risks are of no concern, "a lower bound on acceptable risk levels, no matter what the associated benefits", such as "a cutoff level of 10^{-6} individual lifetime risk [of death]" (Fiksel, 1985; pp. 257–258.). A common argument for this standpoint is that in general we tend to accept risks at that level without worrying much about them. But of course, if we accept some risks of a certain size then that does not commit us to accepting all risks of the same size. Concededly, there may be good reasons why we have previously accepted risks of that size, for instance that they are associated with outweighing benefits or that they are impossible to reduce. However, these reasons do not necessarily apply to the new risks that we are urged to accept. Furthermore, even if each of a large number of small risks may in itself be tolerable, the combination of all of them may add up to a large total risk that we are unwilling to tolerate (Bicevskis, 1982; Hansson, 2004a; Pearce, Russell, & Griffiths, 1981). It follows from this that even a very small risk imposition needs a justification.

If someone claims that all risks below 10^{-6} are negligible, you can easily test the sincerity of that claim by asking her whether she will let you play "seven dice Russian roulette" on her. In this game, a fair die is rolled seven times in a row. If it lands on a six all seven times, then you play Russian roulette on her with a cartridge in one of the six chambers of the revolver. If she answers no, then she has just contradicted the view that all risks below 10^{-6} are negligible (This one is about 0.6×10^{-6}). If she answers yes, then the next question is how many times she will allow the game to be played against her.

In radiation protection, the idea of such a de minimis level is seldom heard. However, another argument has sometimes been put forward that would allow for the acceptance of much higher radiation doses. It has repeatedly been claimed that if no adverse health effects from an exposure have been detected, then that exposure can be accepted. Most statements to that effect have been made by laypersons, but sometimes similar claims have been made by professed experts or by authorities with access to expertise. In 1950, Robert Stone, a radiation expert with the American military, proposed that humans be exposed experimentally to up to 150 R (a dose that can give rise to acute radiation sickness) with the motivation that "it seems unlikely that any particular person would realize that any damage had been done on him by such exposure" (Moreno, 2001, p. 145). Similarly, the influential US-based Health Physics Society wrote in 1996 in a position statement on radiological protection:

"...[E]stimate of risk should be limited to individuals receiving a dose of 5 rem in one year or a lifetime dose of 10 rem in addition to natural background. Below these doses, risk estimates should not be used; expressions of risk should only be qualitative emphasizing the inability to detect any increased health detriment (i.e., zero health effects is the most likely outcome)." (Health Physics Society 1996)

Subsequently, the Society has modified this statement, and now says that "zero health effects is a likely outcome" at exposures where no increased health detriment can be detected (Health Physics Society 2004; Health Physics Society 2010). Both formulations indicate that if a potential risk factor does not give rise to any detectable detrimental effect then that is a good reason to believe that it does not give rise to any risk of concern. In other words, it is assumed that indetectability is in itself a sufficient justification for risk impositions.

In order to evaluate that argument we need to ask the question: How large detrimental effects can go undetected even if competent epidemiological studies are being performed on exposed populations? It turns out that for purely statistical reasons, surprisingly large effects can escape detection. Suppose for instance that a certain exposure increases the lifetime incidence of lung cancer among those exposed from 10.0 to 10.5%. Or suppose that it increases the total lifetime cancer mortality in a population from 25 to 26%, evenly distributed over the different forms of cancer. In both cases chances are small that epidemiological studies would lead to discovery of the increase, since it would probably be indistinguishable from random variations. (As a rough rule of thumb, epidemiological studies cannot reliably detect excess relative risks that are about 10% or smaller. Hansson, 1995, 1997, 2002; Vainio & Tomatis, 1985.)

This means that risks can go undetected that would be considered to be significant public health problems if they were discovered. Presumably the ethical problem is that humans *die* due to preventable exposures, not that it is *known that they die* due to such exposures. Therefore, these examples show that the absence of detected effects from a radiation exposure (or from any other potentially harmful exposure) does not give us sufficient reason to believe that there is no such effect. (There may of course be other reasons to believe that low exposures have no effect. This is the case for some but not all chemical carcinogens. It is not the case for ionizing radiation)

The question of indetectable effects has also been the subject of a parallel discussion in moral philosophy. Parts of the early discussion made use of the "lawn-crossing example" (Harrison, 1953, p. 107; Österberg, 1989). Suppose that there is a lawn in your way between home and work. Each time you approach it, you can choose between crossing it and walking around it. Each time you cross it, you make a perceptible time gain. No single crossing makes a (perceptible) difference in the condition of the lawn. However, if you cross it every time you walk this route then it will be seriously damaged. Now let us assume that you are anxious to have the lawn retained in its original shape. Indeed, you put higher value on this than on all the small time gains, taken together, that you can make by crossing the lawn. This would seem to put you in the following seemingly paradoxical situation:

You prefer crossing the lawn once to not crossing it at all, since that involves a noticeable time gain but no noticeable loss in the condition of the lawn. For the same reason your prefer crossing it twice to crossing it once. Similarly, you prefer crossing it three times to crossing it twice, crossing it four times to crossing it three times,... and indeed crossing it a thousand times to crossing it nine hundred and ninety-nine times. But yet you do not prefer crossing it a thousand times to not crossing it at all, because of the conspicuous difference in the condition of the lawn between these two alternatives.

A more drastic example with essentially the same structure has attracted considerable attention among moral philosophers since it was proposed by Warren S. Quinn (1990). It is usually called "the self-torturer": A physiological device has been in implanted in a person's body. The device has 1001 settings, from 0 (off) to 1000. To begin with it is set at 0. Each week, the self-torturer has two options. He may leave the device as it is, or he can advance its dial one setting. He can only advance it one step per week, and he can never revert to a lower setting. At each advance, he gets a payment of $10,000. The snag is that the device is connected to his sense of pain. As the dial is moved from 0 to 1000, his physiological state progresses from no pain to unbearable pain. However, each of these steps is imperceptibly small. Therefore, each move on the dial gives him a nice sum of money and no perceptible disadvantage. But when he has gone all the way from 0 to 1000 he will probably regret that he did it. The situation is analogous to that of the lawn-crosser, only more tragic.

In summary, we have two parallel discussions referring to the same problem structure: a discussion about indetectable health effects of radioactive (and other) exposures and a discussion about imperceptibly increasing disadvantages in moral philosophy (Hansson, 1999; Shrader-Frechette, 1987, 1988). In both cases, a sensible solution will have to take the potential contributory effects of our actions into account even when their contribution cannot be discovered in each single step (Hansson, 1993, 2010, pp. 591–592). This is yet another example where contacts between a practical discussion in radiation protection and a more theoretical discussion in moral philosophy can be mutually beneficial.

3.5 VALUING FUTURE EFFECTS

Sometimes when summarizing advantages and disadvantages of an option, we find that they materialize at different points in time. For the smoker, the most important positive effect of smoking is immediate: she avoids the nicotine withdrawal syndrome. The most important negative effect is the risk of serious disease that will typically materialize decades later. (About half of the smokers die prematurely due to smoking Boyle, 1997.) In climate and environmental policies, we are often concerned with measures that cost money today but have their positive effects much later. Nuclear waste management provides what is perhaps the most extreme example of such temporal discrepancies: on the one hand energy is produced to be consumed now, and on the other hand the potential damages from nuclear waste may materialize hundreds of thousands of years ahead.

The standard method for evaluating future outcomes is discounting, a method that was originally developed for money. It is based on the assumption of a positive interest rate. For example, suppose that the interest rate in a bank is constantly 3%, and furthermore suppose that we want to have €100.000 in 10 years. Then it is sufficient to deposit €74.400 in the bank. We can therefore say that the "present value" of receiving €100.000 ten years from now is €74.400. With a similar argument, a loss €100.000 ten years from now corresponds to a loss of €74.400 today. More generally, we can "convert" the value of future money into money now using the following formula:

$$v_0(x) = v_t(x) \times 1/(1+r)^t,$$

where x is the object whose value we are converting, $v_0(x)$ its value now, $v_t(x)$ its value after t years, and r the interest rate (in the example: 0.03). In cost-benefit analysis, this formula is used as a standard. Suppose for instance that we discuss measures that would prevent an accident fifteen years into the future in which 31 persons would die. With a 3% interest rate, the formula tells us to value the loss of 31 lives in fifteen years the same way that we would value a loss of 20 lives today (since $31 \times 1/1{,}03^{15} \approx 20$).

A major problem with this approach is that it yields absurd results if we consider very long time periods. Consider, as a simple schematic example, a hypothetical choice between the following two actions:

1. Killing one person now.
2. Now performing an action that will lead to the death of the whole population of the earth, 10 billion people, in the year 2800.

If we apply discounting, and use a discount rate of 3%, then the first of these actions will be worse than the second. The example is unrealistic, but it illustrates that even very large disasters will have almost zero (dis)value if they take place a couple of hundred years ahead from now. (Lowering the discount rate only delays this effect. With a discount rate of 0.5% it will still be worse that one person dies today than that 10 billion people die in 4620 years.) If we applied discounting to radioactive waste management, then we could in practice disregard what happens after the first thousand years or so.

But interestingly enough, discounting has seldom been applied to nuclear waste. Effects far off into the future are treated as equally serious as if they were to take place today (and it is effects in the distant future that are most difficult to prevent). On the other hand, in economic planning for waste depositories, monetary costs are discounted in the usual way. This might appear inconsistent, but in fact it is not. There are strong reasons to discount money, assuming that we will continue to have a monetary economy with positive interest rates. But this argument does not extend for instance to losses in human life or spoliation of the environment. Money can be replaced by other money, but human lives cannot be replaced by other human lives, and neither can species be replaced by other species. And we can deposit money in a bank and hopefully see it grow, but we cannot deposit lives or species. These are strong arguments in favor of restricting discounting to money and that which can be replaced by money, in other words in favor of the approach that is commonly applied in nuclear waste management.

In most other areas, discounting is applied to all values, including values referring to lives, health, and the environment. It is for instance standardly applied in economic analysis of climate change. We learn from nuclear waste management that it is possible to discount money and whatever is monetizable, without discounting nonmonetizable effects. This is yet another case in which ways of thinking from radiation protection may be generalizable.

3.6 PROTECTING THE MOST SENSITIVE PEOPLE

The ICRP has provided a comprehensive summary of the scientific information about differences in sensitivity to the harmful effects of ionizing radiation. They concluded that at any given level of exposure, the cancer risk is about 39% higher for women than for men (ICRP, 2007, p. 210). Furthermore, young children are a "particularly sensitive subgroup" with a risk that may be as high as about three times that of the population as a whole. There are also small minorities of the population (well below 1%) that have very high sensitivity to radiation due to genetic factors, usually a defect in DNA repair genes.

In spite of this information, the ICRP has chosen to adjust the level of protection to the average exposed individual. This means for instance that the occupational exposure limits are calculated with reference to an average worker. It can however be questioned whether this is satisfactory from the viewpoint of the more sensitive subpopulations. If I am exposed to a dose that gives rise to a certain risk, can that exposure be defended by pointing out that the risk from that dose would be smaller for an average person than it is for me? (Hansson, 2009a).

This discussion has an interesting parallel in the ethical discussion on what is the proper distribuendum of justice, i.e. exactly what it is that should be fairly distributed. This has often been called the discussion of "equality of what" since different answers to the question give rise to different strands of egalitarianism. Perhaps the most obvious answer is that the proper distribuendum consists of the redistributable resources that we can use to live our lives. This is essentially the answer given by John Rawls in his *A Theory of Justice* (1971). He used the term "primary social goods" to denote those redistributable goods that almost everyone values. Money, rights, and power are primary social goods. Neither health nor intelligence are primary social goods since they are beyond social control and cannot be redistributed. Bungy jumps are not primary social goods since they are not valued by everyone. According to Rawls, it is the primary social goods that we should attempt to distribute fairly.

Rawls has been criticized for being insensitive to people with special needs. Economist Kenneth Arrow has expressed this criticism as follows:

> *"[C]onsider the haemophiliac who needs about $4000 worth per annum of coagulant therapy to arrive at a state of security from bleeding at all comparable to that of the normal person. Does equal income mean equality?"*
>
> (Arrow, 1973, p. 254)

Presumably, the reason why we care about the distribution of resources is that they to a large extent determine a person's quality of life. If the relation between resources (in the form of primary social goods) and quality of life were constant and well-determined, then it would not make any difference which of them we take to be the proper distribuendum. However, that relation differs between persons (Sen, 1982, p. 353), and we must then ask: which is the ultimate good that should be the distribuendum? Is it the resources or is it the quality of life? In the last two or three decades, this discussion has developed considerably, and a ramifying set of sophisticated answers to the question is available (Matravers, 2002).

The corresponding question in radiation protection is: which is the ultimate evil that should be the evitandum (that which should be avoided)? Is it the dose of ionizing radiation or is it the increased risk of serious disease and perhaps death? The parallel with the "currency of justice" issue in moral philosophy is obvious—of course with the usual exchange between maximizing desirables and minimizing undesirables.

3.7 CONCLUSION

In summary, radiation protection and moral philosophy deal to a large extent with parallel issues such as:

- The choice of a suitable aggregation level for moral assessment: Is it the individual allotments of goods (doses) that should be morally assessed, or their total sum, or perhaps both?
- The choice of a method to give priority to the least advantaged persons: By setting inviolable limits or by giving extra weights to improvements from disadvantaged positions?
- Indetectable effects: How should we morally assess actions (doses) that have no detectable effect but nevertheless contribute to significant effects when combined with other actions (doses) of the same type?
- Future effects: How do we value goods (doses) that will materialize in the future, perhaps the distant future?
- The choice of a distribuendum: What is it that should be distributed in a fair way: doses or risks, social resources or actual welfare?

In spite of all these similarities, contacts have been few between the two disciplines. I hope to have shown that such contacts should increase and that the two areas have much to learn from each other.

REFERENCES

Arrow, K. (1973). Some ordinalist–utilitarian notes on Rawls's theory of justice. *Journal of Philosophy*, *70*, 245–263.

Bentham, J. (1780). *An introduction to the principles of morals and legislation*. London: T. Payne. http://gallica.bnf.fr/ark:/12148/bpt6k93974k/f2.image.r=.langEN.

Bicevskis, A. (1982). Unacceptability of acceptable risk. *Search*, *13*(1–2), 31–34.

Boyle, P. (1997). Cancer, cigarette smoking and premature death in Europe: a review including the Recommendations of European Cancer Experts Consensus Meeting, Helsinki, October 1996. *Lung Cancer*, *17*(1), 1–60.

Cohen, B. L. (2003). Probabilistic risk analysis for a high-level radioactive waste repository. *Risk Analysis*, *23*, 909–915.

Fiksel, J. (1985). Toward a de minimis policy in risk regulation. *Risk Analysis*, *5*, 257–259.

Franklin, B. (1970). Albert Henry Smyth (Ed.), *The writings of Benjamin Franklin* (Vol. 5, pp. 1767–1772). New York: Haskell House.

Guidi, M. E. L. (2008). Everybody to count for one, nobody for more than one. *Revue D'études Benthamiennes*, Vol. 4. http://etudes-benthamiennes.revues.org/182.

Hansson, S. O. (1993). Money-pumps, self-torturers and the demons of real life. *Australasian Journal of Philosophy*, *71*, 476–485.

Hansson, S. O. (1995). The detection level. *Regulatory Toxicology and Pharmacology*, *22*, 103–109.

Hansson, S. O. (1997). Can we reverse the burden of proof? *Toxicology Letters*, *90*, 223–228.
Hansson, S. O. (1999). The moral significance of indetectable effects. *Risk*, *10*, 101–108.
Hansson, S. O. (2002). Replacing the no effect level (NOEL) with bounded effect levels (OBEL and LEBEL). *Statistics in Medicine*, *21*, 3071–3078.
Hansson, S. O. (2003a). Ethical criteria of risk acceptance. *Erkenntnis*, *59*, 291–309.
Hansson, S. O. (2003b). Applying philosophy. *Theoria*, *69*(1–2), 1–3. 2003.
Hansson, S. O. (2004a). Fallacies of risk. *Journal of Risk Research*, *7*, 353–360. 2004.
Hansson, S. O. (2004b). Weighing risks and benefits. *Topoi*, *23*, 145–152.
Hansson, S. O. (2006). Uncertainty and the ethics of clinical trials. *Theoretical Medicine and Bioethics*, *27*, 149–167.
Hansson, S. O. (2007). Ethics and radiation protection. *Journal of Radiological Protection*, *27*, 147–156.
Hansson, S. O. (2009a). Should we protect the most sensitive people? *Journal of Radiological Protection*, *29*, 211–218.
Hansson, S. O. (2009b). Ethics beyond application. In T. Takala, P. Herissone-Kelly & S. Holm (Eds.), *Cutting through the surface: Philosophical approaches to bioethics* (pp. 19–28). Amsterdam and New York: Rodopi.
Hansson, S. O. (2010). The harmful influence of decision theory on ethics. *Ethical Theory and Moral Practice*, *13*, 585–593.
Hansson, S. O. The moral Oracle's test. *Ethical Theory and Moral Practice*. (in press).
Harrison, J. (1953). Utilitarianism, universalisation, and our duty to be just. *Proceedings of the Aristotelian Society*, *53*, 105–134.
Health Physics Society. (1996). Radiation risk in perspective. *Position Statement of the Health Physics Society*. Adopted January 1996. Downloaded in December 1998 from: http://www2.org/hps/rad.htm.
Health Physics Society. (2004). Radiation risk in Perspective. *Position Statement of the Health Physics Society*. Revised August 2004. Downloaded in March 2007 from: http://www.hps.org.
Health Physics Society. (2010). Radiation risk in Perspective. *Position Statement of the Health Physics Society*. Revised July 2010. Downloaded in May 2012 from: http://www.hps.org.
Hermansson, H. (2007). The ethics of NIMBY conflicts. *Ethical Theory and Moral Practice*, *10*, 23–34.
Heyd, D. (1996). Experimenting with embryos: can philosophy help? *Bioethics*, *10*, 292–309.
International Commission on Radiological Protection. (1989). Optimisation and decision-making in radiological protection: ICRP publication No. 55. *Annals of the ICRP*, *20*(1), 1–69.
International Commission on Radiological Protection. (2007). The 2007 recommendations of the International Commission on Radiological Protection: ICRP publication No. 103. *Annals of the ICRP*, *37*(2–4), 1–332.
Kymlicka, W. (1993). Moral philosophy and public policy: the case of the new reproductive technologies. *Bioethics*, *7*, 1–26.
Luloff, A. E., Albrecht, S. L., & Bourke, L. (1998). NIMBY and the hazardous and toxic waste siting dilemma: the need for concept clarification. *Society and Natural Resources*, *11*, 81–89.
Matravers, M. (2002). Responsibilty, luck, and the 'equality of what?' debate. *Political Studies*, *50*, 558–572.
Moreno, J. D. (2001). *Undue risk. Secret state experiments on humans*. New York: Routledge.
Österberg, J. (1989). One more turn on the lawn. In Sten Lindström & Wlodek Rabinowicz (Eds.), *So many words. Philosophical essays dedicated to Sven Danielsson on the occasion of his fiftieth birthday* (pp. 125–133). Uppsala: Uppsala University, Department of Philosophy.
Parfit, D. (1997). Equality and priority. *Ratio*, *10*, 202–221.

Pearce, D. W., Russell, S., & Griffiths, R. F. (1981). Risk assessment: use and misuse. *Proceedings of the Royal Society of London, Series A: Mathematical, Physical and Engineering Sciences, 376*(1764), 181–192.

Quinn, W. S. (1990). The puzzle of the self-torturer. *Philosophical Studies, 59*, 79–90.

Rawls, J. (1971). *A theory of justice*. Cambridge, Massachusetts: Harvard University.

Sen, A. (1982). Equality of what. In A. Sen, *Choice, welfare and measurement* (pp. 353–369). Oxford: Blackwell.

Shrader-Frechette, K. (1987). Parfit and mistakes in moral mathematics. *Ethics, 98*, 50–60.

Shrader-Frechette, K. (1988). Parfit, risk assessment and imperceptible effects. *Public Affairs Quarterly, 2*, 75–96.

Vainio, H., & Tomatis, L. (1985). Exposure to carcinogens: scientific and regulatory aspects. *Annals of the American Conference of Governmental Industrial Hygienists, 12*, 135–143.

Wikman, P. (2004). Trivial risks and the new radiation protection system. *Journal of Radiological Protection, 24*, 3–11.

Wikman-Svahn, P., Peterson, M., & Hansson, S. O. (2006). Principles of protection: a formal approach for evaluating dose distributions. *Journal of Radiological Protection, 26*, 69–84.

Williford, M. (1975). Bentham on the rights of women. *Journal of the History of Ideas, 36*, 167–176.

Chapter 4

A Cross-Cultural Approach to Radiation Ethics[a]

Friedo Zölzer
Department of Radiology, Toxicology and Civil Protection, University of South Bohemia, České Budějovice, Czech Republic
E-mail: zoelzer@zsf.jcu.cz

Chapter Outline

4.1 Background 53
4.2 Ethics of Radiation Protection in a Globalizing World 55
4.3 "Principles of Biomedical Ethics" as an Example 56
4.4 Possible Sources of Cross-Cultural Ethics 57
4.5 The Need for Cross-cultural Discourse 58
4.6 The Relevance of the "Four Principles" for Cross-Cultural Radiation Protection Ethics 59
4.6.1 Respect for Autonomy 59
4.6.2 Non-Maleficence and Beneficence 60
4.6.3 Justice 61
4.7 Further Cross-Culturally Accepted Principles with Relevance for Radiation Protection 62
4.7.1 Concern for the Underprivileged 63
4.7.5 Intergenerational Equity 63
4.7.6 Precaution 64
4.8 Conclusion 66

4.1 BACKGROUND

The recommendations of the International Commission on Radiological Protection (ICRP, Clarke & Valentin, 2009; ICRP, 2007a) obviously presuppose

a. This article is based on a presentation at the Symposium on Ethics of Environmental Health in Prague, September 2011.

Radioactivity in the Environment, Volume 19
ISSN 1569-4860, http://dx.doi.org/10.1016/B978-0-08-045015-5.00004-6

certain elements of moral philosophy, but these are not always made explicit. Individual authors, among them members of the commission itself (Clarke, 2003; González, 2011; Streffer, Witt, Gethmann, Heinloth, and Rumpff, 2005), have identified arguments from utilitarian and deontological, sometimes other kinds of ethics.

Thus, for instance, the principle of justification ("Any decision that alters the radiation exposure situation should do more good than harm") calls for a weighing of positive and negative consequences of radiation exposures and is therefore often thought to be based on utilitarian thinking. However, alternative interpretations have been offered, ranging from Aristotelianism (Hansson, 2007) to Machiavellianism (González, 2011). Without going into any detail, this fact alone shows that the moral basis of the first principle is not unambiguous—to say the least.

Somewhat less controversial is the assignment of the principle of optimization ("The likelihood of exposure, the number of people exposed, and the magnitude of their individual doses shall be kept as low as reasonably achievable, taking into account economic and societal factors"). As it aims to minimize risk, while leaving room for other aspects of human well-being, it is generally considered to be utilitarian in nature. Whereas the justification principle is only looking for a net positive outcome, this second principle is to ensure the widest possible margin between cost and benefit. ICRP has even explicitly recommended cost-benefit analysis as a tool for optimization (ICRP, 1973).

By contrast, the principle of dose limitation ("The total dose to any individual from regulated sources in planned exposure situations…should not exceed the limits specified") stems from the consideration that doing good to some people cannot justify doing harm to others. It is not acceptable, for instance, to expose one individual to a relatively high risk in order to save many from a relatively low one, even if this would lead to a reduction of the collective risk. In the third principle, we therefore see a deontological argument at work, where the emphasis is on the rights of individuals rather than on overall usefulness.

The problem with all this is that in moral philosophy, utilitarian, and deontological theories are considered to be mutually exclusive, because they have different priorities. For the utilitarian, all that counts is the "greatest happiness for the greatest number" (Bentham, 1744), whereas the deontologist will insist that you should "treat humanity, whether in your own person or in the person of any other, never merely as a means to an end" (Kant, 1785). It is not clear how a combination of these two is supposed to work, as we can easily think of situations where one would be completely incompatible with the other. The current philosophical foundation of radiation protection is therefore rather problematic (For further discussion, see Clarke, 2003; Gardiner, 2008; Hansson, 2007; Persson, & Shader-Frechette, 2001; Shrader-Frechette & Persson, 1997).

How this heterogeneous system of principles developed historically is an interesting question as well, but will not be pursued here (Clarke & Valentin, 2009). Suffice it to say that ICRP proposed the ALARA principle ("as low as reasonably achievable") as early as 1958 in its Publication 1 (ICRP, 1959)[b]. The other two principles made their first appearance in Publication 26, which came out in 1977 (ICRP, 1977). Neither at that point nor later, however, has there been much of a critical reflection on the principles' philosophical background, and on what it might mean to combine elements of opposing ethical approaches.

4.2 ETHICS OF RADIATION PROTECTION IN A GLOBALIZING WORLD

My question here goes beyond the incompatibility between utilitarian and deontological arguments. I should like to ask if it is at all appropriate in a globalizing world to base the recommendations of an international advisory body such as ICRP mainly on ethical theories developed in Europe during the era of enlightenment. Less than 30% of the world's population is living in Europe and the Americas, but over 50% in Asia and another 20% in Africa and the Middle East. Can we really expect the majority of mankind to adopt principles of radiation protection developed in a context largely alien to them?

It is true that population numbers do not reflect the relative use of radioactive materials or radiation around the globe, but this situation is gradually changing. According to the World Nuclear Association, there are currently 434 nuclear power reactors in operation, only 115 of them, or 26%, in Asia, Africa, and the Middle East (http://www.world-nuclear.org/info/reactors.html). However, of the 64 reactors worldwide under construction and the 160 reactors planned, 43 and 102, or 67% and 64%, respectively, will be operating outside Europe and the Americas. As for medical radiology equipment, data published by the World Health Organization suggest a similar trend. Computer tomography, for instance, is currently almost exclusively a matter of high income countries, with typically 10–30 units per million population, whereas in low-income countries this figure is 0–2 units per million population (http://www.who.int/gho/publications/world_health_statistics/EN_WHS2012_Full.pdf). Again, in spite of their economical difficulties, many countries in Asia, Africa, and the Middle East are on their way to catching up.

So, is it appropriate to carry on as we have done for the last 50 years? Currently, nobody seems to be complaining. But this may well change and our system of radiation protection may be considered unacceptable in parts of the world. What would we do in such a situation? Would we give up the idea of international recommendations for radiation protection and accept a situation where different principles were applied in different places around the globe? What if there was disagreement on issues that would affect people beyond the local or regional context?

I think that both the geographical distribution of mankind and the changes in the use of radioactive materials and radiation to be expected over the coming

---

b. In 1958, the exact wording was "as low as practicable" (Publication 1, ICRP, 1959), later, in 1966,"as low as readily achievable, economic and social consequences being taken into account" (Publication 9, ICRP, 1966)

years suggest that our system of radiation protection should not be based exclusively on certain theories of "Western" ethics. It seems to me that the discourse needs to be raised to another level. Instead of proceeding from concepts developed in one particular cultural context, we should acquire a truly global perspective, developing arguments that are acceptable to people of different backgrounds. My proposal is therefore to work toward "cross-cultural" ethics of radiation protection (For examples of similar approaches in other areas, see Veatch, 2000). This of course immediately leads to another question: Do different cultures have fundamentally different approaches to moral issues, or is there something like a "common morality" that we can draw upon? (For a sceptical viewpoint on this question, see Turner, 2003).

4.3 "PRINCIPLES OF BIOMEDICAL ETHICS" AS AN EXAMPLE

One of the most widely used frameworks of biomedical ethics is the one developed by Beauchamp and Childress (Beauchamp & Childress, 1979). It is based on four principles:

1. Respect for autonomy,
2. Nonmaleficence,
3. Beneficence, and
4. Justice.

and these are assumed to be rooted in a "common morality", which is "not relative to cultures or individuals, because it transcends both" (Beauchamp & Childress, 2009d).

Initially, Beauchamp and Childress were not speaking about different cultures. They were just trying to find middle-level principles that the former as a utilitarian and the latter as a deontologist could agree to without referring to one single, more fundamental principle, such as usefulness or individual rights. It is not that the utilitarian and the deontologist each contributed one or more principles which the other had to accept in exchange of getting some of his own ideas through. Rather both could fully agree with all four principles, albeit for different reasons (Beauchamp & Childress, 2009f).

The four principles have prima facie validity, which means that they apply as long as there is no conflict between them. If there is, they need "balancing", i.e. their relative importance has to be weighed in each case at hand. The principles also need "specification", i.e. concrete rules or guidelines have to be derived for different areas of application, which can to a certain extent vary between cultural contexts. How exactly to "balance" and to "specify" is the matter of long discussions in Beauchamp and Childress' book (Beauchamp and Childress, 2009e).

My main proposition in the following is that the approach of Beauchamp and Childress could become a model for the ethics of radiation protection, in that we could try to identify relevant principles in the "common morality", and thus avoid the "Western" bias, which has prevailed so far. However, my approach differs from the one proposed by Beauchamp and Childress in two aspects, namely on

- how we find the underlying principles of the "common morality", and
- how we "balance" the principles and "specify" them in different contexts.

4.4 POSSIBLE SOURCES OF CROSS-CULTURAL ETHICS

Beauchamp and Childress are not really interested in the question of where and how the "common morality" can be found. When they introduced the term, they just claimed that "all morally serious persons" (Beauchamp & Childress, 1994), or in the current version "all persons committed to morality", would agree with their four principles. I do not find this convincing. More effort is needed to show that these principles have cross-cultural validity—or to look for other, more widely acceptable ones.

The possibility exists, of course, that we would use empirical research to test the assumption that we have got the underlying principles right (Beauchamp & Childress, 2009a), but I am not convinced that anthropological or cultural studies alone would be meaningful. A universal "opinion poll" that would find out what people around the globe are thinking about the pertinent questions would just reflect current dispositions and would be very much subject to fluctuations. We have to look for something with greater long-term validity.

Orientation has been provided throughout the ages by the religious and philosophical traditions of the different cultures. Although our "Western" society is largely secularized, and fundamentalism, fanaticism, and extremism have brought religion into discredit, we cannot ignore the fact that these traditions continue to be of great influence for people not versed in "Western" secular philosophy. And even in the "West", the importance of Christianity is probably still much greater than the number of people attending Sunday church service would suggest. The views of Europeans and Americans have been shaped at least as much by Christian values passed on from generation to generation for centuries, as by the philosophical traditions of the enlightenment era. An analysis of "common morality" can therefore not pretend that religion has no role to play in the twenty-first century.

My suggestion then is that the most important documents for the construction of a "common morality" are the sacred scriptures of the world's great religions, such as the Vedas and the Bhagavadgita for the Hindus, the Sermons of the Buddha for the Buddhists, the Torah for the Jews, the Gospels for the Christians, the Quran for the Muslims, the Writings of Bahá'u'lláh for the Bahá'ís, and so on. They provide a framework of orientation for the believers (even though there may be some disagreement regarding their exact meaning), because they are considered to be divinely inspired. A nonbeliever will of course have some difficulty with this notion, but can instead perhaps take such writings as crystallizations of hundreds or thousands of years of human experience. Another category of useful documents for our purpose are those produced by way of intra- and interreligious dialogues, because they already reflect a certain cross-cultural agreement. A prominent example would be the "Declaration Toward a Global Ethic" signed at the Parliament of the World's Religions in 1993 (Küng & Kuschel, 1993a).

There are also relevant cultural expressions outside the context of (organized) religion. Thus, we should not ignore oral traditions in the form of proverbs, stories, legends, and myths, especially those of indigenous peoples who have no written records. We should also take into consideration secular texts of various kinds that have had a formative influence over the centuries. The Hippocratic Oath comes to mind—or the works of certain philosophers of ancient Greece and China (even if Confucius' writings are perhaps more appropriately classified as sacred scripture). In addition to these time-honored traditions, some modern documents like the "Universal Declaration of Human Rights" (http://www.un.org/en/documents/udhr/index.shtml) adopted by the United Nations General Assembly in 1948, or the "Universal Declaration on Bioethics and Human Rights" adopted by the UNESCO General Conference in 2005 (http://unesdoc.unesco.org/images/0014/001461/146180e.pdf) have been suggested to already constitute "common heritage of humankind" (ten Have and Gordijn, 2013, Chapter 63).

4.5 THE NEED FOR CROSS-CULTURAL DISCOURSE

Having established certain commonalities across cultural boundaries, however, does not mean the end of our task. A mere collection of principles cannot yet be considered cross-cultural ethics. We need to develop what we have identified as "global ethic" into a coherent system, where in particular it is clear how the principles are to be "balanced" with each other and how they are "specified" in different areas of application. Beauchamp and Childress, as mentioned above, have gone a long way to resolving such questions. In their work, they rely much on the concept of a "reflective equilibrium" (Rawls, 1971; Rawls, 1999): one starts with certain firmly held ethical beliefs, proposes some principles that seem to build up on those beliefs, applies these to new questions, looks for consistency between the original beliefs and the new applications, and modifies either component of the system until everything fits together without contradiction (Beauchamp & Childress, 2009b).

This is well and good, but I think it underestimates the difficulties of cross-cultural understanding. If, as Ludwig Wittgenstein maintains, "the limits of my language mean the limits of my world" (Wittgenstein, 1921), we cannot hope to develop a cross-culturally acceptable set of principles and work with it if we talk only to people from our own "world". If we really want to know what the meaning of those principles in other cultures is, and how they should be "balanced" and "specified", we have to actually talk to each other across cultural borders. The ethic contained in our principles can be developed into real cross-cultural ethics only through a global discourse[c]. In this regard, I even have some

c. The doyen of discourse ethics, Jürgen Habermas, who himself claims to be "religiously unmusical", has nevertheless recently argued that "the liberal state has an interest of its own in unleashing religious voices in the political public sphere, for it cannot know whether secular society would not otherwise cut itself off from key resources for the creation of meaning and identity." Habermas, 2008

sympathy for opinion polls, only I think they should be accompanying the development of cross-cultural ethics rather than driving it.

In the remainder of this paper, I will try to indicate how the established principles of radiation protection are compatible with, or even follow from those proposed by Beauchamp and Childress and can be anchored more deeply in the above-mentioned written and oral traditions of mankind. I will then mention some problematic areas of radiation protection where I see further potential for a cross-cultural approach to provide support for certain positions and against others.

4.6 THE RELEVANCE OF THE "FOUR PRINCIPLES" FOR CROSS-CULTURAL RADIATION PROTECTION ETHICS

4.6.1 Respect for Autonomy

The first of the four principles is perhaps the most problematic. It has been criticized as being "more or less ethno-ethics of American society" (Fox, 1990; Matsuoka, 2007), but of little relevance elsewhere in the world. In particular, some authors claim that people of Asian background would generally not agree with it, or at least define it differently from Beauchamp and Childress (Fagan, 2004; Fan, 1997). In "Principles of biomedical ethics" the role of this principle is to ensure that the patient is the main decision maker in his or her own case. An important corollary therefore is the concept of "informed consent", which means that neither therapy nor research can be carried out without the agreement of a competent patient. This understanding of "autonomy" is certainly common in what we call the "West", but not so much in other parts of the world. There is at least anecdotal evidence that in Latin America, in Muslim Countries, in Africa, in China, and in South East Asia, decision making is not primarily a matter of the individual patient, but very much a matter of the patient's family (Justo & Villarreal, 2003). And it does not appear as if that would have to be considered just a current phenomenon, whereas the written and oral traditions would actually place emphasis on autonomy as it is understood in the "West". So the question is indeed whether autonomy in the individualistic sense of Beauchamp and Childress is justifiable as a cross-cultural principle.

It may therefore be reasonable for our discussion to rather look at a closely related concept, namely "respect for human dignity" (On the relationship between the two concepts, see Andorno, 2009; van Brussel, 2012). It is expressed in different ways around the world, but as a fundamental principle it is virtually ubiquitous. In the Bhagavadgita, Krishna states, "I am the same to all beings…In a Brahma…and an outcast, the wise see the same thing" (The Bhagavadgita, 1973, 9:29 and 32). Similar statements are reported of the Buddha and Confucius (Lepard, 2005). In the Bible, the prophet Malachi asks, "Do we not have one father? Has not one God created us?" (Bible, 2012, Malachi 2:10) The same idea is expressed in the Quranic verse, "We have conferred dignity on the children of Adam…and favoured them far above most of Our creation" (Quran, 1980, 17:70). And a much-cited passage from Bahá'u'lláh's writings reads: "Ye are all the leaves of one tree and the drops of one ocean" (Tablets of Bahá'u'lláh, 1994, Bishárát 37).

These are just short glimpses from different religious sources, but the broad agreement on the notion that all human beings share the same dignity is also reflected in the "Declaration Toward a Global Ethic" of the Parliament of World's Religions in 1993. It says that "every human being without distinction of age, sex, race, skin colour, physical or mental ability, language, religion, political view, or national or social origin possesses an inalienable and untouchable dignity, and everyone, the individual as well as the state, is therefore obliged to honor this dignity and protect it" (Küng & Kuschel, 1993b).

Human dignity has also for centuries been invoked by secular philosophers. This strand of thought begins with Stoicism, continues through the Renaissance, and leads up to Enlightenment (Kretzmer & Klein, 2002). In our time, together with the above-mentioned religious traditions, it has played a very prominent role in the drawing up of the "Universal Declaration of Human Rights" (http://www.un.org/en/documents/udhr/index.shtml) of 1948 and the "Universal Declaration of Bioethics and Human Rights" of 2005 (http://unesdoc.unesco.org/images/0014/001461/146180e.pdf). Incidentally, the latter document mentions autonomy as well, but it accords the first place to human dignity.

There is no space here to discuss in detail what importance "respect for human dignity" could have with regard to radiation protection. I would like to suggest, however, that the third principle of radiation protection, the principle of dose limitation, partly rests on this foundation. To expose somebody to high doses in order to spare others is clearly using him or her merely as a means to an end, and that is—with or without explicit reference to deontological ethics—not in compliance with human dignity.

4.6.2 Non-Maleficence and Beneficence

"To abstain from doing harm" is one of the central features of the Hippocratic Oath (Edelstein, 1943), which was later adopted by Jewish, Christian, and Muslim physicians (Pelligrino, 2008). The principle is also mentioned, albeit indirectly, in similar texts from ancient China (Tsai, 1999, 2005). Of course, it has always been understood that sometimes pain has to be inflicted to achieve healing and thus nonmaleficence has to be balanced with beneficence. To work "for the good of the patient" is part of the Hippocratic Oath as well, and it features quite prominently in the above-mentioned Chinese medical texts (Edelstein, 1943; Tsai, 1999, 2005).

More generally, i.e. outside the context of medicine, both nonmaleficence and beneficence can be seen as core principles in any system of religious ethics. A central concept of both Hinduism and Buddhism is *ahimsa,* which means kindness and nonviolence to all living beings (Peetush, 2011). The Bhagavadgita praises the "gift which is made to one from whom no return is expected" (The Bhagavadgita, 1973, 17:20), whereas the Dhammapada states, "A man is not great because he is a warrior or kills other men, but because he hurts not any living being" (Dhammapada, 1997, 270). Both the Torah and the Gospel express the same thought in a different way by exhorting everybody to "love your neighbour as yourself" (Bible, 2012, Leviticus 19:18 and Matthew 22:39). More concretely, the Talmud observes that to "to save one life is tantamount to saving a whole world" (Talmud, Sanhedrin 37a, quoted in Rosner, 2001), whereas the apostle Paul suggests that "whenever we have the opportunity, let's practice doing good to everyone" (Bible, 2012, Galatians 6:10). The Quran asserts that "Whoever rallies to a good cause shall have a share in its blessings; and whoever rallies to an evil cause shall be answerable for his part in it" (Quran, 1980, 4:85). Nevertheless Islamic jurisprudence has the guideline that "if a less substantial instance of harm and an outweighing benefit are in conflict, the harm is forgiven for the sake of the benefit" (Fadel, 2010).

In the context of radiation protection, nonmaleficence and beneficence together would certainly support the principle of justification as they support very similar principles in medicine. When it comes to the principle of optimization, the matter may be a bit more complicated, as the wording in the ICRP recommendations suggests "taking into account economic and societal factors". Now, the interests of the wider community are certainly one factor that none of our traditions would recommend neglecting, but economical considerations are not usually on the agenda of sacred scriptures. Their emphasis is on the human being, especially his or her spiritual and physical health. But then, of course, economical factors cannot be neglected altogether. Resources are limited and it is simply not possible to invest unlimited money into better living conditions—or better radiation protection, for that matter—when that would mean that other areas of the common weal could not receive attention or even basic needs could not be satisfied. So this question becomes a question of justice.

4.6.3 Justice

The "Golden Rule" is one of the most common ethical guidelines around the world. It is found in every single tradition one may choose to look at, and even its wording is strikingly uniform. A few examples must suffice: "One should never do that to another which one regards as injurious to one's own self" (Hindu, The Mahabharata, 1975, 13:113). "Hurt not others in ways that you yourself would find hurtful" (Buddhist, Udanavarga, 1946, 5:18). "Never impose on others what you would not choose for yourself" (Confucian, Analects XV:24). "That which is hateful to you, do not do to your fellow. That is the whole Torah; the rest is the explanation; go and learn" (Jewish, Talmud, Shabbat 31a, quoted in Kellner, 1993). "Therefore whatever you want people to do for you, do the same for them, because this summarizes the Law and the Prophets" (Christian, Bible, 2012, Matthew 7:12). "None of you [truly] believes until he wishes for his brother what he wishes for himself" (Muslim, Hadiths of an-Nawawi 13). "If thine eyes be turned toward justice, choose thou for thy neighbour that which thou choosest for thyself" (Bahá'í, Tablets of Bahá'u'lláh, 1994, Kalímát-i-Firdawsíyyih 20). Because of its general acceptance, this rule is also foundational to the above-mentioned "Declaration Toward a Global Ethic" of the Parliament of the World's Religions 1993 (Küng & Kuschel, 1993c). It is obvious at least from some of the versions quoted here that the Golden Rule can also serve as support for the principles of nonmaleficence and beneficence. But it seems to me that its greatest importance is for the idea of justice. It asks everyone to consider the interests of the other as if they were his or her own, and thus demands reciprocity.

Justice as such is verifiably an element of "common morality" as well. The Bhagavadgita contains the promise that "He who is equal-minded among friends, companions and foes…among saints and sinners, he excels" (The Bhagavadgita, 2012, 6:9). In the Sermons of Buddha a similar statement is found: "He, whose intentions are righteousness and justice, will meet with no failure" (Mahavagga, quoted in Fozdar, 1973). The Psalms observe that, "He loves righteousness and justice; the world is filled with the gracious love of the Lord," (Bible, 2012, Psalms 33:5) whereas in the introduction to the Proverbs the reader is assured that here he will acquire "the discipline that produces wise behavior, righteousness, justice, and upright living" (Bible, 2012, Proverbs 1:3). Muhammad advises his followers, "whenever you judge between people, to judge with justice" (Quran, 1980, 4:58). And Bahá'u'lláh writes that "No light can compare with the light of justice. The establishment of order in the world and the tranquillity of the nations depend upon it" (Bahá'u'lláh, 1988, Epistle to the Son of the Wolf, 53).

Again, a look at secular philosophy will be instructive, as justice has not only been of prime importance since Antiquity, but has also been systematically studied early on (Johnston, 2011). Aristotle, for instance, distinguished between different forms of justice, and his analysis has exerted decisive influence on later thought. The form that Beauchamp and Childress are talking about (Beauchamp & Childress, 2009c), and that is certainly also implied by the sacred scriptures quoted above, is "distributive justice". It concerns the allocation of goods and burdens, of rights and duties in a society. But even this one form can be viewed from different perspectives. Which allocation of goods and burdens is just? An egalitarian one, one that considers merits, one that considers needs, or one that respects historical developments? All this needs to be made the subject not only of philosophical debate, but also of cross-cultural discourse.

For radiation protection the principle of justice would seem to play the role of the second pillar of the dose limitation principle, because it exhorts to a just allocation of burdens. But as mentioned in the last section, its reach is certainly beyond that and it has implications for the optimization principle as well. It could help with a better explication of exactly how "economic and societal factor" should be "taken into account" when determining what is "as low as reasonably achievable."

4.7 FURTHER CROSS-CULTURALLY ACCEPTED PRINCIPLES WITH RELEVANCE FOR RADIATION PROTECTION

I will discuss only three questions here that have been and are still being discussed in the context of radiation protection ethics, but cannot be solved on the basis of the three main principles mentioned (justification, optimization, dose limitation).

4.7.1 Concern for the Underprivileged

My first example is this: How can we assess a situation in which profits and burdens are distributed unequally, i.e. the good is provided preferentially to one group of individuals and the harm to another? In this case, I think the cross-cultural approach has indeed something to contribute. Many if not all philosophical and religious traditions agree that special attention must be given to the underprivileged. We find a similar way of thinking in John Rawls' "Theory of Justice" (Rawls, 1971; Rawls, 1999), where he states that "social and economic inequalities are to be arranged so that they are to be of the greatest benefit to the least-advantaged members of society." Rawls is generally considered a deontological philosopher, but in this particular instance I do think his theory very much reflects "common morality".

So, let us again have a look at the primary sources. The Rig Veda recommends, "Let the rich satisfy the poor implorer, and bend his eye upon a longer pathway. Riches come now to one, now to another" (RigVeda, 2009, 10:117,5). The Buddha promises, "He who pursues wealth in a lawful way, and having done so gives freely of his wealth thus lawfully obtained - by so giving…he begets much merit" (Sutta Nipata, quoted in Woodward, 2002). Confucius' counsel is: "Exemplary people help the needy and do not add to the wealth of the rich" (Analects VI:4). In the Psalms it is stated that "Blessed is the one who is considerate of the destitute; the Lord will deliver him when the times are evil" (Bible, 2012, Psalms 41:1). Of Jesus Christ we read, "Since you didn't do it for one of the least important of these, you didn't do it for me" (Bible, 2012, Matthew 25:45). And Muhammad says about the "doers of good" that they "[would assign] in all that they possessed a due share unto such as might ask [for help] and such as might suffer privation" (Quran, 1980, 51:16 and 19).

4.7.5 Intergenerational Equity

My second example: How can we take into account the risks for future generations? Different models have been proposed, some of them suggesting a discount rate approach in which future good and harm count less than prompt consequences, (Newell & Pizer, 2003) but the International Atomic Energy Authority has stated as principles, that "Radioactive waste shall be managed in such a way as to assure that predicted impacts on the health of future generations will not be greater than relevant levels of impact that are acceptable today" and "Radioactive waste shall be managed in such a way that will not impose undue burdens on future generations" (IAEA).

This is neither utilitarian nor deontological. At least nothing like this is part of the original versions of those theories. But it is certainly in line with a clear cross-cultural agreement about intergenerational equity. Both Hinduism and Buddhism are very much concerned with the idea of *karma*, which sees each thought or action as part of an ever-continuing cycle of cause and effect (For further information, see Doniger O'Flaherty, 1999). In line with this, a Hindu delegation to the World's Parliament of Religion stated, for instance, that "we must do all that is humanly possible to protect the Earth and her resources for the present as well as future generations" (Hindu Declaration on Climate Change, 2009), and the Dalai Lama made a similar pronouncement: "Now that we are aware of the dangerous factors, it is very important that we examine our responsibilities and our commitment to values, and think of the kind of world we are to bequeath to future generations" (Fox, 1996). The responsibility for those who come after us is expressed somewhat differently in the Torah, where God speaks to Abraham, "I'm establishing my covenant between me and you, and with your descendants who come after you, generation after generation, as an eternal covenant" (Bible, 2012, Genesis 17:7), and this concept of eternal covenant is equally important for Christians and Muslims (For further information, see Covenant). Bahá'u'lláh adds still another component to this by saying, "All men have been created to carry forward an ever-advancing civilization" (Gleanings from the Writings of Bahá'u'lláh, 1990, 109:2), which according to a statement of the Bahá'í International Community "offers hope to a dispirited humanity and the promise that it is truly possible both to meet the needs of present and future generations" (Palmer & Inlay, 2003). Here we can also mention African customary law, which is aptly summarized by a Nigerian chief as follows: "I conceive that land belongs to a vast family of whom many are dead, a few are living, and countless hosts are still unborn"(Mugambwa & Kmankwah, 2002). And as an example of recent international documents, we can look at a passage from the report of the United Nations World Commission on Environment and Development of 1987 ("Brundtland Commission"), which maintains that development must meet "the needs of the present without compromising the ability of future generations to meet their own needs" (Our Common Future).

4.7.6 Precaution

Finally, my last example is this: How should we deal with risks for which there is no direct evidence? For decades, the linear-nonthreshold (LNT) model has been assumed to be the most appropriate way of extrapolating from radiation risks at higher doses to those at lower doses. The International Commission on Radiological Protection in its most recent recommendations still upholds this view and says that although there are no data to support or reject the LNT model in the very low dose range, it "remains a prudent basis for radiological protection" (ICRP, 2007b). Elsewhere, however, the Commission suggests that "The aggregation of very low individual doses over extended time periods is inappropriate, and in particular, the calculation of the number of cancer deaths based on collective effective doses from trivial individual doses should be avoided"

(ICRP, 2007c). This is justified by saying that "the assumptions implicit in the calculation of collective effective dose...conceal large biological and statistical uncertainties"(ICRP, 2007d). There is an element of inconsistency here, as the calculation of collective dose presupposes the very LNT model that is otherwise considered applicable even in the absence of direct evidence. I am ready to admit the possibility that reasons to disregard "trivial doses" may exist in certain situations, but I do not think that we can forgo ethically sound arguments if these reasons are to prevail over established principles. In the particular case here, the fact that the Commission's recommendation is based on "uncertainties" cuts across its own statement that the application of the linear-nonthreshold model is "commensurate with the 'precautionary principle'"(ICRP, 2007b).

This principle was first formulated in the late twentieth century. For instance, the United Nations Conference on Environment and Development in Rio de Janeiro 1992, also called the Earth Summit, proposed the following: "Where there are threats of serious or irreversible damage, lack of full scientific certainty shall not be used as a reason for postponing cost-effective measures to prevent environmental degradation" (Rio Declaration). Another important version is the one drawn up by a group of scientists from different disciplines gathered at the Wingspread Conference: "When an activity raises threats of harm to human health or the environment, precautionary measures should be taken even if some cause and effect relationships are not fully established scientifically"(Wingspread Statement).

Again, the Precautionary Principle is not utilitarian or deontological. Neither can it be expected to appear in its modern form in the written and oral traditions of different cultures. Exhortations to prudence, however, are ubiquitous, and they are generally interpreted, by people referring to those traditions for orientation, as suggesting a precautionary approach. Thus, in the Mahabharata Krishna advises to "act like a person in fear before the cause of fear actually presents itself" (The Mahabharata, 1975, 12:138), whereas Shotoku Taishi, the first Buddhist regent of Japan, puts it this way: "When big things are at stake, the danger of the error is great. Therefore, many should discuss and clarify the matter together, so the correct way may be found" (Greenstein & Thompson, 2007). Confucius simple says that "The cautious seldom err" (Analects IV:23). In the Proverbs, we find the following statement: "Those who are prudent see danger and take refuge, but the naïve continue on and suffer the consequences" (Bible, 2012, Proverbs 27:12), and Muhammad reportedly counseled one of his followers who complained that God had let his camel escape: "Tie up your camel first then put your trust in God" (Htay, Arif, Soualhi, Zaharin, & Shaugee, 2013). As samples of an explicit reference to the Precautionary Principle I will give just two quotes: The North Dakota Conference of Churches issued a statement in 2003, saying "We endorse the Precautionary Principle as a primary guide...We believe that it is a discipline consistent with our Christian calling as stewards of creation and advocates of economic and social justice" (Rural Life Committee, 2003). And, to refer again to a nonliterate culture, a representative of the Australian Aboriginals and Torres Strait Islanders expressed his understanding in this way: "Over the past 60,000 years we, the indigenous people of the world, have successfully managed our natural environment to provide for our cultural and physical needs. We have no need to study the non-indigenous concepts of the precautionary principle [and others]. For us, they are already incorporated within our traditions" (Missi, 1998). Although this sounds a little dismissive of dialogue, indigenous people's perspectives have definitely been appreciated by those who are working on cross-cultural ethics.

Whatever the reader may think of my particular example, the case of the "trivial doses", it seems obvious to me that the Precautionary Principle is of relevance to our system of radiation protection. It will have to be taken seriously if we are interested in cross-culturally valid principles, but of course it can be only one of several components of radiation protection ethics, which needs to be balanced with other principles as well as specified for this particular area of application.

4.8 CONCLUSION

The approach which I have presented here takes account of both religious and philosophical traditions of different cultures. If my examples focus on the sacred scriptures of the major world religions this is because I do think that they have exerted a deeper and more lasting influence on mankind than secular sources. Nevertheless, my suggestion is to develop cross-cultural, not just interreligious ethics.

In discussions about these ideas, the most frequently asked question is whether the reference to religious and philosophical traditions is not too backward oriented and in danger of missing out on developments of the twentieth century such as the emancipation of women. I would like to offer three arguments in response to this: First, my claim is not that every single ethical question can be solved equally well with a cross-cultural approach. We have to try and see how far the "common morality" will get us with respect to questions of radiation protection. Second, I am convinced that the wrongs of the past cannot be attributed to the primary sources referred to above (as is frequently suspected, especially in the case of sacred scriptures). Rather it is people of later ages that have misused them and associated with them ideas alien to the origins. I do not think, for instance, that there is a cross-cultural agreement about women as second class citizens. Third, I am not a fundamentalist. As many others I think that the religious and philosophical traditions have to be viewed in the context of their times, and we may come to the conclusion here and there that certain concepts are outdated and cannot be considered "common morality" for our time. This kind of argument will have to be very cautiously applied to not give way to arbitrariness, but again leading a real discourse with real people across cultural borders will prove helpful.

Finally, I would like to emphasize that I am obviously not advocating a total revamp of the system of radiation protection, but suggest a different approach to its ethical foundation. This approach would seem to be less biased toward "Western" philosophical tradition, and therefore more acceptable for people of different cultural backgrounds. It may also give fresh insight into some of the problems, which are difficult to solve with the current mix of utilitarian and deontological approaches in radiation protection.

REFERENCES

Analects. (D. Hinton, Trans.) (1999) Berkeley CA: Counterpoint Press.

Andorno, R. (2009). Human dignity and human rights as a common ground for a global bioethics. *Journal of Medicine and Philosophy*, *34*, 223–240.

Bahá'u'lláh. (1988). *Epistle to the Son of the Wolf.* Wilmette IL: Bahá'í Publishing Trust.

Gleanings from the Writings of Bahá'u'lláh. (1990). Wilmette IL: Baha'i Publishing Trust.

Tablets of Bahá'u'lláh revealed after the Kitáb-i-Aqdas. (1994). Wilmette IL: Bahá'í Publishing Trust.

Beauchamp, T. L., & Childress, J. F. (1979). *Principles of biomedical ethics.* Oxford University Press.

Beauchamp, T. L., & Childress, J. F. (1994). *Principles of biomedical ethics.* Oxford University Press.

Beauchamp, & Childress. (2009a). *Principles of biomedical ethics.* Oxford University Press. p.392–394.

Beauchamp, & Childress. (2009b). *Principles of biomedical ethics.* Oxford University Press. p.381–387.

Beauchamp, & Childress. (2009c). *Principles of biomedical ethics.* Oxford University Press. p.241.

Beauchamp, T. L., & Childress, J. F. (2009d). *Principles of biomedical ethics.* Oxford University Press. pp.4,387–397.

Beauchamp, T. L., & Childress, J. F. (2009e). *Principles of biomedical ethics.* Oxford University Press. pp.14–25.

Beauchamp, T. L., & Childress, J. F. (2009f). *Principles of biomedical ethics.* Oxford University Press. p.40.

Bentham, J. (1744). *A fragment on government.* J.S. Mill. (1861) Utilitarianism.

The Bhagavadgita. (1973). (S. Radhkrishnan, Trans.) New York: Harper and Row.

The Holy Bible, international standard version. (2012). La Mirada CA: Davidson Press.

van Brussel, L. (2012). Autonomy and dignity: a discussion on contingency and dominance. *Health Care Analysis.* Epub ahead of print.

Clarke, R. H. (2003). Changing philosophy in ICRP: the evolution of protection ethics and principles. *International Journal of Low Radiation*, *1*, 39–49.

Clarke, R. H., & Valentin, J. (2009). The history of the ICRP and the evolution of its policies. ICRP Publication 109. *Annals of the ICRP*, *39*, 75–110.

Cleary, T. (1992). *The essential Confucius.* San Francisco: Harper. p.19 (Analects VI:4).

Covenant, Encyclopaedia Britannica, http://www.britannica.com/EBchecked/topic/141085/covenant.

The Dhammapada. (1997). (S. Radhkrishnan, Trans.) Oxford University Press.

Doniger O'Flaherty, W. (1999). *Karma and rebirth in classical Indian traditions.* Delhi: Motilal Banarsidass.

Edelstein, L. (1943). *The Hippocratic oath: Text, translation, and interpretation.* Johns Hopkins University Press.

Fadel, H. E. (2010). Ethics of clinical research: an Islamic perspective. *The Journal of Islamic Medical Association*, *42*, 59–69.

Fagan, A. (2004). Challenging the bioethical application of the autonomy principle with multicultural societies. *Journal of Applied Philosophy*, *21*, 15–31.

Fan, R. (1997). Self-determination vs family-determination: two incommensurable principles of autonomy. *Bioethics*, *11*, 309–322.

Fox, R. (1990). The evolution of American bioethics: a sociological perspective. In G. Weisz (Ed.), *Social science perspective on medical ethics* (pp. 201–220). Philadelphia: University of Pennsylvania Press.

Fox, M. W. (1996). *The boundless circle: Caring for creatures and creation*. Wheaton IL: Quest Books. p.215.

Fozdar, J. (1973). *The God of Buddha*. New York: Asia Publishing House. p. 56.

Gardiner, S. (2008). Why we need more than justification in the ethics of nuclear protection: a view from outside. In G. Eggermont & B. Feltz (Eds.), *Ethics and radiological protection* (pp. 97–111). Louvain-la-Neuve: Academia.

González, A. J. (2011). The Argentine approach to radiation safety: its ethical basis. *Science and Technology of Nuclear Installations*, 69–83.

Greenstein, D., Thompson D. (2007) Putting "Public" back in public health work, http://www.mediate.com/mobile/article.cfm?id=3207.

Habermas, J. (2008). *Between naturalism and religion: Philosophical essays*. Cambridge: Polity. p.131.

Hadiths of an-Nawawi (Islamic oral tradition), http://www.iium.edu.my/deed/hadith/.

Hansson, S. O. (2007). Ethics and radiation protection. *Journal of Radiological Protection*, *27*, 147–156.

Hindu Declaration on Climate Change. (2009). *Presented at the parliament of the world's religions*. Melbourne, Australia http://home.sandiego.edu/~kaufmann/hnrs379/Singh_et_al._2009.pdf.

Htay, S. N. N., Arif, M., Soualhi, Y., Zaharin, H. R., & Shaugee, I. (2013). *Accounting, auditing and governance for takaful operations*. Singapore: Wiley. p.4.

IAEA safety fundamentals: the principles of radioactive waste management, http://www.oecd-nea.org/rwm/reports/1995/geodisp/annex1.html.

ICRP (1959) Recommendations of the International Commission on Radiological Protection. ICRP Publication 1, Oxford: Pergamon Press.

ICRP. (1966). *Recommendations of the International Commission on Radiological Protection. ICRP Publication 9*. Oxford: Pergamon Press.

ICRP. (1973). *Implications of commission recommendations that doses be kept as low as readily achievable. ICRP Publication 22*. New York: Pergamon Press.

ICRP. (1977). Recommendations of the International Commission on Radiological Protection. ICRP Publication 26. *Annals of the ICRP*, *1*.

ICRP. (2007a). The most recent one is ICRP (2007). The 2007 recommendations of the International Commission on Radiological Protection. ICRP Publication 1003. *Annals of the ICRP*, *37*.

ICRP. (2007b). The most recent one is ICRP (2007). The 2007 recommendations of the International Commission on Radiological Protection. ICRP Publication 1003. *Annals of the ICRP*, *37*, p. 43.

ICRP. (2007c). The most recent one is ICRP (2007). The 2007 recommendations of the International Commission on Radiological Protection. ICRP Publication 1003. *Annals of the ICRP*, *37*, p.13.

ICRP. (2007d). The most recent one is ICRP (2007d). The 2007 recommendations of the International Commission on Radiological Protection. ICRP Publication 1003. *Annals of the ICRP*, *37*, p.76.

Johnston, D. (2011). *A brief history of justice*. Chichester: Wiley-Blackwell.

Justo, L., & Villarreal, J. (2003). Autonomy as a universal expectation: a review and a research proposal. *Eubios Journal of Asian and International Bioethics*, *13*, 53–57.

Kant, I. (1785) Grundlegung zur Metaphysik Der Sitten [Groundwork of the metaphysics of morals].

Kellner, M. (1993). Jewish ethics. In P. Singer (Ed.), *A companion to ethics* (pp. 86–87). Oxford: Blackwell.

Kretzmer, D., & Klein, E. (Eds.), (2002). *The concept of human dignity in human rights discourse.* The Hague: Kluwer Law International.

Küng, H. & Kusche, K.-J. (Eds.), (1993a). *A global ethic. The declaration of the Parliament of the World's Religions.* London/Continuum, New York: SCM Press.

Küng, H., & Kuschel, K.-J. (1993b). In *A global ethic. The declaration of the Parliament of the World's Religions* London/Continuum, New York: SCM Press. p. 23.

Küng, H., & Kuschel, K.-J. (1993c). In *A global ethic. The declaration of the Parliament of the World's Religions* London/Continuum, New York: SCM Press. p. 15.

Lepard, B. D. (2005). *Hope for a global ethic.* Wilmette IL: Bahá'í Publishing Trust. pp. 67–69.

The Mahabharata. (K. M. Ganguli, Trans.) (1975) New Delhi: Orient Book Distributors.

Matsuoka, E. (2007). The issue of particulars and universals in bioethics: some ideas from cultural anthropology. *Journal of Philosophy and Ethics in Health Care and Medicine, 2*, 44–65.

Missi, C. L. (1998). An indigenous perspective on flying fox harvesting. *Australian Biologist, 11*, 72–75.

Mugambwa, J., & Kmankwah, H. (2002). *Land law and policy in Papua New Guinea.* New York: Routledge-Cavendish. p. 129.

Newell, R. G., & Pizer, W. A. (2003). Discounting the distant future: how much do uncertain rates increase valuations? *Journal of Environmental Economics and Management, 46*, 52–71.

Our common future: Report of the World Commission on Environment and Development, http://www.un-documents.net/wced-ocf.htm, chapter 2.

Palmer, M., & Inlay, V. (2003). *Faith in conservation: New approaches to religions and the environment.* Washington D.C.: World Bank Publications. p.72.

Peetush, A. (2011). Ahimsa. In D. K. Chatterjee (Ed.), *Encyclopedia of global justice* (pp. 23–26). New York: Springer.

Pelligrino, E. D. (2008). Some personal reflections on the "appearance" of bioethics today. *Studia Bioetica, 1*, 52–57.

Persson, L., & Shrader-Frechette, K. (2001). An evaluation of the ethical principles of the ICRP's radiation protection standards for workers. *Health Physics, 80*, 225–234.

The message of the Quran. (M. Asad, Trans.) (1980). Gibraltar: Dar al-Andalus.

Rawls, J. (1971). *A theory of justice.* Harvard University Press.

Rawls, J. (1999). *A theory of justice.* Harvard University Press.

The RigVeda. (R.T.H. Griffith, Trans.) (2009). Santa Cruz CA: Evinity Publishing.

Rio declaration on environment and development, http://www.un-documents.net/rio-dec.htm, principle 15.

Rosner, F. (2001). *Biomedical ethics and Jewish law.* Persey City NY: Ktav Publishing House. p.223.

A response to issues and values related to genetically modified organisms, a statement of the Rural Life Committee of the North Dakota Conference of Churches (2003), http://www.sehn.org/ndcc.html.

Shrader-Frechette, K., & Persson, L. (1997). Ethical problems in radiation protection. *Health Physics, 73*, 373–382.

Streffer, C., Witt, A., Gethmann, C.-F., Heinloth, K., & Rumpff, K. (2005). *Ethische Probleme einer langfristigen globalen Energieversorgung.* Berlin, New York: de Gruyter.

ten Have, H., & Gordijn, B. (2013). Global bioethics. In H. ten Have & B. Gordijn (Eds.), *Compendium and atlas of global bioethics.* Dordrecht: Springer. (chapter 63).

Tsai, D. F. -C. (1999). Ancient Chinese medical ethics and the four principles of biomedical ethics. *Journal of Medical Ethics, 25*, 315–321.

Tsai, D. F. -C. (2005). The bioethical principles and Confucius' moral philosophy. *Journal of Medical Ethics*, *31*, 159–163.

Turner, L. (2003). Bioethics in a multicultural world: medicine and morality in pluralistic settings. *Health Care Analysis*, *11*, 99–117.

Udanavarga. (A. Subhadra, Trans.) (1946). London: Universalist Press.

Veatch, R. M. (Ed.), (2000). *Cross-cultural perspectives in medical ethics*. Burlington: Jones and Bartlett.

Wingspread statement on the precautionary principle, http://www.psrast.org/precaut.htm.

Wittgenstein, L. (1921) Tractatus logico-philosophicus 5.6.

Woodward, F. L. (2002). *Some sayings of the Buddha*. New Delhi: Asian Education Services. p.93.

Chapter 5

Ethical Aspects of Ecological Risks from Radiation*

Deborah Oughton
Centre for Environmental Radioactivity, Norwegian University of Life Sciences, ÅS, Norway
E-mail: deborah.oughton@umb.no

Chapter Outline

5.1 Introduction 71
5.2 Valuing the Environment: Philosophical Theories 73
5.2.1 Biocentricsm 73
5.2.2 Ecocentrism 75
5.2.3 Anthropocentrism 76
5.3 Common Ethical Principles 78
5.4 Harms and Values in Practical Radiation Protection 78
5.5 Assigning Monetary Value to the Environment 80
5.6 Conclusion: Relevance of the Value Debate to Ecological Radiological Protection 81

5.1 INTRODUCTION

The new millennium has seen a growing consensus that radiation risk management needs to address the question of effects on the environment (IAEA, 2002; ICRP 2007; IUR, 2002; Pentreath, 1998, 1999; 2009). Prior to this, radiation protection was almost exclusively concerned with humans, under the assumption that that dose limits for exposure of humans will usually entail sufficient protection to the environment (ICRP, 1991 Section 16). Challenges to this approach included that it was not in line with the assessment and management of other environmental stressors, and that there were a number of cases, such as marine or geological disposal, where wildlife and ecosystems could be exposed to high levels of radiation even when human doses were low. Thus a requirement to explicitly address the impacts on nonhuman organisms is now part of international

* Revised from *Journal of Environmental Radioactivity*, 66(1-2), Oughton D, "Protection of the environment from ionizing radiation: ethical issues," 3-18 © 2003 with permission of Elsevier.

Radioactivity in the Environment, Volume 19
ISSN 1569-4860, http://dx.doi.org/10.1016/B978-0-08-045015-5.00005-8

radiation protection recommendations and standards (IAEA, 2011; ICRP, 2007), as well as national strategy in a number of countries (Copplestone et al., 2009).

Many of the international organizations involved in the development of tools and frameworks for assessing environmental risks recognized that producing a practical and coherent system of radiological protection for wildlife raises a number of ethical questions. The International Atomic Energy Agency (IAEA) produced a report on "Ethical Considerations in Protecting the Environment from the effects of Ionizing Radiation" (IAEA, 2002), which also addressed the cultural, scientific, and social influences on environmental worldviews, as well as links to political protection principles such as sustainability and biodiversity (Figure 5.1). Ethical aspects have also been addressed in International Union of Radioecology (IUR) and International Commission on Radiological Protection (ICRP) publications on environmental protection (ICRP, 2008; IUR, 2002, 2012). Common to all approaches is an appreciation of the diversity in ethical and cultural views on valuing the environment, and that this diversity should be respected in environmental protection frameworks.

Building on this work, the present paper summarizes some of the main ethical issues concerning the protection of the environment from radiation, largely based on the IAEA report and previous papers (e.g. Oughton, 2003), and looks at more recent developments on environmental protection in radiation risk assessment. The first part gives an overview of different philosophical worldviews on valuing the environment in a context of radiation risk. This is followed by an evaluation of some of the more recent challenges to the proposed environmental protection frameworks, including practical applications following the Chernobyl and Fukushima accidents. The final part of the paper offers some

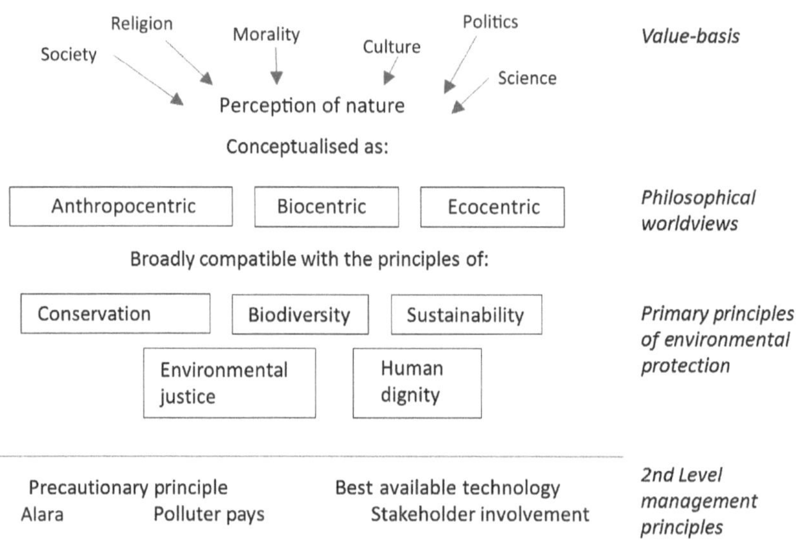

FIGURE 5.1 Links between the value-basis of perception of nature, philosophical worldviews and environmental protection principles (drawn according to concepts discussed in IAEA, 2002).

recommendations for how ethical evaluation can aid in producing a robust and transparent approach to protection of the environment from ionizing radiation.

5.2 VALUING THE ENVIRONMENT: PHILOSOPHICAL THEORIES

Environmental ethicists have been debating the matter of why one attaches value to the environment for a number of decades (Rolston, 1988; Sterba, 1994). Central philosophical issues include the question of moral standing and whether the environment has intrinsic or inherent value (i.e. value in itself) or extrinsic or instrumental value (i.e. value because of human interest). Although environmental ethics is a relatively young field within philosophy, a number of distinct views on this question have emerged. In contemporary environmental philosophy, the most fundamental source of divergence arises between the anthropocentric and the nonanthropocentric view. An anthropocentric ethic (literally human-centered) alleges that only humans have moral standing and that environmental degradation matters only in so far as it influences human interests (Bookchin, 1991; Norton, 1988). Proponents of a nonanthropocentric ethic reject this assumption, and attribute moral standing either to other living organisms or to the ecosystem as a whole, contending that effects on the environment matter irrespective of their consequences for humans. Although a variety of different views can be found in the literature, the biocentric and ecocentric outlooks are arguably the two main contenders.

5.2.1 Biocentricsm

Proponents of biocentrism (literally "life-centered") assert that individual life-forms other than humans can have moral standing, and should be respected for what they are—not only because they affect the interests of humans. Different biocentric views exist as to which criterion forms the basis for moral standing, and what hierarchy (if any) exists between different species. But all views derive moral value from some biological characteristic of individual members of species, such as sentience or the ability to feel pleasure or pain (Singer, 1991), self-consciousness (Regan, 1980) or inherent worth or a "good of their own" of all living things (Goodpaster, 1978; Taylor, 1986).

Biocentric outlooks can be found within supporters of both utilitarian and deontological theories of ethics. Utilitarians can include the welfare, interests, or preference satisfaction of animals in their utility calculations; deontologists can find room for rights of or duties to animals. The Australian utilitarian philosopher, Peter Singer, is an influential proponent of animal ethics, and suggests that sentience represents the fundamental criterion for moral standing (Singer, 1991). Welfare or well-being matters for any life-form with the capacity to feel pleasure and pain. In this, he advances an idea first proposed by Bentham when considering who or what should count in a utilitarian evaluation: "The question is not, Can they reason? Nor, Can they talk? But, Can they suffer?" (Bentham, 1789, Chapter 17). Although the calculation may allow a hierarchical weighting of different species, human interests are not inalienable and can be outweighed if the amount of suffering caused to animals is large enough.

Deontologists might suggest that the notion of rights and duties should be extended to the animal or biological kingdom. One of the strongest proponents of animal rights, Tom Regan argues that like humans, some nonhuman animals have consciousness or self-awareness and a capability for reasoning (Regan, 1980), and some form of rights attribution to animals can be found in national legislation (e.g. New Zealand). However, critics have claimed that the debate around giving "rights" to nonhuman species or indeed to whole ecosystems is a futile response to the increasing tendency of human society toward environmental destruction. They draw parallels with the way that human rights have emerged as a well-meaning, and yet, to date, depressingly ineffective way of counteracting the modern day atrocities of warfare or racism (Bradford, 1993). The critique harks back to Bentham's notorious claim that "natural rights is simple nonsense; natural and imprescriptible rights, rhetorical nonsense–nonsense upon stilts (Bentham, 1824)". Nonetheless, recognition of human rights is a cornerstone of international conventions, and the possible future extension of these rights to other species is not easily dismissed.

Because biocentrism focuses on individuals rather than the diversity of species, the outlook has been described as an "individualistic" environmental ethic (Rolston, 1991; Sagoff, 1984). In practical policy-making, biocentric outlooks have had the greatest influence in issues of animal welfare and the use of animals in research (Sagoff, 1984). The ICRP's Reference Animals and Plants (RAP) approach (Table 5.1) is consistent with a biocentric methodology for assessing radiation effects on individual nonhuman species. Although, as discussed below, this does not necessarily make it a biocentric value-basis for protecting those individuals. The idea of including impacts on animals in radiation protection optimization is also compatible with a broadly utilitarian approach. In this case optimization would include both the direct impacts of radiation on nonhumans, as well as the more general (and often more damaging) consequences for the environment of reducing doses to human (see Oughton, Bay, Forsberg, Kaiser, & Howard, 2004 for examples of the environmental and animal welfare side-effects of accident remediation).

Nevertheless, optimization in radiation protection rarely considers exactly why one is bothered about environmental impacts, and there can of course still be disagreements on which species and which effects matter. For example, Singer's criterion of sentience only encompasses vertebrates, whereas Paul Taylor suggests that all living organisms are equal moral subjects (egalitarian biocentrism) since each has some goal to its existence (Taylor, 1986). Note that for any biocentric view, as soon as the ethically relevant factor for assigning moral standing diverges from the "speciesist" criterion of simply being human to some trait such as rationality, consciousness or sentience, one is faced the problem of how to deal with those members of the human species that, due to some force of circumstance (accidental or otherwise), might be considered to rank lower than the higher animals.

TABLE 5.1 ICRP Reference Animals and Plants (ICRP, 2008)

RAP	General Class	Available standard chemical toxicity tests[a]
Deer	Large mammal	No
Rat	Small mammal	Yes
Duck	Birds	Limited
Frog	Amphibian	Some life-stages
Trout	Freshwater fish/pelagic	Yes
Flatfish	Marine fish/benthic	Yes
Bee	Insect	No
Crab	Marine crustacean	Related species
Earthworm	Annelids/soil invertebrate	Yes
Pine tree	Conifers	Limited
Wild grasses	Grasses	Yes
Brown seaweed	Macro-algae	Related species

[a]*For example, Organization for Economic Co-operation and Development (OECD) standard guidelines for toxicity tests, ISO procedures, etc., that aid in comparing the impact of radiation with chemical stressors.*

5.2.2 Ecocentrism

Supporters of an ecocentric philosophy claim that the diversity of species, ecosystems, rivers, mountains, and landscapes can have value in themselves, irrespective of the consequences on humans or other individuals of nonhuman species. All ecocentrics attach particular value to the diversity, dynamics and interactions within healthy ecosystems, but differ in their views on the causes of, and proper solutions to, modern environmental problems. Callicott (1979, 1989) and Næss (1974) both see the Western, instrumental view of nature as a main source of environmental problems. Ecofeminists suggest the problem lies in the history of male dominance and sexist oppression of females (Warren, 1990); others that it stems from the social and economic structure of society (Bookchin, 1991). Many link problems to the Judeo-Christian tradition, and, more specifically, in the Biblical quotation (e.g. White, 1967):

"Let us make man in our image, after our likeness: and let them have dominion over the fish of the sea and over the fowl of the air, and over the cattle, and over all the earth, and over every creeping thing that creepeth upon the earth"

(Genesis 1: 197326–30).

Although other philosophers have pointed out that the bible also contains examples of human obligations to respect nature (Ariansen, 1996).

Most ecocentrics claim that mankind needs a radical change from an anthropocentric attitude of domination and exploitation of natural resources toward a greater respect for the integrity of nature (deep ecologists like Næss, are perhaps more radical than others). In evaluating actions, Callicot defends the land-ethic maxim of Aldo Leopold, "A thing is right when it tends to preserve the integrity, stability, and beauty of the biotic community; it is wrong when it tends otherwise" (Leopold, 1949). The general concern for the biotic and abiotic community as a whole leads to the alternative classification of the outlook as a "holistic" ethic (Sagoff, 1984). The inclusion of the abiotic components of the environment in ecocentrism, together with the fact that most definitions of the environment in international legislation include man, biota, abiota, and physical surroundings, raises the issue of how to deal with the abiotic (i.e. soil, rocks, water) in environmental protection, particularly since many environmental standards are based on concentrations in media.

In radiation protection, the ecocentric view has been linked to the ecosystem approach of environmental assessment and management (IUR, 2012), and has been presented as an alternative to the reference animals and plants approach forwarded by the ICRP (ICRP, 2008). One of the criticisms of the RAP approach is that the 12 selected species do not permit an ecosystem level assessment. To do this, one needs a broader range of ecologically relevant species covering producers, predators, and decomposers, as well as insights in to differences in the sensitivity of species (Bréchignac et al., 2011); variability in sensitivity is a driving factor for ecosystem change since some species can prosper by the impacts on others. This does not mean that the ICRP approach is not capable of providing relevant information; but there is a need for data on a wider variety of species. And as for biocentrism, the method of carrying out an environmental impact assessment should not be taken as the same as ascribing moral value to those entities. As discussed below, ecosystem service approaches to environmental protection can be blatantly anthropocentric in both their approach and underlying value-system.

5.2.3 Anthropocentrism

In defence of anthropocentrism, both scientists and philosophers have argued that human interests can provide a powerful set of motives for protecting the nature (Sober, 1986; Wilson, 1984). Understanding the economic and social impacts of environmental damage on humans can provide a strong incentive to protect the ecosystem. On a more philosophical defence of anthropocentrism, William Frankena suggests that only humans are capable of "valuing" in an ethical sense (Frankena, 1973, 1979). Although, in recent years, some interesting research has been carried out on the morality of animals (Bekoff & Pierce, 2012). In reply to Leopold and Callicott, anthropocentrics ask: who is to answer the question of when a biotic community is stable and beautiful? Can such counsel ever express more than the ecological interests of humans and the species they most closely identify with? (Fritzell, 1987).

Anthropocentrics are also concerned about impacts of radiation on animals and plants (and even soil and water, should that impact on human use of the resources), but they do not consider these entities to have moral standing or value in themselves, only by virtue of the consequences to humans. As an example, Kant's philosophy was clearly human-centered, but his morality did include restrictions on what harms one might cause to animals. His rationale being that people who mistreat animals are likely to develop a habit that inclines them to treat humans in the same fashion (Kant, 1785; Regan & Singer, 1976).

Interestingly, the anthropocentric and the nonanthropocentric ethic tend to highlight both man's uniqueness and our oneness with nature. Humans are the only ethical animal, the only "valuer"; humans are responsible for environmental destruction unmatched by any other species, population growth is a singularly human problem. On the other hand, biology, evolutionary science, and genetics have shown that humans are continuous with the rest of nature, "yet none of this scientific reasoning can guarantee that we will develop ethical concern or a proper relation to the biosphere, any more than the knowledge that other human beings are our genetic kin will prevent us from annihilating them in war" (Bradford, 1993). We may agree that humans have a responsibility not to damage the environment, but disagree on what measures are needed to correct human behavior, and when intervention to protect the environment is necessary.

The ICRP statement that "if man is adequately protected, then other living species are also likely to be sufficiently protected" (ICRP 1977) is widely perceived to be an anthropocentric approach to environmental protection. This is understandable when combined with the strong historical human focus on the radiation protection. Exposure experiments on animals were carried out largely to provide information on human effects; the majority of studies on environmental transfer concentrated on those food-chains with humans at the top. But whilst the statement is clearly an anthropocentric approach to risk assessment, it does not necessarily mean that radiation protection does not value the environment. In the 1960s, the operators of the Windscale plant took the trouble to evaluate the possible environmental impact of its radioactive discharges (Dunster, Garner, Howells, & Wix, 1964; cited in Kershaw, Pentreath, Woodhead, & Hunt, 1992).

5.3 COMMON ETHICAL PRINCIPLES

Despite the apparent diversity of these ethical theories, it is important to realize that although they may disagree quite strongly over *why*, exactly, certain factors are relevant to ethics, there can still be room for consensus on some common features. For evaluation of any action involving exposure of humans, animals, or plants to radiation, each of the above theories would find it morally relevant to ask: (1) who and what is being affected? (2) what is the relative size of the benefits and the harms arising from the exposure? (3) what is the distribution of the risks and the benefits? and (4) what alternative courses of action are available?

With respect to protection of the environment and nonhuman species, all theories can defend the principle that radiation protection should not be limited to humans. Since regulations already exist for the protection of the environment from other contaminants, all other things being equal, there is no ethically relevant reason why effects caused by radiation exposure should be treated differently. However, the different theories might disagree on which types of effects matter most, depending for example on whether harms are evaluated in terms of sentience, animal rights, consequences for existing humans or effects on future generations. Two examples of the types of challenges in practical radiation protection are discussed in the next sections. First, the question of linking ecological change with risk of harm; second the issue of assigning a monetary value to environmental impacts. Both of these aspects have been the focus of recent discussions in environmental radiation protection.

5.4 HARMS AND VALUES IN PRACTICAL RADIATION PROTECTION

No one disputes that exposure to radiation can cause *changes* in biota and the environment, but what many experts question is the long-term consequences of such changes. It is accepted that deterministic, stochastic, and hereditary effects in plants, insects, and animals, have been seen both in the laboratory and after serious accidents and that species can show large variations in radiological sensitivity (UNSCEAR, 2008). Scientists have documented genetic mutations in a number of species following the Chernobyl accident (Ellegren, Lindgren, Primmer, & Moller, 1997; Pomerantseva, Ramaiya, & Chekhovich, 1997; IAEA, 2006; Møller and Mousseau, 2007; Møller and Mousseau, 2009) and damage to pine trees in the Red Forest resulted in the pine forests being replaced by the more radiation-resistant birch (IAEA, 2006; Kovalchuk, Abramov, Pogribny, & Kovalchuk, 2004, Kovalchuk et al., 2005). Similar ecosystem level effects were also reported after the Kyshtym accident, including a change from coniferous to deciduous forests, and population level effects on some insects and mammals (JNREG, 1997). Immediately after the Fukushima accident, questions were raised about the possible ecosystem effects and studies suggesting possible impacts on butterflies in contaminated areas were widely reported in both the scientific and traditional media (Hiyama et al. 2012).

But scientists disagree over whether or not these changes reflect permanent or serious ecological damage—after all the forests grew back, the wildlife returned, and genetic change is not always a bad thing (Baker et al., 1996). Indeed, some people have suggested that the ecological benefit of removing humans from the Chernobyl area might outweigh any radiation detriments (Mycio 2005). The consequences that are deemed "harmful" depend on the level of protection awarded to the various components of the environment (individual, population, species, ecosystem). This in turn can depend on the moral standing of those components.

The regulation of human exposure to radiation takes effects on individuals very seriously. Management of environmental hazards tends to focus on the risk of harm to *populations*. In this respect, most environmental risk managers make a clear moral distinction between human and nonhuman species: individual humans matter; individual animals tend not to. The types of radiation exposure that result in observable (and probably, therefore, unacceptable) damage on a population level are thought to be far higher than the mGy/yr levels at which intervention to protect humans takes place. While this might be true for mortality, however, it need not be the case for other biological endpoints such as reproductive ability and genetic effects. In some cases, such as for endangered species, effects on the individual are deemed to matter—even if not quite as much as for individual humans. Of course the variety of nonanthropocentric views may offer quite different interpretations and explanations on this last point. Some might be offended by the mere presence of man-made radionuclide in the environment, irrespective of any discernible effect on humans or biota.

To conclude, supporters of both anthropocentric and nonanthropocentric ethics can agree that harms to nonhuman populations should be avoided. They may disagree on the level of population change that can be accepted, and which populations should be considered the most important to protect. Likewise all viewpoints could find specific cases where the individual would be the appropriate level of protection: the anthropocentric and ecocentric may focus on endangered species or habitats; the biocentric on certain individuals as having value in themselves. Both the anthropocentric and ecocentric may find it necessary to also address changes in the abiotic environment, i.e. increased concentrations of radionuclides in soil, water, and air. Anthropocentric support for such views may arise from aesthetics or a wish to "preserve" "pristine" environments such as the Arctic; ecocentric support may arise from considerations of the inherent value of all components of the ecosystem. To conclude, population effects can be an appropriate focus for environmental protection from ionizing radiation, but not at the exclusion of effects on individuals, ecosystems, or even the abiotic environment itself.

5.5 ASSIGNING MONETARY VALUE TO THE ENVIRONMENT

A number of philosophers and politicians are concerned by the tendency of environmental policy to attach monetary value to the environment (Barde & Pearce, 1990; Spash, 2011). A similar issue has been raised in human radiation protection (see Valentine, this issue) since a limit on the amount of money invested to reduce one manSv can be recalculated in terms of "a price on a life" (assuming, of course, that the linear, nonthreshold hypothesis holds). A similar view is provided by social ecologists who suggest that economic and cultural issues lie at the core of the most serious environmental problems we face today (Bookchin, 1991, 1993). Hence, ecological problems cannot be understood, much less, clearly resolved, without dealing with problems within society. "An environmental philosophy that fails to recognize the interrelatedness of the social and natural crisis will fail to uncover and confront the real sources of the ecological meltdown occurring today" (Bradford, 1993). The interrelationship between environment, economy and society is grounded in the principle of sustainable development, and a central component of an ecosystem approach to environmental protection (Costanza et al., 1997; Millennium Assessment, 2005). These approaches focus on the ecosystem, rather than single species, and the sustainable use of resources. They stress the inherent dynamic interactions between system components (including humans), potential feedback loops, indirect effects, and resilience. Similar ideas can be found in the concepts of ecosystem services and ecological economics, which are aimed predominantly at the ultimate benefits of ecosystems for humans, either financially or otherwise, and are arguably more human-centred than while the ecosystem approach. Nevertheless, all approaches share a fundamental recognition of the integration and interdependency of humans and the environment.

Other ecologists, however, suggest that the root of the problem is capitalism itself and, in turn, the reduction of all societal values to profits and losses. In a market economy, nothing can be sacred, since to be sacred means to be "nonexchangeable" (Kovel, 1993; Spash, 2011). The challenge is that if one does not attach monetary value to the environmental consequences of actions, then it makes it difficult to account for those consequences in a cost-benefit evaluation. Honest accounting of the interests of present as well as future generations can make environmentally damaging policies unprofitable.

An analysis of the economic consequences of the Japan tsunami and Fukushima accident on fishing industries offers an interesting perspective on the issue. The ecological economist Shunsuke Managi has pointed out that since Japanese fishing industries were heavily subsided, the government is actually saving money through fishing restrictions. Furthermore, in many areas the traditional fishing was unsustainable and outdated, hence rebuilding after the tsunami offers the opportunity for a rejuvenation of the industry (Pacchioli, 2013). There are also ecological benefits from a ban or restriction in fishing over large areas. On the negative side there can be complex social consequences caused by demographic changes in the, predominately young, people moving out of contaminated areas and not carrying on in family business. This type of holistic analysis is also in line with ecosystem approaches to environmental impact assessment, as proposed by the IUR and other environmentalists (IUR, 2012), as a possible way of reconciling anthropocentric and nonanthropocentric worldviews in practice. Recognizing some of the more fundamental concerns ecocentrics have about the links between ecological damage and monetary valuation of natural resources, perhaps the most important recommendation is that such damage (or changes) are not assessed only in terms of instrumental value, assessors should also respect the idea of intrinsic value of plants, animals, and the environment.

5.6 CONCLUSION: RELEVANCE OF THE VALUE DEBATE TO ECOLOGICAL RADIOLOGICAL PROTECTION

Difficulties in defining the basis for valuation of the environment include fundamental questions such as what exactly constitutes harming the environment and how the environment should be valued. Both of these, typically philosophical, problems arise in assessments of any environmental contaminant. Although philosophers might disagree about the *way* in which the environment should be valued, almost all philosophers would agree that damage to the environment should *matter* in risk assessment. Furthermore, most people would agree that harms caused by exposure of nonhuman species to radiation should carry weight in optimization and justification—either because the species has value in itself and/or because of the potential consequences for future human generations.

In practice, the variety of cultural and religious beliefs in the way humans perceive nature, and the differences in opinions on what has moral standing and why, can have a strong influence on the question of what it is, exactly, that we mean by ecological harm. Environmental policy needs to be able to acknowledge, respect, and protect this diversity in beliefs. It would be näive to expect radiation protection practitioner to resolve such fundamental problems within environmental philosophy, yet it is important that any framework developed should be flexible enough to incorporate both anthropocentric and ecocentric values. Although humans (as do all other organisms) use the environment instrumentally simply to survive, that does not preclude allocation of intrinsic value to biota, the abiotic environment, or ecosystems as a whole. To be successful, and broadly justifiable in practice, environmental policy needs to consider both issues (Rolston, 1991; Shrader-Frechette, 1991).

Because there are no easy answers to the challenges highlighted above, any system of environmental protection should be sufficiently flexible to allow such conflicts to be addressed. In this respect, ethics should be seen as a tool rather than a burden in policy making. Ethical evaluation can be valuable both in identifying controversies and in forcing decision makers to address the issues, and clarify the premises upon which decisions are being made. Showing that decision makers are aware of, and have considered, such conflicts is an important step in making ethical issues transparent in policy making. Ethical evaluations also encourage attempts to find alternative solutions in order to mitigate or avoid the ethical insult, and help to document the assumptions and reasons behind eventual disputes. For instance, it is helpful to know whether experts disagree on ways of managing radiation risks due to a matter of fact (e.g. they might disagree about the environmental consequences or the probable cost of remediation) or a matter of ethics (e.g. they may disagree about the relative importance of human interests against those of nonhuman species).

Ethicists put great weight on "treating like-cases equally". In this respect, protecting the environment from radiation will need to be put into context with the risks from other environmental contaminants and detriments. Unless there are clear, morally relevant grounds, radiation damage should not be treated differently than other hazards. The significant progress made in developing frameworks and tools for assessment of the effects of ionizing radiation over the past two decades (e.g. ICRP, ERICA) mean that decision makers have a much more robust scientific basis for comparison of the ecological impacts of radiation with other environmental stressors.

To conclude, there is a need for a holistic evaluation of the environmental impacts of ionizing radiation that not only considers the direct consequences on the health of humans and nonhuman species, but also the more complex social, ethical, and economic consequences of both human and nonhuman exposures. Ethical risk evaluation for both humans and the environment extends the issue of whether a risk is acceptable, into dimensions that go beyond its probability of harm; ethical risk management asks questions other than those connected simply to the size of the radiation dose and the cost of reducing those doses.

REFERENCES

Ariansen, P. (1996). Miljøfilosofi. Univesitetsforlaget: Oslo.

Baker, R. J., Hamilton, M. J., VandenBussche, R. A., Wiggins, L. E., Sugg, D. W., Smith, M. H., et al. (1996). Small mammals from the most radioactive sites near the Chernobyl nuclear reactor power plant. *Journal of Mammalogy*, *77*, 155–170.

Barde, J.-P., & Pearce, D. W. (1990). *Valuing the environment*. Paris: OCDE.

Bekoff, M., & Pierce, J. (2012). *Wild justice: The moral lives of animals*. Chicago University Press. pp. 192.

Bentham, J. (1789). *Introduction to the principles of morals and legislation*. Sect. 1, (1948 edition), New York: Hafner.

Bentham, J. (1824). Anarchical Fallacies, Art. II. In *Collected works of Jeremy Bentham (1983)*. London: Clarendon Press.

Bookchin, M. (1991). *The ecology of freedom* (2nd ed.). Montreal: Black Rose Books.

Bookchin, M. (1993). What is social ecology? In M. E. Zimmerman, J. B. Callicott, G. Sessions, K. J. Warren & J. Clark (Eds.), *Environmental philosophy* (pp. 354–373). Englewood Cliffs, NJ: Prentice-Hall.

Bréchignac, F., Bradshaw, C., Carroll, S., Jaworska, A., Kapustka, L., Monte, L., et al. (2011). Recommendations from the International Union of Radioecology to improve guidance on radiation protection. *Integrated Environmental Assessment and Management*, *7*, 411–413.

Bradford, G. (1993). What is social ecology. In M. E Zimmerman , J. B. Callicott, G. Sessions, K. J. Warren & J. Clark (Eds.), *Environmental Philosophy* (pp. 418–437). Engelwood Clifts, NJ: Prentice-Hall.

Pentreath, R. J. (1998). Radiological protection for the natural environment. *Radiation Protection Dosimetry*, *75*, 175–179.

Callicott, J. B. (1979). Elements of an environmental ethic: moral considerability and the biotic community. *Environmental Ethics*, *1*, 71–81.

Callicott, B. (1989). *In defense of the land ethic*. Albany: State University Press of New York.

Copplestone, D., Andersson, P., Garnier-Laplace, J., Beresford, N. A., Howard, B. J., Howe, P., et al. (2009). Protection of the environment from ionising radiation in a regulatory context (PROTECT): review of current regulatory approaches to both chemicals and radioactive substances. *Radioprotection*, *44*, 186–188.

Costanza, R., d'Arge, R., de Groot, R., Farberk, S., Grasso, M., Hannon, B., et al. (1997). The value of the world's ecosystem services and natural capital. *Nature*, *387*, 253–260.

Dunster, H. J., Garner, R., Howells, H., & Wix, L. F. U. (1964). Environmental monitoring associated with the discharge of low activity radioactive waste from the Windscale works to the Irish Sea. *Health Physics*, *10*, 353–362.

Ellegren, H., Lindgren, G., Primmer, C. R., & Moller, A. P. (1997). Fitness loss and germline mutations in barn swallows breeding in Chernobyl. *Nature*, *389*, 593–596.

Frankena, W. K. (1973). *Ethics*. Englewood Cliffs, NJ: Prentice-Hall.

Frankena, W. K. (1979). Ethics and the environment. In *Ethics and problems of the environment*. Notre Dame: University of Notre Dame Press.

Fritzell, P. A. (1987). The conflicts of ecological conscience. In J. B. Callicott (Ed.), *Companion to a Sand County Almanac*. Madison: University of Winsconsin Press.

Goodpaster, K. (1978). On being morally considerable. *Journal of Philosophy*, *75*, 308–325.

Hiyama., Nohara, C., Kinjo, S., Taira, W., Gima, S., Tanahara, A., et al., (2012). The biological impacts of the Fukushima nuclear accident on the pale grass blue butterfly. *Nature Scientific Reports*, *570*, 1–10.

IAEA. (2002). *Ethical considerations in protecting the environment from the effects of ionizing radiation*. TECDOC 1270, Vienna, Austria: International Atomic Energy Agency.

IAEA. (2006). *Chernobyl's Legacy: Health, Environmental and Socio-Economic Impacts*. Vienna, Austria: International Atomic Energy Agency.

IAEA. (2011). *Basic safety standards*. GSR Part 3, Vienna, Austria: International Atomic Energy Agency.

ICRP. (1977). Recommendations of the International Commission on Radiological Protection. ICRP Publication 26. *Annals of the ICRP*, *1*(3).

ICRP. (1991). 1990 recommendations of the International Commission on Radiological Protection. ICRP Publication 60. *Annals of the ICRP*, *21*(1–3).

ICRP. (2007). The 2007 recommendations of the International Commission on Radiological Protection. ICRP Publication 103. *Annals of the ICRP*, *37*(2–4).

ICRP. (2008). Environmental protection: the concept and use of Reference Animals and Plants. ICRP Publication 108. *Annals of the ICRP*, *38*(4–6).

IUR. (2002). Protection of the environment. Current status and future work. International Union of Radioecology, IUR Report 03, Osteras, Norway. (www.iur-uir.org)

IUR. (2012). *Towards an ecosystem approach for protection with emphasis on radiological hazards*. IUR Report 7 (2nd ed.).

JNREG. (1997). *Sources contributing to radioactive contamination of the Techa River and areas surrounding the Mayak Production Association, Urals, Russia*. Østerås: Joint Norwegian Russian Expert Group (JNREG).

Kant, E. (1785). *Groundwork of the metaphysic of morals*. (M. Gregor, Trans.), *Cambridge texts in the history of philosophy*. Cambridge: Cambridge University Press.

Kershaw, P. J., Pentreath, R. J., Woodhead, D. S., & Hunt, G. J. (1992). A review of radioactivity in the Irish Sea: a report prepared for the Marine Monitoring Management Group. In *Aquatic environment monitoring report 32*. Lowestoft: MAFF Directorate of Fisheries Research.

Kovalchuk, I., Abramov, V., Pogribny, I., & Kovalchuk, O. (2004). Molecular aspects of plant adaptation to life in the Chernobyl zone. *Plant Physiology*, *135*(1), 357–363.

Kovalchuk, O., Burke, P., Arkhipov, A., Kuchma, N., James, S. J., Kovalchuk, I., et al. (2005). Genome hypermethylation in *Pinus silvestris* of Chernobyl—a mechanism for radiation adaptation? *Mutation Research*, *529*(1–2), 13–20.

Kovel, J. (1993). The marriage of radical ecologies. In M. E. Zimmerman, J. B. Callicott, G. Sessions, K. J. Warren & J. Clark (Eds.), *Environmental philosophy* (pp. 406–417). Englewood Cliffs, NJ: Prentice-Hall.

Leopold, A. (1949). *A Sand County Almanac* 1997, (2nd ed.). Oxford: Oxford University Press. (See also J. B. Callicott, (Ed.) 1987. A Companion to A Sand County Almanac, University of Wisconsin, Madison).

Millennium Ecosystem Assessment. (2005). *Ecosystems and human well-being: Synthesis*. Washington, DC: Island Press. (p. 160) http://www.maweb.org/en/index.aspx.

Mycio, M. (2005). *Wormwood Forest: A Natural History of Chernobyl*. Joseph Henry Press.

Møller, A. P., & Mousseau, T. A. (2007). Species richness and abundance of birds in relation to radiation at Chernobyl. *Biology Letters*, *3*, 483–486.

Møller, A. P., & Mousseau, T. A. (2009). Reduced abundance of insects and spiders linked to radiation at Chernobyl 20 years after the accident. *Biology Letters*, *5*, 356–359.

Norton, B. (1988). *Why preserve natural variety?* Princeton: Princeton University Press.

Næss, A. (1974). *Økologi, sammfunn og livstil: Utkast til en økosofi*. Oslo: Universitetsforlaget. (Ecology, Community and Lifestyle: Outline of an Ecosophy (D. Rothenburg, Trans.), (1990), Cambridge University Press, Cambridge).

Oughton, D. H. (2003). Ethical issues in protection of the environment from ionising radiation. *Journal of Environmental Radioactivity*, *66*, 3–18.

Oughton, D. H. (2011). Social and ethical issues in environmental risk management. *Integrated Environmental Assessment and Management*, *7*, 404–405.

Oughton, D. H., Bay, I., Forsberg, E.-M., Kaiser, M., & Howard, B. (2004). An ethical dimension to sustainable resoration and long-term management of contaminated areas. *Journal of Environmental Radioactivity*, *74*, 171–183.

Pacchioli, D. (2013). Seafood Safety and Policy: what's safe to eat? How can we know? *Oceanus*, *50*, 16–19.

Pentreath, R. J. (1999). A system for radiological protection of the environment: some initial thoughts and ideas. *Journal of Radiological Protection*, *19*, 117–128.

Pentreath, R. J. (2009). Radioecology, radiobiology and radiological protection: frameworks and fractures. *Journal of Environmental Radioactivity*, *100* (12), 1019–1026.

Pomerantseva, M. D., Ramaiya, L. K., & Chekhovich, A. V. (1997). Genetic disorders in house mouse germ cells after the Chernobyl catastrophe. *Mutation Research—Fundamental and Molecular Mechanisms of Mutagenesis*, *381*, 97–103.

Regan, T. (1980). Animal rights, human wrongs. *Environmental Ethics*, *2*, 99–120.

Regan, T., & Singer, P. (1976). *Animal rights and human obligations*. Englewood Cliffs, NJ: Prentice-Hall.

Rolston, H. (1988). *Environmental ethics*. Philadelphia: Temple University Press.

Rolston, H. I., III (1991). *Challenges in environmental ethics. ecology, economics, ethics: The broken circle*. London: Yale University Press.

Sagoff, M. (1984). Animal liberation and environmental ethics: bad marriage quick divorce. *Ossgood Hall Law Journal*, *22*, 297–307.

Shrader-Frechette, K. S. (1991). *Risk and rationality*. Berkeley: University of California Press.

Singer, P. (1991). *Animal liberation: A new ethics for our treatment of animals* (2nd ed.). London: Thorsons.

Sober, E. (1986). Philosophical problems for environmentalism. In *The preservation of species: The value of biological diversity* (pp. 180–188). Princeton: Princeton University Press.

Spash, C. L. (2011). Social ecological economics: Understanding the past to see the future. *American Journal of Economics and Sociology*, *70*, 340–375.

Sterba, J. T. (1994). Reconciling anthropogenic and nonanthropogenic environmental ethics. *Environmental Values*, *3*, 229–244.

Taylor, P. (1986). *Respect for nature*. Princeton: Princeton University Press.

United Nations. (1992). Rio-declaration, Article 15.

UNSCEAR. (2008). *United Nations Scientific Committee on the Effects of Atomic Radiation. Report to the general assembly, Annex e*. New York: UNSCEAR.

Warren, K. J. (1990). The power and the promise of ecological feminism. *Environmental Ethics*, *12*, 125–146.

White, L. (1967). Science, 155, 1203–1207.

Wilson, E. O. (1984). *Biophilia*. Cambridge: Cambridge University Press.

Chapter 6

Why Chemical Risk Assessment can Learn from Radiation Risk Assessment

Carl F. Cranor
Environmental Toxicology, University of California, Riverside, CA, USA
E-mail: carl.cranor@ucr.edu

Chapter Outline

6.1 Introduction 87
6.2 Some Principles and Presumptions of Radiation Protection 88
6.3 Contamination 90
6.4 The Developmental Basis of Disease 92
6.5 Contamination of Developing Children 92
6.6 Adverse Health Effects 93
6.7 Particular Substances have No Obvious Thresholds 98
6.8 A Unified Approach to Dose-response Assessment 99
6.9 Conclusion 100

6.1 INTRODUCTION

The biological models for radiation and many chemicals have been seen as having substantial differences. Radiation has no lowest exposure at which adverse health effects result and the dose–response curve has been assumed to be linear from levels at which it can be measured to the no-exposure level—lower doses are less likely to cause adverse health effects, but they are never nonexistent. In contrast, at least some chemical exposures have been assumed to have threshold effects for individuals, exposure levels below which for a chemical taken in isolation no adverse health effects occur. Thus, although there can be exposures above the threshold at which humans are adversely affected, once exposure is less than the threshold level, there are no adverse effects. However, thresholds might be different for different individuals and identifying thresholds for populations is more difficult (more on this below).

Radioactivity in the Environment, Volume 19
ISSN 1569-4860, http://dx.doi.org/10.1016/B978-0-08-045015-5.00006-X

The picture just described no longer seems applicable or at best quite misleading. The emerging science of the developmental origins of disease reveals that very low, even tiny, doses that would not contribute to harm in adults can contribute to harm in developing children. At odds with the chemical threshold model, researchers have identified some substances that appear to have no lowest safe dose (at least, to date), much like radiation. More importantly, since we now live in a world in which people are surrounded and permeated by chemical substances and we have different individual thresholds for adverse effects, even if particular chemicals act by means of threshold mechanisms, these conditions together suggest that responses to chemical contamination should begin to incorporate policy responses similar to those of radiation in order to properly protect the general public. The upshot is that protection from chemical exposures should begin to incorporate presumptions of no threshold in order to protect the public. This presumption could be overridden only if there were good evidence contrary to the background conditions of the presumption.

6.2 SOME PRINCIPLES AND PRESUMPTIONS OF RADIATION PROTECTION

Radiation can produce two different kinds of effects on humans: tissue reactions and stochastic effects. "*Tissue reactions*...are characterized by a threshold dose, above which the effects always occur ... Tissue reactions are caused by the extensive damage or killing of living cells in organs and are generally limited to accidents or controlled medical circumstances" (Wikman-Svahn, 2012). In contrast, *stochastic effects* "do not necessarily occur in an exposed individual, but with a certain probability. Stochastic effects are caused by modification of cells (e.g. damage to the DNA), which may lead to the development of cancer and hereditary diseases" (Wikman-Svahn, 2012).

Somewhat oversimplifying, the different biological reactions lead to two different models: tissue damage effects are based on the idea that the threshold dose represents a cutoff between damage and no damage. Below the threshold no damage is presumed to occur, but above the threshold the tissue effects occur in an exposed individual. In contrast, "the risk of stochastic effects is best represented by a linear dose–response relationship—the so-called linear no-threshold (LNT) model" (Wikman-Svahn, 2012).

In what follows in order to contrast adverse effects from chemicals with adverse effects from radiation I focus on stochastic effects and the linear no-threshold (LNT) model. The stochastic model seems appropriate for both cancer risks and for hereditary risks, or risks to the germ cells of person that are passed from one generation to another. Per Wikman-Svahn, in a recent doctoral dissertation from the Royal Institute of Technology, summarizes the main conclusions concerning these effects. "The mainstream scientific view on these matters ... is that a threshold for stochastic effects is not likely and that a linear dose–response relationship for small doses is more credible than other alternatives. This linear relationship does not necessarily apply to each individual or each cancer type, but is seen as representing the average response in a population and over all cancer types." The LNT model does generate some disagreement; some scientists believe that it underrepresents risks from radiation, because at low doses radiation causes more damage than the linear model suggests, while others believe that it overestimates risks because of its linearity (Wikman-Svahn, 2012).

Despite some degree of disagreement within the literature, for what follows below I assume that for cancers and adverse hereditary effects from radiation exposure the LNT model best describes the biology. For exposures to chemicals early in the history of chemical carcinogenesis scientists appeared to believe that chemicals similarly contributed to harm by means of a simple linear model. Subsequent research for carcinogenesis has shown that there are various mechanisms for cancer, not all occurring by means of LNT effects and not caused by a single hit from a chemical.

More recently, there has been more emphasis on threshold models for noncancer harms caused by chemicals. Certainly, in regulatory or tort law contexts in the U. S. industries subject to regulation or to personal injury suits often emphasize the threshold models to explain adverse effects from chemical substances. The reason seems clear: if there are exposures to the chemical that are below the presumed threshold, then there is no case for reducing exposures to the substance and there is no legal case to be made in the tort or personal injury law that an individual exposed to chemicals below the threshold has been harmed by the exposure. A clear biological border between harm and safety makes certain legal arguments much easier, and if a threshold has not been exceeded, this tends to remove the legal rationale for regulatory action and to exonerate a company from tort suits.

Recent scientific research and some subtleties about mechanisms for harm from chemical exposures throw this overly simple assumption into question. In what follows I summarize some recent scientific findings that suggest myriad exposure circumstances support an argument for a policy and legal approach to chemical exposures that more closely resembles the legal and policy response to radiation than the threshold model. In short, despite biological evidence for threshold effects from exposures to some individual chemicals, a general approach that emphasizes the threshold model seems to be misplaced. Wise policy to protect the public from harm from chemical exposures should shift this presumption. It seems much better to presume that chemical exposures contribute to harm by means of something like a no-threshold model than a threshold model. Only if there is good evidence for a threshold approach, given all the exposure conditions and all that is known about the biology of the chemical in human bodies as we find them, should a threshold approach be followed.

6.3 CONTAMINATION

The U.S. Centers for Disease Control (CDC) has a biomonitoring program that has testing protocols for measuring the amounts of industrial chemicals in a person's blood or urine in order to determine concentrations in his or her body. Such measurements identify the concentrations of substances in one's body from, "all routes of exposure—inhalation, absorption through the skin and ingestion, including hand-to-mouth transfer by children." More importantly, biomonitoring reveals the integrated effect of different exposures to the same substance (Sexton, Needham, & Perkle, 2004).

Moreover, the CDC chose the particular substances for investigation because they either constitute substantial exposures or are known or suspected toxic hazards or both. They are called chemical *hazards* because they have intrinsic toxic properties or a "built-in ability to cause an adverse effect" (Faustman & Omenn, 2001; Heinzow, 2009).

The CDC's research is revealing the extent to which U.S. citizens are contaminated by substances of concern. In 2005, the CDC had reliable protocols to identify 148 industrial chemicals in citizen's blood and urine (U.S. Department of Health and Human Services, Centers for Disease Control and Prevention, Third National Report on Human Exposure to Environmental Chemicals, 2005). In 2009, it had protocols for 212 substances (U.S. Department of Health and Human Services, Centers for Disease Control and Prevention, 2009). Currently, it lists more than 300 environmental chemicals or their metabolites in the U.S. citizens. (U.S. Department of Health and Human Services, Centers for Disease Control and Prevention, Environmental Chemicals, 2013).

The significance? Each person is contaminated to a greater or lesser degree, as various studies have shown (more below). Humans are not just exposed to industrial chemicals external to their bodies, but the substances enter our bodies via inhalation, ingestion, or skin absorption. Beyond this, they invade our internal tissues and biological processes. According to Larry Needham, Director of the program, all but the very largest macromolecules will invade our bodily tissues and be processed by various metabolic routes (Needham, 2007).

As Environmental Defense puts the point based on a small study of Canadians, "No matter where people live, how old they are or what they do for a living, they are contaminated with measurable levels of chemicals that can cause cancer and respiratory problems, disrupt hormones, and affect reproduction and neurological development" (Environmental Defense, 2005).

Moreover, since all of the substances identified to date are known or suspected toxicants, these findings are worrisome. Of special concern is that industrial chemicals can penetrate deep into a person's body. For example, when a woman is pregnant, most industrial chemicals, pesticides, and pharmaceuticals can cross the placenta and enter the womb, depending upon such properties as size, electric charge, fat solubility, and so on. As one of the leading experts puts the point, "It is clearly evident that there really is no placental barrier per se: The vast majority of chemicals given the pregnant animal (or woman) reach the fetus in significant concentrations soon after administration." (Schardein, 2000) Such substances can even contaminate the very tissues that go into creating a child before parents ever decide to have a child. This includes women's eggs and men's sperm, genetic sources of children. In addition, many other tissues in their bodies have intimate contact with industrial chemicals.

Once a child is born and begins nursing most substances can similarly enter the breast milk, be conveyed to the child and transfer some of a mother's body burden of industrial chemicals to the child (Heinzow, 2009). The consequence is that even the youngest, most innocent, and seemingly the most pristine of humans experiences intimate contamination of their tissues and bodily organs from conception onwards.

Unlike nuclear radiation, for which many or most sources tend to be associated with workplaces, chemical contaminants are all around us and very close to home. When we use cosmetics or sunblock, we absorb some phthalates through the skin. Some lipsticks can add to the lead in one's body that is present from past exposures to leaded gasoline, lead paint or deposited in the environment. Tap water or vegetables contain small amounts of a component of rocket fuel, fireworks, or munitions, perchlorate. Furniture, drapes, electronic equipment, including television sets and computers, contain some brominated fire retardants, polybrominated diphenyl ethers (PBDEs). They are not chemically bound to the fabrics or plastics, but are merely mixed in, so over time they can disperse into our homes, house dust and ultimately into our bodies. In the U.S., concentrations of PBDEs in citizens' bodies are rapidly increasing even though some steps have been taken to reduce the production and use of some of these chemical products. Recently created chemicals in domestic and international markets are not the only concern; legacy chemicals such as PCBs and DDT have been in the environment and in our bodies for decades. PCBs and the more recent PBDEs travel around the world, enter the ocean, and contaminate ecosystems and animals (Cone, 2005). Indeed, PBDEs have been found in Tasmanian devils, hundreds of miles from any industrialized society (Denholm, 2008).

Phthalates appear to contribute to premature breast development, sex organ problems in males and some reproductive and developmental risks (Rawlins, 2009; Swan et al., 2005). Lead is a well-known neurotoxicant, adversely affecting learning, IQ, and behavioral controls. It also contributes to cardiovascular disease. Adverse effects can occur at surprisingly low concentrations and for some no known safe level has been identified (Navas-Acien, Guallar, Silbergeld, & Rothenberg, 2007; Wigle & Lanphear, 2005). Perchlorate in water can be a special problem for pregnant women, children developing in utero or even newborns. Perchlorate can interfere with thyroid hormones needed for brain development. Pregnant women who have too little circulating thyroid hormone may adversely affect their children's brain development; chemical exposures can contribute to this problem. When young children have too little thyroid hormone, this can interfere with brain development (more below) (Woodruff et al., 2008).

6.4 THE DEVELOPMENTAL BASIS OF DISEASE

Major scientific developments associated with what is now being called the "developmental origins of health and disease" (which I will largely refer as the developmental origins of disease) are leading to a reassessment of the sensitivity of humans to toxic substances. Several considerations support this research: the placenta that had been seen as protecting a developing fetus is no longer considered a barrier to many toxic substances; scientists now understand that humans are exposed to many more substances and exposed earlier in life than previously; during in utero and postnatal development, humans (and mammals more generally) are quite sensitive to toxic influences; and, finally, these effects are exacerbated by a number of other factors. I consider each of these in turn.

This research does not necessarily show that a threshold model of toxicity is not correct at least for quite limited circumstances, but it strongly suggests that any thresholds can be quite low and much lower for developing children than for adults. However, once this information is combined with data about exposures to myriad substances as well as the additive and sometimes-synergistic effects between substances, this supports a presumption for adopting a nonthreshold model for chemical toxicants.

6.5 CONTAMINATION OF DEVELOPING CHILDREN

As introduced above, James Schardein points out, "there really is no placental barrier per se ... " (Schardein, 2000). Toxicologists Rogers and Kavlock (2001) concur: "virtually any substance present in the maternal plasma [blood] will be transported to some extent by the placenta." These findings reject an older view of the womb as a safe, protected capsule within which a child develops, following its own genetic program. In contrast, it is probably better to understand the womb within a woman's body as an internal *environment* that provides food, fluids, and sound (Soto, 2007). However, if this environment contains toxicants, as we now know that all human bodies do, a developing child is exposed to those substances as well. This internal "environment" can expose a child to toxicants by the same routes that provide nourishment and fluids.

Because the placenta constitutes no, or is at best a limited, barrier to chemicals, any contamination of a pregnant woman is likely shared with the children developing in utero. For instance, despite the sound advice for mothers to nurse their newborns, nursing does not protect infants from toxicants. A nursing child begins to ingest toxicants from its mother's body from its first drink. In effect, this transfers some of a mother's body burden of industrial chemicals to the child (Heinzow, 2009).

Consequently, for the above reasons children are not protected from chemical substances until they are born and enter the world as independent living beings; they are contaminated in utero and are born already tainted by industrial chemicals, many of them known toxicants. A news article reported that newborns were tainted with up to 200 industrial chemicals (Fimrite, 2009).

A scientific study in Minneapolis found a significant proportion of children from poor sections of the city have been found contaminated with more than 75 substances or their metabolites: phthalates, metals [lead, mercury], organophosphate pesticides, organochlorine pesticides, polychlorinated biphenyls (PCBs), volatile organic compounds, cotinine (an ingredient in cigarette smoke), environmental tobacco smoke (ETS) (Sexton, Ryan, Adgate, Barr, & Needham, 2011). All the contaminants include known or suspected carcinogens, endocrine disrupters, neurotoxicants, and developmental and respiratory toxicants. At some concentration level all of these will pose risks of disease; some acting alone may cause harm by threshold mechanisms, while others may contribute to harm by linear mechanisms. One small study found 232 industrial chemicals in the umbilical cords of newborns (Environmental Working Group, 2009). As a consequence, scientists now understand that humans are exposed to many more substances and exposed earlier in life than previously.

6.6 ADVERSE HEALTH EFFECTS

Developing children are especially vulnerable to adverse health effects and typically much more susceptible to them than adults because they are in one of the most sensitive life stages. Whatever organ system one considers—the brain, the immune system, reproductive system, or the lungs—each is typically much more vulnerable to toxic harm than the same system in adults. While not all exposures during development will contribute to adverse effects, the fact that developing children are especially sensitive to toxicants is quite worrisome.

Moreover, developing children are typically subject to greater exposures than adults on a body weight basis. According to the consensus statement of first conference on the developmental origins of disease, "the mother's chemical body burden will be shared with her fetus or neonate, and the child may, in some instances, be exposed to larger doses relative to the body weight" (Grandjean et al., 2008). Methylmercury concentrations in the fetal brain can be as much as five times greater than concentrations in the mother's blood (Grandjean et al., 2008; Honda, Hylander, & Sakamoto, 2006) Breast-fed infants may have greater concentrations of lipophilic (fat soluble) toxicants, since breast milk contains considerable fat. For instance, a nursing child's daily dose of PCBs in the breast milk "may be 100-fold higher" than the concentration of the PCBs in the mother's blood "resulting in much greater toxic concentrations in the child than in the mother" (Grandjean et al., 2008). Not all lipophilic toxicants will show similar increases in breast milk, but this seems to be the case for PCBs.

In addition, during development children have lesser defenses than adults. A child's immune system is not developed in utero or at birth. A mother's immune system offers some protection for the child in utero, but her immune system offers less protection for each of them of them considered separately than it would for the mother alone (Talbot, 2009). The blood–brain barrier, which evolved to protect the brain from some toxicants, does not develop until about six months after birth. Once developed it imparts protection against some chemicals entering the brain. Similarly many enzymes that can detoxify toxic substances are often poorly developed in young children, resulting in greater toxic insults to children than adults from industrial contamination. (Interestingly, some enzymes that *increase* the toxicity of comparatively less toxic substances may not have matured, so sometimes children can have greater protection than adults.)

The points above represent a few of the general or typical biological tendencies of developing children that can increase their vulnerability to toxic insults. However, when genetic variability and diversity are considered, the range of adverse effects increases.

For instance, vulnerability to organophosphate pesticides can "vary by age and genotype." Children as well as adults with a variant of a particular gene have lower levels of an enzyme that assists in metabolizing organophosphate pesticides. Having less of this particular enzyme puts them "at higher risk of health effects from organophosphate exposure." (Eskenazi et al., 2008) Potential effects include neurotoxic effects as well as some cardiovascular endpoints (Ecobichon, 2001).

For another example, polycyclic aromatic hydrocarbons (PAHs), formed during incomplete combustion of organic compounds from the combustion of coal, gas and oil, and from side stream and secondhand tobacco smoke can cross the placenta and bind to (or create adducts on) DNA (Perera, Jedrychowski, Rauh, & Whyatt, 1999). This typically alters the DNA's function and causes mutations or incorrect repair leading to cancers or other diseases. Subpopulations of fetuses with more PAH-DNA adducts show increased sensitivity to genetic damage compared with the mother and compared to others (Miller et al., 2002; Perera et al., 1999). This can lead to smaller head circumference, associated with other adverse effects, as well as genetic damage in the newborn (Perera et al., 1999).

As a consequence, while an average or typical child might not be susceptible to a particular contaminant at a particular concentration, human genetic variability can increase or decrease the extent of sensitivity. This fact of biology increases the range of susceptibility of developing children to adverse effects compared with adults.

The greater vulnerability of developing children to disease has a further consequence less typical of adults. Because young children have more years of future life ahead of them than adults, if children are contaminated with toxicants before they are born or in early childhood, and disease processes are quickly initiated, there is more time for diseases or dysfunctions to fully develop so they can be clinically detected during a lifetime. A disease process might require one, two or three critical steps to occur before the disease is fully initiated. However, if one or two steps occur in utero, as DES likely did, or in early childhood, as occurs with lead, then fewer steps would need to occur later in life for full-fledged disease or dysfunction to appear (Heindel, 2008). Miller et al. (2002)

point out, "Cancer is a multistage process and the occurrence of the first stages in childhood increases the chance that the entire process will be completed, and a cancer produced, within an individual's lifetime."

The general vulnerability of children plus greater exposures and (generally) lesser biological defenses than adults have resulted in risks of diseases for developing children. All of these processes are exacerbated by genetic variation that can increase vulnerability to toxicants. Moreover, toxic effects in developing children usually occur at much lower concentrations than those that cause adverse effects in adults.

Adult humans who ingested fish contaminated with methylmercury from Minimata Bay in Japan suffered adverse effects, but children who were contaminated in utero experienced quite catastrophic effects (Honda et al., 2006). Children contracted cerebral palsy at 10 times the rate of unexposed children and a number died (Weiss, 1994). In part, this occurred because they had much greater exposures to methylmercury in the brain, which has a selective affinity for it, and, of course, they were in general much more susceptible to adverse effects than adults (Honda et al., 2006).

In utero exposure to the synthetic estrogen diethylstilbestrol (DES) caused dramatic rates of early life vaginal cancer in young women (about 20 years of age) and also increased breast cancer in DES daughters as they reached middle age (Kortenkamp, 2008). DES mothers do not appear to have suffered cancer of the reproductive tract, but have subsequently experienced an elevated rate of breast cancer because of DES they took decades earlier (Titus-Ernstoff et al., 2001). Similarly, while Thalidomide caused some peripheral neuropathy in some women who took it, this sedative generally seemed to have benefited them. However, developing children exposed in utero to the Thalidomide their mothers' ingested suffered terrible physical abnormalities and birth defects along with neurological problems (Landrigan, Kimmel, Correa, & Eskenazi, 2004). Some anticonvulsive drugs can reduce convulsions in women prone to them (for example, because of epilepsy), but can cause birth defects in children exposed to them in utero (Landrigan et al., 2004).

Children have higher rates of leukemia and thyroid cancer from radiation exposure than adults at similar exposures. Teenage women exposed to radiation tend to have higher rates of breast cancer than older women similarly exposed (Miller et al., 2002). In addition, women younger than 14 who were exposed to greater concentrations of DDT when it was in widespread use in the U.S. contracted breast cancer at a fivefold higher rate than older women with similar exposures (Cohn, Wolff, Cirillo, & Sholtz, 2007). For developing children whose blood–brain barriers have not developed, cadmium and monosodium glutamate can "enter the developing brain freely" (Rodier, 1995). Some hormones can have adverse effects at exceedingly low levels. For instance, Tamoxifen, which is now used to treat breast cancer, promotes cancer at two or more orders of magnitude below therapeutic levels (Vandenberg et al., 2012).

To this point I have mentioned adverse effects in humans who were exposed to one substance at a time. However, as presented above, a more realistic understanding of exposures is that we are all routinely contaminated by multiple industrial chemicals, many of them toxic.

Some substances add to the toxic effects of other compounds. This is seen with estrogen mimicking compounds, dioxin-like compounds and androgen antagonists. That is, some substances add their toxic effects together because toxicants and naturally occurring biochemicals in the body attach to the same cellular receptor (Simon, Britt, & James, 2007). For estrogens, it appears that a woman exposed to more estrogen from endogenous or exogenous sources over a lifetime is at greater risk for breast cancer (Kortenkamp, 2008). Thus, when a person is contaminated by toxicants that attach to the same receptor, this can increase risks of any diseases they cause.

In addition to this point, there are more general additive effects that pose concerns. Woodruff et al. (2008) have discussed several compounds that can function via different biological pathways but that cause the same adverse effects. For example, pregnant women need sufficient levels of thyroid hormones to facilitate proper neurological, including brain, development of their children. If circulating thyroid hormones are too low in a pregnant woman, a developing child can experience poor brain development. Women could have insufficient thyroid hormones because of their circumstances, e.g. too little iodine in their diets. However, even if this were not the case, Woodruff et al. have shown that one class of substances, e.g. dioxins, dibenzofurans and dioxin-like PCBs, adversely affect one group of liver enzymes reducing thyroid hormones, while another class of compounds, e.g. nondioxin-like PCBs, affect different liver enzymes that also reduce circulating thyroid hormones. It also appears that the brominated fire retardants (PBDEs) along with perchlorate, a discarded rocket fuel and fireworks component, can also contribute to similar adverse effects, but by two additional and different pathways (Woodruff et al., 2008). Thus, although the four classes of substances act by four different biological pathways, the substances produce "a dose-additive effect on [thyroid hormones] at *environmentally relevant* doses ... demonstrating exposures to chemicals acting on different [biological] pathways can have cumulative effects..." Consequently, "It is appropriate to presume cumulative effects unless there is evidence to the contrary, and it is important for risk assessments to consider real-life exposure mixtures" (Woodruff et al., 2008).

When the above research is combined with the data indicating that humans are contaminated by a number of substances, this shows that people can be much more vulnerable to toxic insults than had previously been considered. The conceptual point is that if a population had no exposures to other substances that could contribute to the same adverse endpoint and no special biological susceptibility, then exposures from a single substance might cause no adverse effects in the population. However, when co-exposures are considered, even without any biological susceptibility to the exposures, the co-exposures plus a new exposure that could contribute to the same adverse effects could be sufficient to push some portion of the population into the range at which adverse effects occur. Finally, if the population has co-exposures and some factors that increase biological susceptibility, then a larger percentage of the population would be at risk from a new exposure (Woodruff et al., 2008).

The factors that can increase biological susceptibility include prominently (1) genetic variation among adults, (2) the special susceptibility of children during development, and (3) the genetic variation of children during their vulnerable developmental period. This research strongly supports the idea that new exposures could induce adverse effects at much lower levels than a single exposure taken by itself and in absence of any particular biological susceptibility. Thus, a risk assessment for exposure to a new substance is not properly considered in isolation from co-exposures and the much greater susceptibility of some subpopulations. The proper level of a new exposure against the background of a heterogeneous and already contaminated population very likely may require reducing it to a level as low as practically possible to prevent harm to the most susceptible subpopulations. Of course, this would need to be addressed case by case.

This point can be illustrated in another way by considering substances that are assumed to act by means of threshold mechanisms (introduced above), and thus contrary to the LNT model of radiation. As a preliminary point, we should note that thresholds are appropriate for individuals and much more difficult to identify for populations because of interindividual variability (Lutz, 1990). Once genetic variation within subpopulations of humans is taken into account, a threshold, nonlinear model with a comparatively high threshold can be transformed into a linear model.

The argument proceeds as follows. Assume that the substance in question is revealed by animal data to act by means of a mechanism that produces adverse effects in animal population A by means of a threshold and nonlinear mechanism. This tends to be what is seen in many animal studies as a result of exposure to a single chemical.

However, when a second homogeneous but genetically somewhat different animal population B is similarly dosed and the adverse effects are combined into a single graph the shape of the dose–response curve has changed—it has two thresholds at which diseases can be induced. The reason is that the two somewhat genetically different populations are affected at different thresholds by the exposure.

Going a step further and assuming that a larger number of genetically different populations in which there exists the so-called polygenic determinants of sensitivity are similarly dosed with a single substance, the dose–response curve would change again. It would resemble a shallow and rounded step function reflecting the different thresholds at which the substance triggers the same cancers at different concentrations because of the different genetic susceptibilities in the populations. Finally, when both population heterogeneity and life style contributions in human populations add to adverse effects, this results in a *linear* dose–response relationship from the highest doses to the tumor rate of the control group (Lutz, 1990).

Lutz (1990) summarizes his conclusions about assessing the risks from carcinogens this way: There are a large number of toxicological mechanisms "that generate nonlinear parts in the dose–response curve in a *homogeneous* population." (Emphasis added) However, at the lowest doses, scientists typically see linear dose–response. This is particularly true for genotoxic carcinogens, substances that form adducts on DNA. (At higher doses, he also points out that there are many possibilities for nonthreshold reactions, but typically the main concern is what occurs at low doses.) However, even when there are nonthreshold mechanisms at work in the biological processes leading to cancer, a sufficient number of those in a heterogeneous population can result in a linear process.

Overall, Lutz concludes, "For risk assessment in a heterogeneous [human] population, therefore, linear extrapolation from the high-dose incidence to the control rate has to be taken into consideration even if the mechanism of action would result in a nonlinear shape of the dose–response curve in a homogeneous population" (Lutz, 1990). The reason? "In a heterogeneous population such as humans, nonlinear shapes of the dose–response curve are linearized by the presence of genetic and life-style factors that affect the sensitivity for the development of cancer." Consequently, even though studies in animals administered a single substance in isolation and subject to no other external carcinogens typically shows a threshold mechanism, epidemiological studies support the linear, no threshold view. "In human studies, significant deviation from linearity are more difficult to find…" and these are found in only two reports (Lutz, 1990).

The Lutz argument importantly augments the Woodruff et al. (2008) findings. Multiple hits by a carcinogen in a genetically heterogeneous population not only lowers the risk level from additional substances, it also tends to make the dose–response curve linear. Thus, for carcinogens in a heterogeneous population, the dose–response curve tends to be linear even though many particular mechanisms contributing to cancer tend to act by thresholds and tend to be nonlinear. More subtle research may reveal similar patterns for multiple hits from noncancer-causing substances. Moreover, mutagenic carcinogens—cancer-causing substances that cause genetic mutations—independent of the considerations Lutz adduces, appear to have no threshold (Eastmond, 2010).

6.7 PARTICULAR SUBSTANCES HAVE NO OBVIOUS THRESHOLDS

Beyond the discussion above about the exquisite sensitivity of developing children, the Woodruff et al. discussion of multiple chemical exposures, and the Lutz arguments about the linearity of dose–response to carcinogens, several noncarcinogenic chemical agents either have no known lowest dose or have caused adverse effects at such low concentrations that they might be considered to have no safe dose for all practical purposes.

According to epidemiological studies, there appears to be no threshold for lead toxicity during development, early childhood, or even adulthood. This appears to be the case whether one considers neurological or cardiovascular effects (Bellinger & Needleman, 2003; Canfield et al., 2003; Lanphear, 2005; Silbergeld & Rothenberg, 2007).

For at least one thalidomide baby a single dose of one 50 mg (or perhaps one 100 mg) pill caused malformations (Claudio, Kwa, Russell, & Wallinga, 2000). Thus, for at least some of the most susceptible children, there would seem to be no practical safe dose. Similarly, a single dose of valproic acid (anti-epileptic drug) in animal studies has been shown to cause autism-like behavior (Dufour-Rainfray et al., 2011). Scientists conducting research into estrogens have found that a single 8 part per billion dose of various synthetic estrogens modify the epigenome of mice and cause obesity (Vom Saal, 2011).

6.8 A UNIFIED APPROACH TO DOSE-RESPONSE ASSESSMENT

To conclude this argument, a committee of the National Academy of Sciences has recommended that in order "to evaluate risks in ways that are consistent among chemicals, that account adequately for variability and uncertainty, and that provide information that is timely, efficient, and maximally useful for risk characterization and risk management," the U.S. Environmental Protection Agency must address the challenges revealed by the above science.

The committee notes, "For cancer it has generally been assumed that there is no dose threshold of effect, and dose–response assessments have focused on quantifying risk at low doses and estimating a population risk for a given magnitude of exposure. For noncancer effects, a dose threshold (low-dose nonlinearity) has been assumed, below which effects are not expected to occur or are extremely unlikely in an exposed population …" However, "Noncancer effects do not necessarily have a threshold, or low-dose nonlinearity, and the mode of action of carcinogens varies. Background exposures and underlying disease processes contribute to population background risk and can lead to linearity at the population doses of concern." And because reference dose cutoffs that are typically used for substances that act by threshold mechanisms, "do not quantify risk for different magnitudes of exposure but rather provide a bright line between possible harm and safety, their use in risk–risk and risk-benefit comparisons and in risk-management decision-making is limited" (National Research Council, 2009).

Consequently, "Scientific and risk-management considerations both support unification of cancer and noncancer dose–response assessment approaches. The committee therefore recommends a consistent, unified approach for dose–response modeling that includes formal, systematic assessment of background disease processes and exposures, possible vulnerable populations, and modes of action that may affect a chemical's dose–response relationship in humans." This unified approach appears to treat both processes as being essentially linear until there is evidence to the contrary (National Research Council, 2009). However, this approach will require EPA in its risk assessments to judge what percentage of a population below a reference dose cutoff if there is one may still be at risk and to assess the benefits and costs of protecting that group.

6.9 CONCLUSION

The above research suggests that scientists must reconceptualize how they approach the assessment of potential adverse health effects from industrial chemicals. They should imaginatively conduct research before exposure to the extent this is possible in order to prevent health problems from arising in the first place (Cranor, 2011). However, to the extent this may not be permitted by existing laws, wise policy based on recent science seems to recommend that even after the fact risk assessments should shift the presumption toward a no threshold model, much like that utilized in assessing the risks from radiation exposures, in order to protect the public from harm. Only if there is good evidence for a threshold approach, given all the exposure conditions and all that is known about the biology of the chemical in human bodies as we find them, should a population threshold approach be followed.

From above, the reasons for this are several. Heterogeneous human populations are much more vulnerable to harm that heretofore have been considered. This is especially true for developing children. There is both a wide range of genetic and other biological heterogeneity. In addition, most humans are already contaminated by a hundreds of industrial chemicals as part of everyday living. Biological heterogeneity and existing contamination are likely to shift larger portions of the population into a range of vulnerability to disease, even at low levels of exposure. One might put this point another way. It is a mistake to infer that because a single substance tested in a homogeneous population of rodents shows threshold mechanisms of action, when a heterogeneous human population is already exposed to hundreds of chemicals, there will be a population threshold. It might turn out that a population threshold could be identified, but the emerging body of scientific evidence suggests that the presumption should be against it and a good evidence for a threshold would be needed to overcome a nonthreshold presumption.

Risk assessment for radiation is based on radiation reaching target cells and causing cancer, whereas chemicals typically must be metabolized by human bodies before they reach cells and do damage. And, radiation appears to cause harm directly as a result of exposure. A presumption in favor of a no threshold model for assessing the risks from industrial chemicals in part results from the fact that our world and the people and animals in it have been so contaminated by myriad chemicals, that even when an individual substance might act by means of a threshold mechanism, in the actual world with biologically heterogeneous and contaminated populations, the much better scientific and public health presumption should be that in the actual world as we know it substances will act by nonthreshold mechanisms against this background.

This above research strongly suggests that when a substance taken in isolation acts by means of a threshold mechanism in individuals, any population thresholds can be extremely low and much lower for developing children than for adults. However, once this information is combined with data about exposures to myriad substances as well as the additive and sometimes-synergistic effects between substances, this supports a presumption for adopting as a starting point for risk assessment a nonthreshold model for chemical toxicants. This presumption could be overridden, but a threshold model for actual human exposures now may well be the exception rather than the rule as had been previously believed.

REFERENCES

Bellinger, D., & Needleman, H. L. (2003). Intellectual impairment and blood lead levels. *New England Journal of Medicine*, *349*(5), 500–502.

Canfield, R. I., Henderson, C. R., Cory-Slechta, D. A., Cox, C., Jusko, T. A., & Lanphear, B. P. (2003). Intellectual impairment in children with blood lead concentrations below 10 micrograms per deciliter. *New England Journal of Medicine*, *348*, 1517–1526.

Claudio, L., Kwa, W. C., Russell, A. L., & Wallinga, D. (2000). Testing methods for developmental neurotoxicity of environmental chemicals. *Toxicology and Applied Pharmacology*, *164*, 1–14.

Cohn, B. A., Wolff, M. S., Cirillo, P. M., & Sholtz, R. I. (2007). DDT and breast cancer in young women: new data on the significance of age at exposure. *Environmental Health Perspectives*, *115*, 1406–1414.

Cone, M. (2005). *Silent snow: The slow poisoning of the arctic*. New York: Grove Press.

Cranor, C. F. (2011). *Legally poisoned: How the law puts us at risk from toxicants*. Cambridge, MA: Harvard University Press.

Denholm, M. (January 22, 2008). *Cancer agents found in Tasmanian devils*. News.Com.AU, www.theaustralian.com.au/news/cancer-agents-in-tassie-devils/story-e6frg6ox-1111115. (Last visited on 13.05.08).

Dufour-Rainfraya, D., Vourc'ha, P., Tourleta, S., Guilloteaua, D., Chalona, S., & Andres, C. R. (2011). Fetal exposure to teratogens: evidence of genes involved in autism. *Neuroscience and Biobehavioral Reviews*, *35*, 1254–1265.

Eastmond, D. A. (2010). Personal communication.

Ecobichon, D. J. (2001). Toxic effects of pesticides. In Curtis D. Klaassen (Ed.), *Casarett and Doull's toxicology* (6th ed., pp. 763–810). New York: Pergamon Press.

Environmental Defense. (2005). *Toxic nation: a report on pollution in Canadians*. (pp. 1–42), Located at: www.toxicnation.ca. Accessed 29.09.08.

Environmental Working Group. (2009). *Pollution in people: Cord blood contaminants in minority newborns*. Located at: www.ewg.org/minoritycordblood/fullreport. Accessed 12.03.11.

Eskenazi, B., Rosas, L. G., Marks, A. R., Bradman, A., Harley, K., Holland, N., et al. (2008). Pesticide toxicity and the developing brain. *Basic and Clinical Pharmacology and Toxicology*, *102*, 228–236.

Faustman, E. M., & Omenn, G. S. (2001). Risk assessment. In Curtis D. Klaassen (Ed.), *Casarett and Doull's toxicology* (6th ed., pp. 83–104). New York: Pergamon Press.

Fimrite, P. (December 3, 2009). *Study: Chemicals, pollutants found in newborns*. San Francisco Chronicle, www.sfgate.com. Accessed 03.12.09.

Grandjean, P., Bellinger, D., Bergman, A., Cordier, S., Davey-Smith, G., Eskenazi, B., et al. (2008). The Faroes statement: human health effects of developmental exposure to chemicals in our environment. *Basic and Clinical Pharmacology and Toxicology*, *102*, 73–75.

Heindel, J. J. (2008). Animal models for probing the developmental basis of disease and dysfunction paradigm. *Basic and Clinical Pharmacology and Toxicology*, *102*, 76–81.

Heinzow, B. G. J. (2009). Endocrine disruptors in human breast milk and the health-related issues of breastfeeding. In I. Ian Shaw (Ed.), *Endocrine-disrupting chemicals in food* (pp. 322–355). Cambridge: Woodhead Publishing.

Honda, S., Hylander, L., & Sakamoto, M. (2006). Recent advances in evaluation of health effects on mercury with special reference to methylmercury—a minireview. *Environmental Health and Preventive Medicine*; *11*, 171–176.

Kortenkamp, A. (2008). *Breast cancer and exposure to hormonally active chemicals: an appraisal of the scientific evidence*. Background paper published by the Health and Environment Alliance and CHEM Trust. Located at: www.chemtrust.org.uk. 29 Accessed 19.08.08.

Landrigan, P. J., Kimmel, C. A., Correa, A., & Eskenazi, B. (2004). Children's health and the environment: public health Issues and challenges for risk assessment. *Environmental Health Perspectives*, *112*, 257–265.

Lanphear, B. P. (2005). "Origins and evolution of children's environmental health," in "Essays on the Future of Environmental Health Research: a tribute to Kenneth Olden," special issue. *Environmental Health Perspectives*, 24–31.

Lutz, W. K. (1990). Dose-response relationship and low dose extrapolation in chemical carcinogensis. *Carcinogenesis*, *11*(8), 1243–1247.

Miller, M. D., Marty, M. A., Arcus, A., Brown, J., Morry, D., & Sandy, M. (2002). Differences between children and adults: Implications for risk assessment at California EPA. *International Journal of Toxicology*, *21*, 403–418.

National Research Council. (2009). *Committee on improving risk analysis approaches used by the U.S. EPA. "Science and decisions: Advancing risk assessment" (free executive summary)*. (pp. 7–8). www.nap.edu/catalog/12209.html. Accessed 27.07.09.

Navas-Acien, A., Guallar, E., Silbergeld, E. K., & Rothenberg, S. J. (2007). Lead exposure and cardiovascular disease—a systematic review. *Environmental Health Perspectives*, *115*, 472–482.

Needham, L. L. (2007). Personal communication, Faroes Islands.

Perera, F., Jedrychowski, W., Rauh, V., & Whyatt, R. M. (1999). Molecular epidemiologic research on the effects of environmental pollutants on the fetus. *Environmental Health Perspectives*, *107*(Suppl. 3), 451–460.

Rawlins, R. (2009). Teething on toxins: in search of regulatory solutions for toys and cosmetics. *Fordham Environmental Law Review*, *20*, 1–50.

Rodier, P. M. (1995). Developing brain as a target of toxicity. *Environmental Health Perspectives*, *103*(6), 73–76.

Rogers, J. M., & Kavlock, R. J. (2001). Developmental toxicology. In Curtis Klaassen (Ed.), *Casarett and Doull's toxicology* (6th ed., pp. 351–386). New York: Pergamon Press.

Schardein, J. L. (2000). *Chemically induced birth defects* (3rd ed. revised and expanded). New York: Marcel Dekker.

Sexton, K., Needham, L. L., & Perkle, J. L. (2004). Human biomonitoring of environmental chemicals: measuring chemicals in human tissues is the 'Gold Standard' for assessing people's exposure to pollution. *American Scientist*, *92*, 38–45.

Sexton, K., Ryan, A. D., Adgate, J. L., Barr, D. B., & Needham, L. L. (2011). Biomarker measurements of concurrent exposure to multiple environmental chemicals and chemical classes in children. *Journal of Toxicology and Environmental Health, Part A*, *74*(14), 927–942.

Silbergeld, E. K., & Rothenberg, S. J. (2007). Lead exposure and cardiovascular disease—a systematic review. *Environmental Health Perspectives*, *115*, 472–482.

Simon, T., Britt, J. K., & James, R. C. (2007). Development of a neurotoxic equivalence scheme of relative potency for assessing the risk of PCB mixtures. *Regulatory Toxicology and Pharmacology*, *48*, 148–170.

Soto, A. (2007). Does breast cancer begin in the womb? In *Presentation at International Conference of Fetal Programming and Developmental Toxicity, Torshavn, Faroe Islands, May 20–24, 2007.*

Swan, A, Main, K. M., Liu, F., Stewart, S. L., Kruse, R. L., et al., (2005). "Decrease in Anogenital Distance Among Male Infants with Prenatal Phthalate Exposure," *Environmental Health Perspectives*, *113*, (8) 1056–1061.

Talbot, P. (2009). Department of Cell Biology & Neuroscience at the University of California, Riverside, and Developmental Biologist, Personal communication.

Titus-Ernstoff, L., Hatch, E. E., Hoover, R. N., Palmer, J. R., Greenberg, E. R., Ricker, W., et al. (2001). Long-term cancer risk in women given diethylstilbestrol (DES) during pregnancy. *British Journal of Cancer*, *84*, 126–133.

U.S. Department of Health and Human Services, Centers for Disease Control and Prevention, National Center for Environmental Health. (September, 2005). *Third National Report on Human Exposure to Environmental Chemicals*. www.cdc.gov. Accessed 20.08.08.

U.S. Department of Health and Human Services, Centers for Disease Control and Prevention, National Center for Environmental Health. (December, 2009). *Fourth National Report on Human Exposure to Environmental Chemicals*. www.cdc.gov/exposurereport. Accessed 13.01.10.

U.S. Department of Health and Human Services, Centers for Disease Control and Prevention, National Center for Environmental Health. (2013). *National Biomonitoring Program, Environmental Chemicals*. Located at: http://www.cdc.gov/biomonitoring/environmental_chemicals.html. Accessed 20.01.13.

Vandenberg, L. N., Colborn, T., Hayes, T. B., Heindel, J. J., Jacobs, D. R., Jr., Lee, D.-H., et al. (June, 2012). Hormones and endocrine-disrupting chemicals: low-dose effects and nonmonotonic dose responses. *Endocrine Reviews*, *33*(3), 378–455.

Vom Saal, F. S. (2011). Personal communication.

Weiss, B. (1994). The developmental neurotoxicity of methylmercury. In Herbert L. Needleman & David Bellinger (Eds.), *Prenatal exposure to toxicants: developmental consequences* (pp. 112–129). Baltimore and London: The Johns Hopkins University Press.

Wigle, D. T., & Lanphear, B. P. (2005). Human health risks from low-level environmental exposures. *PLoS Medicine*, *2*, 1232–1234.

Wikman-Svahn, P. (2012). *Ethical aspects of radiation risk management*. Doctoral Thesis in Philosophy, Royal Institute of Technology, Stockholm, Sweden.

Woodruff, T. J., Zeise, L., Axelrad, D. A., Guyton, K. Z., Janssen, S., Miller, M., et al. (2008). Meeting report: moving upstream-evaluating adverse upstream end points for improved risk assessment and decision-making. *Environmental Health Perspectives*, *16*, 1568–1575.

Part II

Putting Protection to Practice

Chapter 7

Ethical Issues in Clinical Radiology

Jim Malone
Medical Physics, Trinity College Centre for Health Sciences, Dublin, Ireland
E-mail: jfmalone@tcd.ie or jifmal@gmail.com

Chapter Outline

7.1 Introduction 107
7.2 Ethics in Medical Radiology 109
7.2.1 Core Principles in Medical Ethics 109
7.2.2 A Basic Analysis 111
7.3 Medical, Social, and Legal Context for Radiology 114
7.3.1 Overutilization and Health Economics 116
7.4 Risk, Uncertainty, Communication, and Skeptical Doctors 116
7.5 Justification Issues 118
7.5.1 Awareness 119
7.5.2 Appropriateness and Referral Guidelines 120
7.5.3 Audit (Clinical) 121
7.5.4 Some Reasons for Failure of Justification 121
7.6 Some Special Concerns 121
7.6.1 Nonmedical Exposures 121
7.6.2 Regulatory Framework 122
7.6.3 Some Pregnancy Issues 123
7.7 Conclusions 124

7.1 INTRODUCTION

There are many approaches to introducing the ethical issues that arise in diagnostic radiology. Perhaps a good starting point is that the benefits of medical radiology are not in doubt; it has greatly enhanced the effectiveness of medical practice. Radiology has been successful in overseeing and delivering into the healthcare system a technology transfer on an exceptional scale in a relatively short time. However, the benefits come at a price, and radiology now accounts for over 98% of manmade human radiation exposure. Every day more than 10 million diagnostic procedures are performed giving 3–4 billion annually (ICRP, 2007a; UNSCEAR, 2008).

Radioactivity in the Environment, Volume 19
ISSN 1569-4860, http://dx.doi.org/10.1016/B978-0-08-045015-5.00007-1

Radiation dosage in diagnostic radiology was, for many years, regarded as a nonissue. Whether or not this was the case is debatable. However, during the last 10–15 years, dose has increased to worrying levels, to an extent that its containment has become a competitive marketing tool. The largest component of the dose involved arises from computerized tomography (CT), which is widely deployed (NCRP, 2009). Both the number of examinations and the dose per examination have relentlessly increased. Each examination should, in theory, provide a diagnostic benefit, whether performed in the west, the developing world, the public sector or the private domain. Each examination also represents a monetary cost and a risk, which the patients and doctors may, or may not, be aware of (IAEA, 2011a). While dose will continue to be an issue, recent developments in CT promise reductions that remain to be fully evaluated.

Radiation protection in medicine is, as in other areas, underpinned by the concepts of *justification*, *optimization* and *dose limits* (Hansson, 2007; ICRP, 2007a, 2007b). Most countries have strict radiation dose limits for the general population (1 milliSievert (mSv) per year) and for professionally exposed workers (20 mSv per year). However, dose limits are not applied to patient medical exposures (IAEA, 2011b; ICRP, 2007a). Thus, paradoxically, a citizen upon becoming a patient loses the protection of a dose limit, and entrusts their care to physicians who seldom know the dose or risk to which they are exposed. The patient can and often does receive significant exposures, larger than the annual dose limit for the public or workers, without receiving any information and, in the case of inappropriate examinations, without commensurate benefit (Fazel et al., 2009; Hansson, 2007; IAEA, 2009; Malone, 2011a; Malone et al., 2011; Picano, 2004a, 2004b). This, in a nutshell, identifies a central practical ethical issue in radiology.

This chapter surveys some of the related social, legal, and medical/radiological issues and reviews related unresolved ethical questions. Initially, background ethical considerations, in general and in medicine, are reviewed to the extent that may guide good practice. The approach taken is exploratory and tentative and will need further work to consolidate and integrate it with the radiation protection value system. This is followed by sections in which the medical, social and legal contexts for clinical radiology are briefly explored and the manner in which dose and risk are deployed is considered. One of the major ethical problems in radiology is justification of medical exposures in practice. Recent work in this area has uncovered deficiencies in practice, in compliance with the law, and in attentiveness to fundamental ethical considerations.

Finally three areas in which special concerns arise are briefly introduced. First some frankly nonmedical deliberate human exposures occur, and often happen in a clinical context. These need special attention. Second there is significant lack of alignment between the ethical framework for radiation protection and its medical counterparts. Finally situations involving pregnancy or possible pregnancy of patients to be radiated need special attention. None of these is treated comprehensively. They are introduced to a level that will allow those interested to explore them as required.[1]

7.2 ETHICS IN MEDICAL RADIOLOGY

7.2.1 Core Principles in Medical Ethics

The thinking behind the current framework for radiation protection in medicine is to be found in earlier publications of the International Commission on Radiological Protection (ICRP, 1977, 1994). These general principles, originating decades ago, still define the approach of ICRP and are repeated in both their current general and special medical recommendations (ICRP, 2007a, 2007b). The core principles/values identified are:

- Justification (of the activity)
- Optimization (performing the task with dose as low as reasonably achievable (ALARA))
- Dose limitation (application of dose limits and dose limitation strategies; dose limits do not apply to patient exposures)

While these have served radiation protection well, they carry much intellectual baggage from an earlier era. They would benefit from being brought into line with contemporary philosophical, social and legal thinking. In addition, the relationship between the ICRP principles and the principles of medical ethics is not self-evident and requires exploration. There is a disconnect between the way values are currently presented and prioritized for medicine/radiology, on the one hand, and radiation protection, on the other.

The practice of medicine and radiology is essentially a global activity, and its ethical content should aspire to be based on a system of principles/values that all cultures can subscribe to. Beauchamp and Childress have, since 1979, pursued this objective for medical ethics and their findings appear to be transferable to radiology with limited modifications. Their basic principles are in Table 7.1 (Beauchamp & Childress, 2012). The first three are presented as four by Beauchamp and Childress; they separate Non Maleficence and Beneficence. For our purpose I have assumed they can be grouped. These three principles/values are found to be relatively culture-independent and can be used to underpin a system of medical ethics that is global in its reach. It is reasonable to assume that this system can be transferred to radiology and there are proposals to this effect[2] (IRPA, 2012; Malone, 2014).

1. Key references are provided, from which the primary literature can be readily identified, rather than several hundred primary references which would be less helpful to the reader.
2. This has been proposed by Zolzer (in press) at the International IRPA Congress in Glasgow 2012. Related presentations can be downloaded from IRPA (2012).

TABLE 7.1 Core Principle for a System of Ethics for Clinical Radiology

Core Principles/Values	Comment(s)
Three Core Values	
1. Autonomy and dignity of individual 2. Non maleficence (Do no harm) including beneficence, (Do good) 3. Justice, access, etc.	Beauchamp and Childress (2012) and Zolzer (2013). Zolzer et al. at IRPA (2012)
Two Additional Values Required	
4. Prudential/Precautionary principle 5. Openness, transparency and accountability	Lochard in (IRPA 12), Malone (2014).
For Asymptomatic Patients	
6. Utilitarian principle	See text

However, there are special problems in radiology, particularly those arising from communicating and managing incomplete knowledge and the uncertainty of risk to patients and the public. These problems need to be addressed in the context of a clear set of values with an ethical content. This gives rise to two additional values which are widely, but not universally, subscribed to in the practice of medicine (IRPA, 2012; Malone, 2014)[1]:

- The Precautionary Principle, and
- Openness, Transparency, and Accountability

as indicated in Table 7.1.

Finally, when screening asymptomatic patients for possible disease, there is a need for additional guidance and in this context the idea of the

- Utilitarian Principle may be deployed.

Values 4–6 are not as culture-free as the three basic principles. However, they need to achieve a high profile of awareness among practitioners to guide currently problematic aspects of radiology practice, and are explicitly included in the list on this basis.

Radiology is a latecomer to explicitly considering these values although this is not uniform throughout the world. Some of the values contrast with the post-WW2 paternalism of the professions, which no longer provides an acceptable guide to appropriate action in radiation protection in radiology (Malone et al., 2011; Malone, O'Connor, & Faulkner, 2009).

7.2.2 A Basic Analysis

A useful general analytic framework for reviewing radiological procedures and their performance is provided in an IAEA overview of the justification issue (IAEA, 2009; Sia, 2009, 2010). This provides an abstract generally applicable approach using categories and terminology frequently employed in ethical analysis of an action. The generality of the approach has the advantage of avoiding pointing an accusing finger at particular groups such as doctors, technologists, physicists, or administrators. The categories are:

The agent: i.e. the individual (or group) performing an act
The act: i.e. the radiological procedure
The recipient: i.e. the individual on whom the radiological procedure is performed.

The analysis applied here retains the terminology used by the IAEA, although it will, in due course, have to be related to the principles set out in the previous section.

7.2.2.1 The Agent

Ethical responsibility ultimately rests on the agent. As used here, the "agent" not only refers to the doctor or technologist performing the action, but can also include the professional and regulatory framework surrounding it (e.g. referring physician, the radiologist, the technologist(s), and corporate/regulatory bodies) (IAEA, 2009; Sia, 2009, 2010).

It is essential that the agent undertakes the act in the best interest of the recipient. Current experience and the published literature suggest that in many clinical settings, the referring physician may have limited awareness of the actual dose and risks involved, and the performing practitioner may not be aware of the patient's condition (Brenner & Hall, 2007; IAEA, 2009; Shiralkar et al., 2003). It is essential that those referring for and performing procedures acquire and maintain a fluent knowledge of what is involved and the consequent risks. The knowledge required includes:

- how the technique is best executed in practice,
- potential benefits of the action,
- awareness of both short- and long-term consequences and risks,
- knowledge of available alternatives, and
- the consequences of not taking action.

In practice, the agent is often well trained and familiar with some of these and lacks familiarity and fluency with others. When the agent's action is based on best practice *and* appropriate evidence-based guidelines, all five bullet points are likely to be satisfied. It is also important that the reasoning behind the agent's decisions is such that it can be made accessible to those to whom he/she is accountable—particularly the recipient, and not just the peer group.

7.2.2.2 The Act

The Act, for the purpose of this paper, is any diagnostic procedure requiring radiation (e.g. a chest X-ray, or an abdominal CT scan) (IAEA, 2009, 2011a). Whether or not the act is performed is influenced by many considerations including:

- The potential benefits,
- The risks of potential harm,
- The judgment of the agent,
- The attitude of the recipient,
- Considerations arising from need for public accountability, and
- Considerations arising from health economics and equity.

With respect to the attitude of society to the field involved or the procedure (e.g. mammography), it is important that the agent be sensitive to societal developments. However, while he/she should be guided by this, it should not be the sole consideration.

With respect to risks or potential harm, there must be a reasonable expectation that the act will lead to benefits that outweigh the risks (ICRP, 2007a). The fact that harm may arise does not mean it should not be done. There is a long tradition of recognizing and accepting the inevitability of concurrent harm arising from interventions, medication, or acts undertaken for the good of the patient. However, the potential for harm must be recognized and fully accounted for in the decision to undertake a procedure. It is also necessary to be prudent and cautious, not least because the tradition of accepting concurrent harm has primarily been applied with therapeutic interventions. It is less well tested in respect of diagnostic procedures, and the area requires further discussion, reflection, and ethical/legal consideration.

Each person is entitled to a reasonable expectation of health and equal access to health care (Durand-Zaleski, 2009; Sutherland, Fisher, & Skinner, 2009; Wennberg et al., 2008). There are many factors that complicate availability and access to radiological procedures that result in favor of one region, one individual, or one group over others. In addition, the professions involved, including regulators, can be self-serving and distort the decision-making.

7.2.2.3 The Recipient and the Consent Issue

The recipient is the individual on whom the radiological procedure, or the act, is performed. Central issues for the recipient are that the procedure be necessary and that it be conducted in a competent way (IAEA, 2009, 2011a; Malone, 2011a). The former is a central ethical issue and, as will be seen in the section on justification, is not always the case. In countries with reasonable education and training for radiologists and technologists, the latter tends not to be a major problem and will not be further discussed here.

A key issue in practice is respect for the autonomy and dignity of the recipient as an individual. This implies that the individual's consent is necessary before being exposed to the act.[3] This, in turn, means that the individual is entitled to know what is to happen. Without knowing what is happening, an individual's consent is unlikely to be real and valid (IAEA, 2011a; Malone et al. 2011; Picano, 2004a, 2004b; Semelka et al., 2012).

For a consent to be valid, the individual must be informed, before the procedure, of:

- what is going to be done,
- why it is being done,
- what will happen as a result (including risks),
- what will happen if it is not done, and
- what else, if anything, can be done instead, i.e. what the alternatives are?

The disclosure of information should:

- be full, frank and open,
- include all material risks, which a reasonable person would be likely to attach significance to,
- be presented in a way that the individual can understand and assimilate, and
- be clarified by encouraging questions, which are answered honestly and completely.

Consultations undertaken by the IAEA identified an undue level of paternalism in the practices of medical imaging (IAEA, 2009, 2011a). The intent of the bulleted actions above is to provide information that counters paternalism and allows the individual to make a good decision. This requires a nonnegotiable explicit or implicit valid informed consent. The bullets will also help ensure that both the agent and the act will always be directed toward the recipient's best interest.

It is never a matter solely for the agent to make a decision for another individual, except in those circumstances where it is neither practicable nor feasible to obtain consent, or where the risk is very small and the consent is clearly implied by the circumstances of the exchange between the agent (practitioner) and the recipient (patient). This advice is at variance with the situation in practice in radiology today, where consent for examinations is seldom obtained, and when it is, patients are often not properly informed, even when facing considerable levels of exposure (Brink, Goske, & Patti, 2012; IAEA, 2009; Picano, 2004a, 2004b; Semelka et al., 2012).

The need to provide information and/or obtain the patient's consent is underpinned in a number of legal instruments and many judicial decisions (IAEA, 2011a). Most legal systems now tend to encourage and enable patients to make

3. Consent and the information required for the patient are used throughout in the ordinary colloquial sense of these words. The precision of the varying legal definitions used in different territories may not always be intended. A more detailed discussion of the legal issues surrounding the information and consent is required (Brink et al., 2012; Semelka et al., 2012).

decisions for themselves about matters that intimately affect their own lives and bodies. The INTERIM revised IAEA BSS requires that a procedure not be carried out unless "the patient has been informed, as appropriate, of the potential benefit of the radiological procedure as well as radiation risks" (EC, 2007; IAEA, 2011b).

7.3 MEDICAL, SOCIAL, AND LEGAL CONTEXT FOR RADIOLOGY

The defining characteristic of medicine, in recent times, has been its immense scientific and technological success coupled with an iconic repositioning in public consciousness. This has been accompanied by growth in expectations from hospitals and medical institutions, to a level where they are unrealistic and place an undue burden on the healthcare system and those working in it. This also, inevitably, creates public disappointment and anger when expectations are not met (Malone, 2009; Malone et al., 2011).

The model for provision of medical services continues to harbor strong paternalist leanings. The health professions have frequently failed to recognize the growth in recognition of individual autonomy. Consumerist culture, transparency, and accountability are dominant influences in the way transactions are expected to take place. Failures in these areas have led to distrust of the authority of health professionals. Examples of failures can be found in the history of various medical scandals, such as the blood products problems, the infant organ retention issues, and many others. The reports of investigations of these scandals often suggest that contributory factors include both paternalism and desensitization of professions to the concerns of the public (Malone, 2009; Malone et al., 2011).

There has been profound social change since the current ICRP system of radiation protection was laid down in ICRP 26. A short list of areas where this is seen includes: euthanasia, assisted suicide, marriage, divorce, single parents, disability, gender, distrust of authority/professions, the right to life and the autonomy of the individual. In many cases, the changes are reflected in the law, social policy and practices of society, including medicine. However, radiology has been a reluctant participant in these developments (Amis et al., 2007; Malone, 2009, 2011).

There is also evidence of a changing model of access to hospital facilities, sometimes almost on a consumerist basis, as can be the case with medical tourism. Medical tourism is encouraged by some governments, industry, and the professions. In radiology, the growth throughout the world of essentially commercial imaging clinics is widespread; the feeling among "customers" of these clinics may be that, if they want an examination, they should be allowed to have it. This feeling is encouraged by promotional websites, leaflets, and brochures.

Arising from this, two types of patient referral or presentation, not traditionally encountered in radiology, now occur:

- Patients may refer themselves for a procedure and appeal to a radiology service to have it undertaken. This is referred to as *self-presentation*.
- A physician (e.g. a cardiologist) who has radiological facilities within his/her own clinic may perform a procedure on a patient instead of referring on to a third party, such as a radiologist. This is called *self-referral*.

Both tend to increase the use of ionizing radiation over and above that which prevails in the traditional approach (Emmanuel & Fuchs, 2008; Holmberg et al., 2010; Malone et al., 2011).[4]

In practice, the service provider can inadvertently, or otherwise, be diverted from his main focus, i.e. the well-being of the recipient. In particular, financial interest in maximizing use of a clinic's resources may interfere with an objective risk–benefit evaluation. When a physician has such a financial interest, it must be disclosed to the patient (IAEA, 2009). In addition, where the procedure cannot be medically justified, the patient should be advised that this is the case.

The prevailing social environment has raised the level of openness, accountability, and transparency expected of professionals and institutions. Also, the manner in which medical imaging centers are organized can render individual accountability difficult. In larger institutions, imaging departments can be very large enterprises, with several hundred staff undertaking 500 to 1000 examinations a day, possibly several hundred thousand per year. This is radiology on an industrial scale and the skills to manage it effectively are not always available. The well-being of the individual patient may be lost in such large systems. The funding and referral arrangements in both public and private systems can make it difficult for radiologists to refuse inappropriate referrals.

When things go wrong in medicine, inquiry into serious problems may initially use a peer review-like process. If this fails, inquiry by a professional body often yielded acceptable results in former times (e.g. The Medical Council in the UK). However, it is now not uncommon for the findings of such a group to be regarded as unsatisfactory and self-serving. When this is the case formal Tribunals of Inquiry follow to determine matters of fact. Sometimes, this progresses to the law courts, which determine both facts and guilt/punishment. This has become a much more common background feature to the lives of most health professionals and is an important context for the practice of radiology (IAEA, 2009; Malone et al., 2011).

---

4. For radiation-protection purposes, medical practitioners are defined as follows. *Radiological medical practitioner*: A health professional with specialist education and training in the medical uses of radiation, who is competent to perform independently or to oversee procedures involving medical exposure in a given specialty. *Referring medical practitioner*: A health professional who, in accordance with national requirements, may refer individuals to a radiological medical practitioner for medical exposure (IAEA, 2011b).

7.3.1 Overutilization and Health Economics

Concern about overutilization of medical imaging services is now well established in health economics and in health technology assessment (HTA). The economic cost and the loss of benefit to those who really need the services were well articulated during the U.S. health reform debate (Amis et al., 2007; Durand-Zaleski, 2009; Malone et al., 2011; Wernberg et al., 2008). A number of initiatives from the radiological community responding to these pressures, and public concern about high doses, have emerged. These include the IMAGE GENTLY and IMAGE WISELY campaigns, which are concerned with children and adults respectively (Gently and Wisely, 2012). These have achieved high profiles. In parallel with these specialist physicians and surgeons have examined their patterns of prescribing diagnostic tests and treatments, and initiated a campaign known as CHOOSE WISELY. Nine U.S. specialty societies each produced an evidence-based list of five tests or treatments that should be more carefully prescribed. Thus, the nine societies have identified a total of 45 tests or treatments that require special attention and are prone to overutilization (Gently and Wisely, 2012). Of the 45, 60% or 27 are imaging tests.

7.4 RISK, UNCERTAINTY, COMMUNICATION, AND SKEPTICAL DOCTORS

When an investigation involves ionizing radiation, the risk–benefit assessment should include the possible long-term risk of malignancy. From the precautionary principle, it is reasonable to take the view that patients have the right to know of possible risk, and that physicians/radiologists have a duty to inform them (IRPA, 2012; Malone, 2014; Malone et al., 2011; Shah, Sachs, & Wilson, 2012). In practice, achieving this will require the development of new operational approaches and an adjustment of the culture of radiation protection as it is practiced in medicine. With high-dose procedures, like CT, this will be best facilitated by open discussion and shared decision-making, something the organizational approach in radiology is not well equipped to deliver.

Radiation is a known carcinogen. The American College of Radiology (ACR) 2007 “White paper on Radiation dose in Medicines” suggests current imaging rates may result in an increased incidence of radiation related cancer in the near future. Some estimate the increase could be up to 1.5 or 2% (Amis et al., 2007; Brenner & Hall, 2007; NRC-BEIR, 2006). The ACR White Paper notes that some physicians are very knowledgeable on these issues and incorporate such information into their decisions, but others do not routinely do so (Amis et al., 2007; Brenner & Hall, 2007; Horton, 2011; ICRP, 2007a; NCRP-BEIR, 2006).

There is both lack of conviction and debate about cancer incidence after the doses that prevail in diagnostic imaging. This is compounded by failure of the radiobiological and medical physics communities to find effective, transparent ways of communicating about dose and risk to health professionals and patients.

Confusion has been added to uncertainty through the arcane esoteric units for radiation dose and the obsessions of radiation metrology (Horton, 2011; IAEA, 2009; Malone, 2008, 2009, 2014). Nevertheless, the data from Japanese A-bomb survivors continues to be the best epidemiological source for the relationship between attributable cancer risk and radiation dose. The problem at low doses, such as those in diagnostic examinations, is that because of a dearth of direct evidence estimates are derived from extrapolating the dose effect curve linearly from higher doses. However, the relationship is now convincingly seen down to about 35 mSv (2–3 CT scans), and probably lower (Horton, 2011; Shah et al., 2012). While it is not possible to select between competing models for this relationship, the best available authorities conclude that a linear no threshold model remains a valid conservative choice for calculating risks at low radiation doses (Brenner & Hall, 2007; IAEA, 2011a; ICRP, 2007a; NRC-BEIR, 2006; Shah et al., 2012). More recently, new data on cardiovascular and other "noncancer" long-term effects from radiation have been noted.

Many radiologists, cardiologists, and others are radiation damage skeptics, and move seamlessly from the view that there is no definitive evidence of damage, to the position that there is no damage, and behave accordingly. This is not a logical and, in addition, is inconsistent with the precautionary principle. Thus, they do not advise patients about risk, are skeptical, and generally disregard it. The position is of course a nonsequitur. In this context, an AAPM statement,[5] of 2011, must be viewed as unbalanced, inconsistent with the precautionary principle and of little service to public and patients (AAPM). Indeed, Shah et al. have taken the AAPM position as a counterexample to the precautionary principle, i.e. benefits are emphasized without reference to risks (Shah et al., 2012). It would be easy to read the AAPM statement, and some of its predecessors, as statement dismissing the risks entirely for most radiology, with little regard for the BEIR Committee and ICRP positions that there may be some risk.

The message should be that there may or may not be a risk; we don't actually know. Using the precautionary principle, the best conservative consensus of the scientific community for future deaths from the radiation dose typically associated with the scan should be used in communication with staff or patients and qualified with an explanation of the uncertainties involved.

Following this approach, the cancer risk from a 64-slice CT coronary angiography may be as high as 1 in 100 for a young woman or in a child. Of course,

5. The AAPM statement reads: "Risks of medical imaging at effective doses below 50 mSv for single procedures or 100 mSv for multiple procedures over short time periods are too low to be detectable and may be nonexistent. Predictions of hypothetical cancer incidence and deaths in patient populations exposed to such low doses are highly speculative and should be discouraged. These predictions are harmful because they lead to sensationalistic articles in the public media that cause some patients and parents to refuse medical imaging procedures, placing them at substantial risk by not receiving the clinical benefits of the prescribed procedures." [See: http://www.aapm.org/org/policies/details.asp?id=318&type=PPe Consulted 18 Jan 2013].

if the scan is necessary in serious or life-threatening situations it must be done, preferably with the consent of the patient after they have been counseled through the various risk issues. The validity of consents or patient information must be in doubt, if the possibility of such risks is absent from the communication of practitioners with the patient (Brink, et al., 2012 IAEA, 2011a; Malone et al., 2011; Picano, 2004a, 2004b; Semelka et al., 2012). The ethical problem is even deeper if, as will be seen in the next section, the examination is inappropriate. Such practices are a source of detriment to society, patients, and medical practice. Use and retention of mandatory consent forms, particularly for higher dose examinations, has been suggested. It may promote the patient's understanding and reminds the physician of his or her responsibilities (Malone et al., 2011).

A joint IAEA/EC workshop recognized that successive approaches to communication of radiation risks to various groups, including patients, practicing physicians, surgeons, radiologists, and allied professionals, have not been effective (IAEA, 2011a). In addition, it identified the need to break out and differentiate the communication tasks facing the professions involved to enable new programs be developed. In addition, a primer publication reviewing the contribution of ethics, law, health economics, and communication to radiation protection was felt to be necessary (IAEA, 2011a; Malone et al., 2011). The importance of a more effective approach in this regard has been given additional weight by recent communications and debate both in the specialist literature and in the public press (IAEA, 2011a; Malone et al., 2011; Marchione, 2010).

Thus, the current situation in radiology is the one in which communication is incomplete and/or unsuccessful. Simple matters of fact and probability are not transmitted in an effective, enduring way to those who need to be aware of and have confidence in them (Picano, 2004a, 2004b; Semelka et al., 2012). There is little nuanced dialogue on balancing benefits and risks with patients whose perceptions and requirements would surely, on many occasions, contribute to and alter the equation. This situation, inevitably and with time, may undermine the social acceptability of current practice and needs to be remedied. Philosophers/ethicists can help with this. They can't tell us what to do but they can introduce us to guiding principles and precedents.

Finally, it is now well recognized that communication of accurate comprehensible information, while essential, does not fully deal with the issues involved. Failure to recognize this leads to social amplification of risk. It is import to attend to social context and emotional response as well as the empirical content of the message (Malone et al., 2011). In addition, one must not loose sight of an overhead of distrust of radiation and radiation-protection professionals, which renders these tasks even more challenging.

7.5 JUSTIFICATION ISSUES

The extent of use of radiology has become a matter of concern for many reasons, including population dose, individual dose, financial and health economics issues, and the appropriateness or justification of the examinations. The focus here is primarily on justification, a cornerstone of the ICRP system of protection, particularly in the absence of dose limits for patients (IAEA, 2009, 2011a; ICRP, 2007a, 2007b; Holmberg et al., 2010; Malone et al., 2011).

Justification of exposure of individuals to radiation in medicine has well accepted differences from other situations. These include:

- justification is (in theory at least) employed individually with each patient,
- consent (explicit or implied) is required for every radiation procedure.

These differences acknowledge that medical exposures are used to help the patient and the justification process ensures patient benefits substantially outweigh any short- or long-term risks. Thus, key to effective implementation of justification in radiology is ensuring that those referred for examinations actually need them (Fazel et al., 2009; Malone et al., 2011). Justification is, in most countries, part of the legal system. However, the impact of regulatory authorities on enforcement has been marginal, at best, as the areas involved are within the competence of health professionals (EC, 1996, 1997; Janssens, 2011).

A joint IAEA/EC workshop, with participation from 40 countries and the relevant international organizations, found in the workshop summary/conclusions that "*There is a significant and systemic practice of inappropriate examination in radiology.*" It also found three practical approaches key to effective implementation of justification. They are

- Awareness: *improving the effectiveness of communicating about radiation risk to patients, public, physicians, surgeons, radiologists, and the allied professionals,*
- Appropriateness: *ensuring that those referred for radiological examinations really need them,* and
- Audit: *clinical audit of the effectiveness of the referral and related processes.*

These are referred to as the three A's and the IAEA proposes a way forward based on a global campaign to embrace them (IAEA, 2011a; Malone et al., 2011). It has been generally well received and more formally adopted by the Heads or European Radiological protection Competent Authorities (HERCA) and the Nordic Radiation Protection Authorities.

7.5.1 Awareness

Data published during the last few years found that doctors/health professionals generally have poor *awareness* of the doses/risks involved in medical imaging and consistently underestimate them (Brenner & Hall, 2007; Horton, 2011; IAEA, 2011a; Malone et al., 2009, 2011; Marchione, 2010; Pallone, 2010; Picano, 2004b; Shiralkar et al., 2003; Sia, 2010). The results of surveys performed on British physicians, Israeli orthopedists, Italian, Brazilian and Australian cardiologists, Canadian and Turkish paediatricians, U.S. emergency room physicians/radiologists, and others show the majority of doctors seriously underestimate the doses and corresponding risks, for most commonly requested procedures. From the published data, it is reasonable to infer the problem is global and applies to junior and senior doctors, whether or not they have attended special courses.

With respect to patient awareness, a limited number of studies are available, which demonstrate they are poorly informed about radiation risks, often due to lack of awareness amongst referring physicians, on whom they rely for such information (IAEA, 2011a).

An awareness campaign to provide a fluent knowledge of dose and risk is essential. It will have to deal with the special problems radiation protection professionals have in coping with the current requirements for transparency. This is absolutely essential where imaging services are marketed directly to the public and to the worried well (IAEA, 2011a).

7.5.2 Appropriateness and Referral Guidelines

The literature on inappropriate examinations and failure of justification is relatively small compared with its importance. Key studies come from the Nordic countries, one from Sweden and one from Finland (Almen et al., 2009; Oikarinen, 2011; Oikarinen et al., 2009). In the Swedish study, the extent of justification was assessed for all CT examinations performed in a typical day (22 March 2006). About 20% of them were judged not to be justified. This disguised notable variations. Less than 2% of trauma examinations were deemed unjustified, whereas 49% of colon examinations were found to be so. The largest difference was for spine examinations, where 23% of those prescribed in hospitals and 68% of those from primary care were unjustified (IAEA, 2011a).

The Finnish study shows the overall rate of unjustified examinations was 30%, which is higher than the Swedish study. Again the rate for trauma was low and that for lumbar spine high at 77%. Other studies, from many countries including the US, Italy, Brazil, Australia, and Portugal, confirm this pattern, but generally have higher rates.

Referral guidelines for diagnostic and interventional radiology have been in existence for 20 years and have been published in the UK, EU, US, Australia, New Zealand, Hong Kong, Canada and elsewhere (ACR, 2013; EC, 2008; IAEA, 2001; RCR, 2013; WHO, 2008; Remedios, 2011). Those published by the EC have been withdrawn due to intellectual property issues. Today's guidelines are increasingly evidence based and are intended to support decision-making rather than be prescriptive. The evidence on which guidelines are based will have to be strengthened and incorporate findings of effectiveness research based on health outcomes.

Guidelines assist in avoiding inappropriate examinations. IT-based order-entry decision support systems for requesting radiological procedures will greatly assist the application of appropriateness criteria in practice (Sistrom et al., 2009; Vartan et al., 2010). This will not only help the justification campaign, which is essentially ethically driven, it will also positively influence health economics and HTA indicators (Amis et al., 2007; Sutherland et al., 2009; Wennberg et al., 2008).

7.5.3 Audit (Clinical)

Most countries seek to establish transparent, tangible procedures for managing quality in health care. A key element of this is *clinical audit*. This, though widely applied to many health-care practices, and required by the EC Directive 97/43/EURATOM on radiation protection of the patient, has been slow to find its place in radiology (EC, 1997). To assist states with implementation of these requirements, the EC prepared guidance on clinical audit in radiology. The approach is flexible and enables the Member States to adopt a form of clinical audit consistent with their national arrangements. Useful advice and practical recipes are available from the IAEA, the EC, and The Royal College of Radiologists (EC, 2009; IAEA, 2011a; RCR, 2009). Justification is a cornerstone of radiation protection and should be among the top priorities in the audit program (Malone et al., 2011).

7.5.4 Some Reasons for Failure of Justification

Several factors contribute to overutilization of medical-imaging services. These include the referring physicians' and practitioners' lack of awareness already mentioned. This can arise from inadequate training. It can also include insufficient knowledge of the patient presenting, foregoing clinical examination in favor of imaging, duplicating examinations already undertaken, inadequate knowledge, experience and confidence in balancing the benefits/risks of the procedure or alternatives, variations in practice based on local preference rather than evidence, and pressure/expectations from patients to undertake possibly unnecessary examinations. "Defensive medicine" gives rise to imaging whose purpose is to protect the physician rather than benefit the patient and this is also an issue (IAEA, 2011a; Lysdahl & Hofmann, 2009; Malone, 2009, 2011a, 2011b).

There is evidence of a growing practice of self-referral, where referring physicians have a financial interest in the imaging arrangements. Increasingly, imaging services are being marketed directly to the public and this is reflected in a growing popularity for medical/dental tourism. Population screening is an established approach in Public Health, but if it involves radiation, it must be justified in both public health and radiation protection terms. Finally, the near-industrial scale of some imaging departments may also be a factor (IAEA, 2011a).

7.6 SOME SPECIAL CONCERNS

7.6.1 Nonmedical Exposures

Medical exposures generally confer benefit on the exposed persons, and are conducted under the supervision of medical practitioners with training and authorization to do so. Some borderline situations exist there where there is a doubt that all the requirements for bona fide medical exposures are met. Examples include: lifestyle radiology, self-presentation, and unapproved screening programs. However, there are also frankly nonmedical exposures, such as those undertaken for security, crime prevention, emigration, and immigration, detection of smuggling and those arising from litigation. These are of a different matter and require additional attention (EC, 2003, 2011; Malone, 2011a).

How do these frankly nonmedical exposures differ from medical ones? Perhaps most important is that medical exposures are generally for the benefit of the patient. This will often not be the case with frankly nonmedical ones. There may be a social gain with no benefit to the individual, who may in fact be disadvantaged (Malone, 2011b).

Medical and nonmedical exposures can also be differentiated by the level of consent required; the confidentiality of all aspects of the process, the framework to ensure this is respected, and finally the governance system within which the exposure takes place. With bona fide medical exposures, consent is always nonnegotiable and within medicine there is also an exceptionally high threshold for confidentiality.

The governance framework is quite different, for example, in migration assessment centers, customs investigation units, or the security service in airports. The problems in ensuring good practice with nonmedical exposures include the diversity and wide distribution of the governance arrangements for them. The exposures may be undertaken and output from them may be used at a distance from the governance arrangements for both medicine and radiation protection.

Discussion within the EU favored redefining the medicolegal exposures as medical. They also favored extending the meaning of "medical benefit" to include health and well-being, so that it includes "benefit" in the sense that siblings of children in *nonaccidental injury* cases, athletes, and others might benefit from being examined even when they are apparently symptom free. This has the advantage of removing a problem by redefinition.

Finally, biomedical research with human volunteers may involve radiation exposure of individuals without direct benefits to them. A wide range of statutory provisions and ethical advice is available for this (EC, 1997, 1998b).

7.6.2 Regulatory Framework

In the EU, it is anecdotally reported that the need for MED Directive grew, at least in part, form a perceived lack of compliance with general radiation protection provisions among the medical community. Even with the MED, regulators feel ineffective and powerless in the face of issues of medical practice and many inspectorates have had little success with the implementation of justification (Janssens, 2011).

The value system informing the MED, and the national regulations deriving from it, is essentially that of ICRP. This system was primarily developed outside medicine for regulating nuclear activities. Paradoxically, it is used primarily within medicine, an area to which it is at least arguable, that it is not well adapted, and hence there is some level of mismatch. Given this, it is surprising that an in-house regulatory system for medicine has not evolved and replaced it (Malone, 2009, 2011a).

The strains of operating the medical system are exacerbated by the need to host nonmedical exposures whose purpose is, for example, essentially social or crime prevention. In addition, there are pressures arising from medical consumerism and asymptomatic exposures. These are not without implications, and may be less socially acceptable than a bona fide radiology for the patient benefit. Ultimately, they may put the medical exposure dose limit exemption at risk.

Thus, there is the question of ensuring whatever solutions are adopted will not ultimately damage the capacity of medicine to function. Continuing failure of justification and/or the presence of questionable nonmedical exposures raise concerns about the medical system as a whole, and complicate the situation. Finally, there is some risk that using precedents from the US for some security practices in Europe may not prove socially acceptable (EC, 2011; Malone, 2011b).

7.6.3 Some Pregnancy Issues

Medical exposure of a pregnant patient gives rise to additional ethical considerations. ICRP advises that when considering irradiation of pregnant patients (or staff) that the fetus be considered as a member of the general public, i.e. an individual with dose limits applicable in their own right (EC, 1998a; ICRP, 2000). Thus, in evaluating risks and benefits from medical exposures during pregnancy, two individuals' needs must be considered. The mother may receive direct benefit while the fetus may be exposed without direct benefit some of the time. On the other hand if, for example, the mother's problem is life threatening, a procedure that involves radiation exposure of the mother, which leads to her survival, may in turn directly benefit the fetus. In this setting, the mother has a role-related responsibility to care for her unborn child as well as to make decisions for herself. The pregnant patient has a right to know the magnitude and type of potential radiation effects that might result from in-utero exposure, and her consent assumes additional important features because of this (EC, 1998a). With diagnostic radiation, the dose involved is seldom, if ever, large enough to warrant considering a termination on grounds of radiation dose alone.

Many countries have policies to avoid irradiation involving the fetus during pregnancy. Where it proves necessary to use radiation, a risk assessment should generally be conducted. Each case must be assessed according to the gestational age at the time of exposure, the extent to which the fetus is included in the radiation field, and the radiation levels received (IAEA, 2011a; Schreiner-Karoussou, 2008).

The situation is more difficult when the woman is not sure or does not know whether or not she is pregnant. In such situations, the EC advises that where pregnancy cannot be excluded it should be assumed (EC, 1997, 1998a). The degree of concern and form(s) of consent that should be obtained vary greatly from country to country in the EU and the US (Schreiner-Karoussou, 2008).[6] They are sometimes dependent upon the extent of the risk to the fetus and mother. However, they are at least as often, dependent on a paternalistic administrative and operational convenience that seldom involve the patient in the decision and probably derive from a view that there is no real risk. The practices adopted frequently do not protect the first few weeks after conception and/or use pregnancy tests that may be unreliable at this time. This is often "justified" on the assumption that no harm will arise as embryos that are radiation damaged will not implant or survive. It is doubtful if this rationale would be knowingly accepted by a significant cohort of women, particularly those with fertility difficulties. Ideally where the mother's irradiation will impact on the fetus, a risk assessment should be undertaken. Based on this the patient should be counseled accordingly by a knowledgeable, experienced person that she will not have difficulty in believing (EC, 1998a). These problems may be reduced by reverting to use of the "10-day rule",[7] which has been reintroduced in some countries for high-dose procedures in which the fetus is in, or proximate to, the direct beam (HPA, 2009).

7.7 CONCLUSIONS

In conclusion, the factors contributing outcomes and ethical behavior are, as we have seen, many faceted. There are many areas of radiology in need of attention from an ethical point of view. The clinical discipline would gain from embracing basic values like individual autonomy, nonmaleficence, beneficence, and justice in its prevailing culture. There are other areas in radiology, like dose management and optimization, in which clear guidance for practitioners on what needs to be done is available. The issue in these areas is implementation of a well-defined approach, rather than resolving uncertainty on what needs to be achieved. These former ones have not been addressed here.

Of the many areas identified as in need of attention and ethical reflection justification and communication of risk are probably the most central and important. The weakness of justification in practice breeches the basic principles of medical ethics, and is also lacking in openness, transparency, and accountability. In addition, screening programs for asymptomatic subjects will often fail on the grounds of their usefulness to individuals and society.

6. With respect to exposure during pregnancy in the US, it was reported at the IRPA 12 Congress that 15% of radiologists or technologists disclose radiation risk, and that 1% of pregnant women receive abdominal radiation. There is great variability in the US in practice and in the type of screening questions asked. The 10-day rule has been abandoned, apparently because it is inconvenient.
7. The "10 day rule" limits performance of diagnostic procedures to which it applies to the first 10 days after the commencement of the menstrual period (HPA, 2009).

There are major issues around how the professions relate to communication of possible risk and consent to clinical radiological procedures. The precautionary principle and prudence offer a clear path forward for how to behave in the absence of full knowledge. Yet practice is dominated by much more extreme positions ranging from skepticism (which asserts there is no risk) to fundamentalism (which asserts the risk is as extrapolated from the Japanese bomb survivors data). Here we have suggested a revised approach, based on the precautionary principle, which is both reasonable and realistic.

There are many other areas that might have been treated. On a few we have offered some observations, which may be helpful in dealing with difficult or sensitive areas. These include possible incompatibilities between the system of medical ethics and the system of radiation protection. The fact that the ethical principles underlying the practice of medicine and the core values of the ICRP system have not been fully explored and aligned is a problem for both systems. In fact the relationship has seldom been explored, let alone aligned. Finally, sensitive areas like irradiation of women during pregnancy and frankly nonmedical exposures have also been briefly raised and are in need of some further attention from an ethical point of view.

ACKNOWLEDGMENTS

I am grateful to the staff at the RPoP Section of the IAEA, particularly Ola Holmberg and Renate Czarwinski, for many discussions on the justification part of this work. I also gratefully acknowledge the support of the Robert Boyle Foundation, its chairman and trustees.

REFERENCES

American College of Radiology (ACR). (2013). *Appropriateness criteria*. http://www.acr.org/Quality-Safety/Appropriateness-Criteria. Accessed 01.24.13.

Almen, A., Leitz, W., Ririchter, S., et al. (2009). *National survey on justification of the CT-examinations in Sweden*. Stockholm Swedish Radiation Safety Authority (SSM)http://www.stralsakerhetsmyndigheten.se/Global/Publikationer/Rapport/Stralskydd/2009/SSM-Rapport-2009-03.pdf. Accessed 01.24.13.

Amis, E. S., Butler, P. F., Applegate, K. E., et al. (2007). American College of Radiology White paper on radiation dose in medicine. *Journal of the American College of Radiology*, *4*, 272–284.

Beauchamp, T. L., & Childress, J. F. (2012). *Principles of biomedical ethics* (7th ed.). Oxford and New York: Oxford University Press0-19-514332-9.

Brenner, D. J., & Hall, E. J. (2007). Current concepts—computed tomography—an increasing source of radiation exposure. *New England Journal of Medicine*, *357*, 2277–2284.

Brink, J. A., Goske, M. J., & Patti, J. A. (2012). Informed decision making trumps informed consent for medical imaging with ionizing radiation. *Radiology*, *262*(1), 11–14. 10.1148/radiol.11111421.

Durand-Zaleski, I. (2009). Organization and delivery of imaging services: the contributions of ethics and political economy. *Radiation Protection Dosimetry*, *135*, 134–136. 10.1093/rpd/ncp043. (Cited 10 Aug 2009).

European Commission (EC). (1996). *Council Directive 96/29/EURATOM laying down basic safety standards for the protection of the health of workers and the general public against the dangers arising from ionizing radiation*. Luxembourg: EC. http://ec.europa.eu/energy/nuclear/radioprotection/doc/legislation/9629_en.pdf. (Cited 21 Jan 2013), This directive is being recast with others including the European BSS, and will shortly be superseded by the recast version.

European Commission (EC). (1997). *Council Directive 97/43/EURATOM on health protection of individuals against the dangers of ionizing radiation in relation to medicine*. Luxembourg: EC. http://ec.europa.eu/energy/nuclear/radioprotection/doc/legislation/9743_en.pdf. (Cited 21 Jan 2013), This directive is being recast with others including the European BSS, and will shortly be superseded by the recast version.

European Commission (EC). (1998a). *Radiation protection. RP 100. Guidance for protection of unborn children and infants irradiated due to parental medical exposures. DG environment, nuclear safety and civil protection*. Luxembourg: EC. http://ec.europa.eu/energy/nuclear/radiation_protection/doc/publication/100_en.pdf. (Cited 21 Jan 2013).

European Commission (EC). (1998b). *Radiation protection. RP 99. Guidance on medical exposures in medical and biomedical research*. Luxembourg: DG Environment, Nuclear Safety and Civil. http://ec.europa.eu/energy/nuclear/radiation_protection/doc/publication/099_en.pdf. (Cited 21 Jan 2013).

European Commission (EC). (2003). Radiation protection. RP 130. Medico-legal exposures, exposures with ionizing radiation without medical indication. In *Proceedings of the International Symposium*. Luxembourg: European Commission, DG TREN. http://ec.europa.eu/energy/nuclear/radiation_protection/doc/publication/130.pdf. (Cited 21 Jan 2013).

European Commission (EC). (2008). *Radiation protection. RP118. Referral guidelines for imaging (2000 updated 2008)*. Luxembourg: DG TREN, EC. http://ec.europa.eu/energy/nuclear/radioprotection/publication/doc/118_en.pdf. (Cited 10 Aug 2009), This publication is no longer available.

European Commission (EC). (2009). *Radiation protection. RP159. European Commission guidelines on clinical audit for medical radiological practices (Diagnostic Radiology, Nuclear Medicine and Radiotherapy)*. Luxembourg: DG TREN, EC. http://ec.europa.eu/energy/nuclear/radiation_protection/doc/publication/159.pdf. (Cited 21 Jan 2013).

European Commission (EC). (2011). Radiation protection. RP 167. International Symposium on non-medical imaging exposures. In *Proceedings of the Symposium, Dublin, 8–9 Oct 2009*. Luxembourg: DG Energy. http://ec.europa.eu/energy/nuclear/radiation_protection/doc/publication/167.pdf. (Cited 21 Jan 2013).

Emmanuel, E. J., & Fuchs, V. R. (2008). The perfect storm of overutilization. *Journal of the American Medical Association*, *299*, 2789–2791.

Fazel, R., Harlan, M., Krumholz, H. M., et al. (2009). Exposure to low-dose ionizing radiation from medical imaging procedures. *New England Journal of Medicine*, *361*(9), 849–857.

Gently and Wisely. (2012). *IMAGE GENTLY; IMAGE WISELY and CHOOSE WISELY*. Available at: http://www.pedrad.org/associations/5364/ig/; http://www.imagewisely.org/ and http://choosingwisely.org/?page_id=13 respectively, Accessed 21.01.13.

Hansson, S. O. (2007). Ethics and radiation protection. *Journal of Radiological Protection*, *27*, 147–156. http://dx.doi.org/10.1088/0952-4746/27/2/002. See also (2011). "Radiation protection—sorting out the arguments". *Philosophy and Technology*, *24*, 363–368. http://dx.doi.org/10.1007/s13347-011-0036-5.

Holmberg, O., Malone, J., Rehani, M., et al. (2010). Current issues and actions in radiation protection of patients. *European Journal of Radiology*, *76*, 15–19.

Horton, P. W. (2011). Dose and risk: the hard facts. In J. Malone, O. Holmberg & R. Czarwinski (Eds.), *Justification of medical exposure in diagnostic imaging: Proceedings of an International Workshop Brussels, 2–4 September 2009* (pp. 83–90). Vienna: IAEA. Accessed 21.01.13 http://www-pub.iaea.org/MTCD/Publications/PDF/Pub1532_web.pdf.

Health Protection Agency (UK) (HPA). (2009). *Protection of pregnant patients during diagnostic medical exposures to ionising radiation RCE 9*. UK: HPA, RCR and COR. Accessed 21.03.13 http://www.hpa.org.uk/Publications/Radiation/DocumentsOf TheHPA/RCE09Protection-PregnantPatientsduringDiagnosticRCE9/.

International Atomic Energy Agency (IAEA). (2009). Malone, J. (Scientific Secretary). Report of a consultation on justification of patient exposures in medical imaging. *Radiation Protection Dosimetry, 135*, 137–144. 10.1093/rpd/ncp107.

International Atomic Energy Agency (IAEA). (2011a). *Radiation protection and safety of radiation sources: International basic safety standards (INTERIM EDITION)*. (This is the revised version of the 1996 BSS), Vienna: IAEA. http://www-pub.iaea.org/MTCD/publications/PDF/p1531interim_web.pdf. Accessed 22.01.13.

International Atomic Energy Agency (IAEA). (2011b). In J. Malone, O. Holmberg & R. Czarwinski (Eds.), *Justification of medical exposure in diagnostic imaging: Proceedings of an International Workshop Brussels, 2–4 2009*. Vienna: IAEA. http://www-pub.iaea.org/MTCD/Publications/PDF/Pub1532_web.pdf. Accessed 22.01.13.

International Commission on Radiological Protection (ICRP). (1994). *Recommendations of the ICRP. Publication 60*. Annals of the ICRP 20/21, Oxford: Pergamon Press.

International Commission on Radiological Protection (ICRP). (1977). *Recommendations of the ICRP. Publication 26*. Annals of the ICRP 1(3) (1977), Oxford: Pergamon Press.

International Commission on Radiological Protection (ICRP). (2000). *Pregnancy and Medical Radiation. Publication 84*. Annals of the ICRP 30(1), Oxford: Pergamon Press.

International Commission on Radiological Protection (ICRP). (2007a). *Radiological Protection in Medicine. Publication 105*. Annals of the ICRP 37(6): (pp. 1–63), Oxford: Pergamon Press.

International Commission on Radiological Protection (ICRP). (2007b). *The 2007 Recommendations of the ICRP. Publication 103*. Annals of the ICRP 37: (pp. 1–332), Oxford: Pergamon Press.

International Radiation Protection Association (IRPA). (2012). International Congress Glasgow. Papers presented on relevant ethics issues by Zölzer, F. "A Cross-Cultural Approach to Questions of Ethics in Radiation Protection"; Lochard, J. "The Precautionary Principle and the Ethical Foundation of the Radiological Protection System" and Cousins, C. "Ethics in Medical Radiological Protection". Presentations Available at: http://www.irpa13glasgow.com/information/downloads/presentations/dochart/ See Thu, Accessed 22.01.13.

Janssens, A. (2011). Opening address—European Commission. In J. Malone, O. Holmberg & R. Czarwinski (Eds.), *Justification of Medical Exposure in Diagnostic Imaging: Proceedings of an International Workshop Brussels, 2–4 September 2009* (pp. 9–10). Vienna: IAEA. http://www-pub.iaea.org/MTCD/Publications/PDF/Pub1532_web.pdf. Accessed 22.01.13.

Lysdahl, K. B., & Hofmann, B. M. (2009). What causes increasing and unnecessary use of radiological investigations? A survey of radiologists' perceptions. *BMC Health Services Research, 9*, 155–164. 10.1186/1472-6963-9-155.

Malone, J. (2008). New ethical issues for radiation protection in diagnostic radiology. *Radiation Protection Dosimetry, 129*, 6–12. 10.1093/rpd/ncn012.

Malone, J. (2009). Radiation protection in medicine: ethical framework revisited. *Radiation Protection Dosimetry, 135*, 71–78. 10.1093/rpd/ncp010.

Malone, J. (2011a). Justification and tools for change: scene setting. In J. Malone, O. Holmberg & R. Czarwinski (Eds.), *Justification of medical exposure in diagnostic imaging: Proceedings of an International Workshop Brussels, 2–4 September 2009* (pp. 17–24). Vienna: IAEA. http://www-pub.iaea.org/MTCD/Publications/PDF/Pub1532_web.pdf. Accessed 21.01.13.

Malone, J. (2011b). Justification, dose limits and dilemmas. In European Commission (EC) (Ed.), *Radiation protection. RP 167 International Symposium on Non-Medical Imaging Exposures. Proceedings of the Symposium. Dublin, 8–9 Oct 2009* (pp. 101–112). Luxembourg: DG Energy http://ec.europa.eu/energy/nuclear/radiation_protection/doc/publication/167.pdf. Accessed 21.01.13.

Malone, J. (2014). Strategies for improving justification—radiation protection of patients. International Conference on Radiation Protection in Medicine—Setting the scene for the next decade. Vienna: IAEA. Presentations Available at: https://rpop.iaea.org/RPOP/RPoP/Content/Documents/Whitepapers/conference/S1-Malone-Strategies-for-improving.pdf Accessed 21.01.13, in press.

Malone, J., O'Connor, U., & Faulkner, K. (Eds.), (2009). *Ethical and justification issues in medical radiation protection. SENTINEL project special initiative: Ethical and justification issues in medical radiation protection. Radiation Protection Dosimetry (Special Issue).* (Vol. 135: Issue 2). http://rpd.oxfordjournals.org/content/135/2.toc:(Cited 13 Jan 2013).

Malone, J., Guleria, R., Craven, C., et al. (2011). Justification of Diagnostic Medical Exposures, some practical issues: report of and International Atomic Energy Agency Consultation. *British Journal of Radiology.* 10.1259/bjr/42893576. http://bjr.birjournals.org/content/early/2011/02/22/bjr.42893576.1.full.pdf+html. Accessed 21.01.13.

Marchione, M. (2010). *Biggest radiation threat is due to medical scans. Americans get most medical radiation in world; dose has grown sixfold.* : The Associated Press. http://rss.msnbc.msn.com/id/37623994/. Accessed 05.09.10.

National Council on Radiation Protection and Measurements (NCRP). (2009). *Ionizing radiation exposure of the population of the United States.* NCRP Report No 160, Bethesda: MD: NCRP.

National Research Council (NRC-BEIR). (2006). *Health risks from exposure to low levels of ionizing radiation: BEIR VII Phase 2".* Washington DC: The National Academies Press.

Oikarinen, H., Merilainen, S., Paakko, E., et al. (2009). Unjustified CT examinations in young patients. *European Journal of Radiology, 19,* 1161–1165.

Oikarinen, H. (2011). Unjustified CT examinations in young patients. In J. Malone, O. Holmberg & R. Czarwinski (Eds.), *Justification of medical exposure in diagnostic imaging: Proceedings of an International Workshop Brussels, 2–4 September 2009* (pp. 155–160). Vienna: IAEA. http://www-pub.iaea.org/MTCD/Publications/PDF/Pub1532_web.pdf. (Cited 13 Jan 2013).

Pallone, F. (February 26, 2010). *Medical radiation: An overview of the issues. Opening statement.* US House of Representatives. Health Subcommittee Hearing. http://docs.google.com/viewer?a=v&q=cache:3gnn4p05RmkJ:energycommerce.house.gov/Press_111/20100226/Pallone.Statement.2010.02.26.pdf+Medical+Radiation:+An+Overview&hl=en&gl=us&pid=bl&srcid=ADGEESic5Vwp9_MLB9W8lNaZa7p2v_AOigIRKDhIjay8D1VULS9-v673RuHG4C-ZJCwG1FkA4ZBxNwHQQHYWIg9RSyvkJFlnmQr46QdsNJu1tAhvlfsSSUtx6ASwZ3bxHpsGwdV3gxGMB&sig=AHIEtbSqszDzE5QRSo310iaEda0rFnJwHQ. Accessed 26.09.10.

Picano, E. (2004a). Informed consent and communication of risk from radiological and nuclear medicine examinations: how to escape from a communication inferno. *British Medical Journal, 329*(7470), 849–851. 10.1136/bmj.329.7470.849.

Picano, E. (2004b). Sustainability of medical imaging, education and debate. *British Medical Journal, 328,* 578–580.

Remedios, D. (2011). Referral guidelines: why, how and for whom? In J. Malone, O. Holmberg & R. Czarwinski (Eds.), *Justification of medical exposure in diagnostic imaging: Proceedings of an International Workshop Brussels, 2–4 September 2009* (pp. 37–42). Vienna: IAEA. http://www-pub.iaea.org/MTCD/Publications/PDF/Pub1532_web.pdf. Accessed 21.01.13.

Royal College of Radiologists (RCR). (2009). *Audit Live*. London: RCR. http://www.rcr.ac.uk/audittemplate.aspx?PageID=1016. Accessed 25.01.13.

Royal College of Radiologists (RCR). (2013). *iRefer. The latest version of the RCR referral guidelines*. London: RCR. http://www.rcr.ac.uk/content.aspx?PageID=995. Accessed 21.01.13.

Semelka, R. C., Armao, D. M., Elias, , Jr., et al. (2012). The information imperative: is it time for an informed consent process explaining the risks of medical radiation? *Radiology*, *262*(1), 15–18. 10.1148/radiol.11110616.

Schreiner-Karoussou, A. (2008). Review of existing issues and practices with respect to irradiation of patients and staff during pregnancy. *Radiation Protection Dosimetry*, *129*(1–3), 299–302. 10.1093/rpd/. and (2009). "A preliminary study of issues and practices concerning pregnancy and ionising radiation" Op. Cit. *135*(2): 79–82.

Shah, D. J., Sachs, R. K., & Wilson, D. J. (2012). Radiation-induced cancer: a modern view. *The British Journal of Radiology*, *85*, e1166–e1173. 10.1259/bjr/25026140.

Shiralkar, S., Rennie, A., Snow, M., et al. (2003). Doctors' knowledge of radiation exposure: questionnaire study. *British Medical Journal*, *327*(7411), 371–372. 10.1136/bmj.327.7411.371.

Sia, S. (2010). *Ethical contexts and theoretical issues: Essays in ethical thinking*. Newcastle-upon-Tyne, UK: Cambridge Scholars Publishing.

Sia, S. (2009). Ethical issues in radiology: a philosophical perspective. *Radiation Protection Dosimetry*, *135*(2), 102–105. 10.1093/rpd/ncp042.

Sistrom, C. L., Pragya, A., Dang, J. B., et al. (2009). Effect of computerized order entry with integrated decision support on the growth of outpatient procedure volumes: seven-year time series analysis. *Radiology*, *251*, 147–155. http://radiology.rsna.org/content/251/1/147.full. Accessed 26.09.10.

Sutherland, J. M., Fisher, E. S., & Skinner, J. S. (2009). Getting past denial—The high cost of health care in the United States. *New England Journal of Medicine*, *361*, 1227–1230.

United Nations Scientific Committee on the Effects of Atomic Radiation (UNSCEAR). (2008). *Report to the General Assembly: Annex on medical exposures. With Annexe A: Medical Radiation Exposures and corrigendum 2011*. New York: United Nations. http://www.unscear.org/unscear/en/publications/2008_1.html. Accessed 23.01.13.

Vartan, M., Vartanians, C. L., Sistrom, J. B., et al. (2010). Increasing the appropriateness of outpatient imaging: effects of a barrier to ordering low-yield examinations. *Radiology*, *255*, 842–849. 10.1148/radiol.10091228.

Wennberg, J. E., Brownlee, S., Fisher, E. S., et al. (2008). *Improving quality and curbing health care spending: Opportunities for the Congress and the Obama administration. A Dartmouth Atlas White Paper*. New Hampshire: Dartmouth Institute for Health Policy and Clinical Practice. http://www.dartmouthatlas.org/downloads/reports/agenda_for_change.pdf. (Cited 5 Sept 2010).

World Health Organization (WHO). (2008). *Global initiative on radiation safety in health care settings*. Geneva http://www.who.int/ionizing_radiation/about/med_exposure/en/index1.html. (Cited 24.01.13).

Zolzer, F. A (2013). cross-cultural approach to radiation ethics. In S. O. Hannson & D. Oughton (Eds.), *Social and ethical aspects of radiation risk management*. Elsevier, in press.

Chapter 8

Ethics in Practice—Protecting Workers

Christopher J. Kalman
Occupational Health, Morelands, Astley Ainslie Hospital, Grange Loan, Edinburgh, UK
E-mail: Chris.Kalman@nhslothian.scot.nhs.uk

Chapter Outline

8.1 Introduction 131
8.2 Radiation Protection in Workers 132
8.3 Ethical Requirements for Occupational Radiation Health Research 137
8.4 Radiation Science and Workers 139
8.5 Summary 140

8.1 INTRODUCTION

As an occupational physician, most of my early recognition of ethics affecting workers focused on aspects of "professional practice" for those workers involved in areas such as health and social care. These aim at the establishment of norms of behavior for the appropriate management of patients or clients. In my own specialty, for example, for many years there has been published guidance of this type in the UK (The Faculty of Occupational Medicine, in press).

It is possible to extend these considerations, both for individual workers and their employing organizations as a whole, anywhere there is a customer or stake-holder focus. Customer relations can therefore be defined by "ethical standards". Other sections of this book take forward examples of these sorts of standards, such as those dealing with local communities.

Separate from this customer focus, it is possible to look at ethical employment based on treatment and management of the workers themselves. Areas such as appropriate reward and recognition, training and development, absence of discrimination or harassment, maintenance of confidentiality, and organizational accountability all form the basis of many human resource policies, playing an important role in achieving staff engagement and shaping the culture of

Radioactivity in the Environment, Volume 19
ISSN 1569-4860, http://dx.doi.org/10.1016/B978-0-08-045015-5.00008-3

an organization. In addition to arguments about ethical performance, in some of these areas there is an increasing focus on statutory and regulatory requirements, such as in the field of potential discrimination, in terms of disability legislation (The Equality Act, 2010). These requirements extend considerations of ethics and culture into a more formal duty of care between an employer and their workforce.

One such area covered both by ethics and legislation is the field of health and safety. The current system of UK legislation goes back to 1974 (The Health and Safety at Work Act, 1974) although earlier systems were in place before that. The system places a general duty on employers to ensure, so far as reasonably practicable, the health, safety, and welfare at work of their employees. Within this general duty, consideration is given to the need for provision of information, instruction, training, and supervision. These overarching requirements are established in enabling legislation and extended in single-subject statutory instruments dealing with specific hazards such as ionizing radiation (The Ionising Radiation Regulations, 1999). The need for information extends not only to precautions and controls, but also to information on the risks to health. Ethically, it seems sound to state that workers should have knowledge of the hazards of their employment, as part of "informed consent" in their contract to work. On a practical level too, from my own experience, workers' compliance with precautions and safety systems are much greater if they are fully aware of the hazards associated with noncompliance. It also provides an effective mechanism to avoid issues based on the traditional machismo of some occupational groups.

Worker ethics of radiation exposure thus should include the recognition that the employer owes the workforce a duty of care in terms of their protection from the radiation hazard, and also in terms of the provision of information about that hazard.

8.2 RADIATION PROTECTION IN WORKERS

Ionizing radiation science is a young but extensive field. While much of its biology and physics are highly complex, and indeed growing in complexity, the resultant systems of radiation protection, which enjoy a significant consensus across the political and scientific spectrum are fundamentally very simple. This simplicity determines that they can be readily explained to radiation workers. At the basic level, concepts of whole body dose equivalence and committed dose, allow particularly complex issues of partial irradiation and contamination with radioactive material all to be dealt with in terms of a single numeric figure of "dose". This figure is then the "cause" of deterministic and stochastic radiation effects that are perhaps more easily thought of in terms of radiation killing cells or leaving them damaged in some way. The thresholds for deterministic (cell killing) effects mean that these do not occur below certain radiation doses, and the linear nonthreshold dose-response relationship for stochastic (cell damage) effects determines that there is no safe dose, and that all doses contribute equally to risk. Armed with this basic information, coupled with some knowledge of the existence of natural background radiation, what is meant by the terms radiation and contamination, the extent of medical exposure and basic cancer statistics, most workers are well able to suggest principles of radiation protection for themselves. Unsurprisingly, these are not normally too dissimilar from those promulgated by the International Commission on radiological protection, as justification, optimization, and limitation (ICRP, 1977).

Having been involved in the initial health surveillance of thousands of workers, I have always believed that the passage of this sort of knowledge is a prerequisite to allow a worker's informed consent to accept "classified-worker" status (The Ionising Radiation Regulations, 1999).

While most radiation incident victims and fatalities in the world today are patients following medical exposure (IAEA, 2001), there remains a steady continuous stream of workers subject to overexposure resulting from irradiation of product or site radiography (Lloyd et al., 1994). Justification that these practices are of benefit remains a key consideration. Justification can, of course, change with time and understanding. I am old enough to remember the routine irradiation of my feet as a child, to assess the need for and fit of new shoes. This practice is most definitely not justified now, based not only on dose to children but also occupational dose to shoe-shop staff. Similarly, I well recall the fluoroscopic mass radiography of the population for tuberculosis and brucellosis, again clearly not justified now, based on the current prevalence of both diseases. Worker dose may well be the only consideration in the justification of practice or it may be a component of a much more complex calculation of risk. Looking, for example, at the disposal of the decommissioned nuclear power submarines: after the vessel has been defuelled, the vast majority of radioactive material remaining in the reactor compartment is ^{60}Co from activation. With the half life of only about 8 years, storage as opposed to early dismantlement could therefore eliminate the major contribution to worker dose and hence radiation risk.

Limitation and more recently optimization have had a significant effect on worker protection. As an occupational health physician, it is interesting to track these changes in terms of changing requirements for statutory health surveillance, as improvements were made.

In the early days of radiology, there was little or no dose control, save perhaps "if your hands were red, you should stop work." With the use of a deterministic (cell killing) effect for dosimetry, it is not surprising that further deterministic effects occurred in the hands and, more severely, causing aplastic anemia in these workers. There are pictures of cancers developing on the stumps of fingers lost to deterministic damage in these early radiologists (Figure 8.1), and this, together with increased risk of leukemia, confirmed stochastic (cell damage) effects too. At the same time, bone cancers in the predominantly female workers pointing their brushes with their lips as they painted radium dials (Figure 8.2) provided further evidence of worker risk (Baverstock & Papworth, 1989), as well as the need to consider radiation protection as a whole, other than a

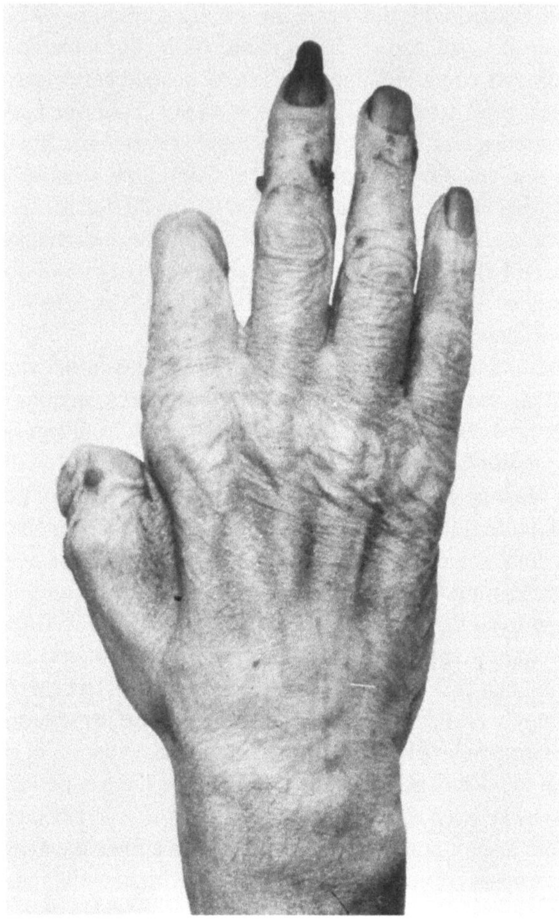

FIGURE 8.1 Early radiologist's hand showing malignant change superimposed on earlier deterministic damage.

specific concentration on a particular high focus area such as radiology. With both deterministic and stochastic effects, health surveillance based on the early detection of disease was clearly indicated, including blood tests for deterministic changes and regular screening for malignant disease. The advent of dose controls brought the elimination of deterministic effects, but ongoing concern about stochastic risk. Health surveillance was modified with initial blood testing to ensure that the worker did not have a hematological condition prior to work and ongoing periodic screening for disease. Optimization in 1976 (ICRP, 1977) brought the concepts of as low as reasonably achievable (alara) and as low as reasonably practicable (alarp). Personal dose was no longer a resource that could be used, but something which needed to be minimized. This resulted in major reductions in both collective and personal dose levels. In parallel,

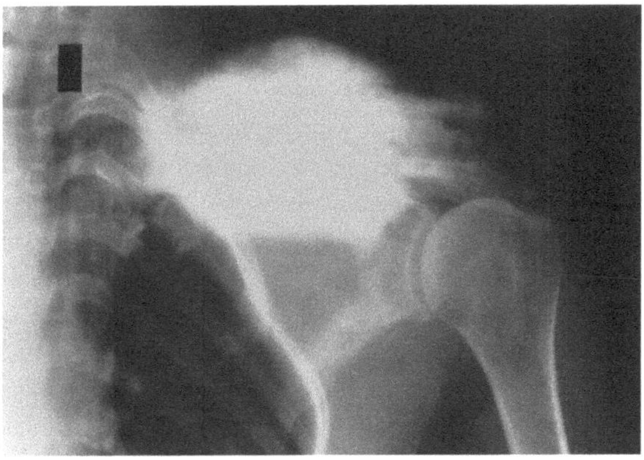

FIGURE 8.2 Primary bone cancer in a radium dial painter.

large-scale epidemiological studies were published (Kendall et al., 1992) demonstrating that, while there may be some evidence for the expected increased cancer risk with the increased dose, cancer morbidity and mortality amongst radiation workers in the UK was significantly lower than for the population as a whole. Against this backdrop, health surveillance for the early detection of disease, self-evidently, was meaningless. Health surveillance however has been retained as a mandatory requirement in Europe (96/29/Euratom, 1996) based on the basic principles of occupational medicine, and as a fit for work assessment, to ensure that the worker is fit to undertake their role. A further indication of the duty of care owed by the employer to the radiation worker.

Effective dose control measures in radiology within the UK were introduced in the early 1920s and their benefit in terms of risk reduction is well demonstrated by epidemiological study (Smith & Doll, 1981). Radium dial painting commenced on a large scale before World War I and bone sarcomas were recognized by 1929 (Rowland & Lucas, 1984). Risk however continued until the 1950s, emphasizing the need for common and consistent radiation protection to cover all practices and uses.

In more recent times, most public, media, and even perhaps regulatory focus has been on "nuclear". There has been very real reduction in individual dose in this area. Against this backdrop, however, in other areas of worker exposure, shockingly, we have seen even the recurrence of deterministic (cell killing) effects or suggestions of dose sharing as a radiation protection tool. As medical use continues to expand, different workers have been involved. In particular, in intervention radiology, deterministic skin effects have been reported since 1992 in doctors and other clinical staff where significant dose can be received for the hands (Streckelhuber et al., 2005). More recently, there are reports of cataracts, another deterministic effect, in the same groups (Vano et al., 2010). These occurrences in new clinical groups have led the need for specific guidance from the International Commission on Radiological Protection (ICRP, 2013) which, not surprisingly, in addition to protection standard, emphasizes the need for training and knowledge about radiation risk for these workers.

Although clearly not so dramatic as the reappearance of deterministic health effects, the arguments for consistency throughout all areas of radiation practice extend into other issues hugely relevant to worker dose, and worker understanding. Against a backdrop of continued planned reduction in most radioactive discharges, nuclear medicine continues to develop and discharges from the medical field seem likely to continue to rise. These result in critical group doses acquired occupationally to groups such as sewage workers, estimated to rival and even exceed dose from designated nuclear sites (UK Strategy for Radioactive Discharges, 2009). In the fields of Naturally Occurring Radioactive Material (norm) critical groups are workers, such as those in commercial kitchens (cooking with natural gas), whose doses are similarly comparable with those from industry (UK Department of Environment Food and Rural Affairs, 2002). While there may be major differences in what is "reasonably practicable" to achieve in terms of dose reduction between processes, if worker understanding and confidence is to be maintained, this must not extend into short hand incorrect simplifications and statements that doses in one area are trivial or "no danger" while other documentation decrees that similar doses elsewhere represent a continued risk that must be acted upon (UK Department of Environment Food and Rural Affairs, 2002).

Accident response and contingency planning provide another important area where there is a need to consider ethical issues in the worker role. The fatalities and acute health effects in the Chernobyl liquidators (Ivanov, Gorski, Makutov, Tsyb, & Souchkevitch, 2001) provide high-profile evidence of the need to extend the employers' duty of care in the areas of worker actions to control or mitigate the consequences of accidents, as well as to implement actions to protect fellow workers or the public. Just as in the earlier paragraphs, however, these considerations are not confined to issues around nuclear power plants, but extend throughout all areas where ionizing radiations are used.

Basic statutory requirements (The Ionising Radiation Regulations, 1999) include the need to have contingency plans for reasonably foreseeable events, as well as specifying the need for workers to be given suitable and sufficient instruction. Worker considerations of this type are also included in statutory requirements for plans required in terms of public protection (Radiation (Emergency Preparedness and Public Information), 1996). As in the earlier paragraphs, the issues of worker information must center around the hazard, and within that context, the knowledge of the events, which are planned for, as well as safety controls and protection. Risk assessment must not be something that is considered solely in terms of statutory approval or licensing, but must form the basis of worker information on foreseeable scenarios. Information on the plan and the worker's role within it is also crucial in terms of what needs to be done, how it can be done, and how monitoring and safety are to be maintained. Knowledge of the equipment and systems that are required is also of importance. Rehearsal adds greatly to understanding and is also a statutory requirement.

8.3 ETHICAL REQUIREMENTS FOR OCCUPATIONAL RADIATION HEALTH RESEARCH

Research has an accepted place in occupational medicine, based on the standard justification of any medically related study, the increase of human knowledge to the benefit of mankind (Royal College of Physicians, 1996). In the occupational setting, however, it is clear that there is an additional justification in support of the health and safety for a specific group of workers upon whom the study is conducted. These two aims may be closely aligned but are not necessarily synonymous.

An earlier paragraph of this chapter mentioned the UK epidemiological study of radiation workers, the so-called National Registry for Radiation Workers (NRRW) (Kendall et al., 1992). This was established with the clear desire to define numeric radiation risk estimates for workers, based on actual worker data. This aim is being achieved by the so-called internal analysis of the study, which involves dividing the workers into dose bands, and looking for the expected increase in cancers as the dose increases. As an individual who has been professionally involved in talking to radiation workers over many years, I have never felt any great worker interest in such numeric dose estimates. Workers, in the main, appear interested in safety of their employment in that particular situation. As such, workers have traditionally paid much greater interest in the so-called external analysis of the NRRW, looking at mortality and morbidity in the radiation worker cohorts, compared with the population at large. Together, the internal and external analyses provide a useful complete study that the two halves taken in isolation fail to achieve.

The identification of the worker as a justification for, and also a customer of, occupational radiation research, requires consideration of an ethical duty of care that the researchers must owe to these workers. Without worker agreement to participate, of course, in many cases no study could take place.

The paragraphs that follow are equally applicable to all occupationally related studies. Radiation, however, is one of those areas that is capable of attracting massive media and public interest; it is therefore not surprising that it is the radiation workers that have suffered great distress, and it is those who have professionally attempted to provide service to the radiation workers that have attempted to take forward an ethical stance to protect them and others in the future.

With the internet, and e-publication, systems of promulgation of the results of health research are currently going through major revolution. The basic principles, however, remain that a paper is submitted to a journal and, followed by a varying level of peer review, achieves publication. It is a publication that allows wider academic review and hence the process by which a consensus on the merits of the work is taken forward. In the modern world, however, additionally, if the work has some potential for media impact, it may be subject to a prepublication embargo. Following press conference launch, there is an immediate general population knowledge through the full range of modern media. Most of this occurs before the journal's normal readership gets their copy through the post.

In the 1990s, in particular, there were a number of studies involving radiation workers that attracted such media attention. Perhaps the most famous, and certainly the most distressing, was what became known as the Gardner Study, and its associated hypothesis (Gardner et al., 1990), that workers and their families were told through headlines of mass circulation newspapers, radio, and television that radiation exposure at work to fathers was causally associated with the incidence of leukemia in their children. With no, or limited, access to actual copies of the paper, few occupational health practitioners involved in the aftermath of the release of this study will ever forget the despair and anguish it caused to real people, to whom we sought to provide occupational health advice.

My own personal experience centers on a telephone call well before 7 a.m to be asked if I had read the morning paper. This was followed by a mad dash to the airport and a flight of approximately 90 min during which I did have sight of a newspaper. Onto a car ride of 1 h, where for the first time I could read the scientific paper, before immediate entry into a large lecture theater holding hundreds of radiation workers and their wives, and my attempts to answer their questions. It makes no difference whether the paper's conclusions of hypotheses proved to be correct or not, the situation is there and then, and workers require care and sensitivity. I hope all would agree that this is no way to treat workers who are real people with families and concerns, who in many cases have consented to participate in the study in question. In fact, of course, the Gardner hypothesis, years later, has not been supported by the evidence (Slovak, Kalman, Davies, & Pilling, 1994).

It is interesting to note that even before the publication of Gardner, consideration had already been given to some ethical aspects of occupational radiation research in the UK. The NRRW from its outset had been an individual consent study, in which workers had to opt in unless their data could not be included. Consent, therefore, is limited to the study as recorded within the NRRW protocol and other researchers wishing to use NRRW data for their own purposes must revisit consent at least at the level of worker representatives. This sort of system of consent recognizes the duty of care, which the researcher owes the worker, acknowledges the worker as a customer of the research, and is an issue that still needs to be addressed by those who periodically argue for increased accessibility and interchange of data (Smith, 1994) between researchers.

After the Gardner paper, and other studies with similar media headlines, the occupational physicians working in the nuclear industry in the UK agreed on a short ethical policy on future involvement in radiation health research. This consisted of four basic principles of *medical confidentiality*, *worker information*, *worker consent* and *worker access to study results prior to publication*. Medical confidentiality, of course, needs no further discussion, where actual medical data are involved in protocol. These requirements for worker knowledge and consent however extend the established requirements for informed consent for invasive medical trials, and recognize workforce rights as customers of, and participants in, occupational health research. The level of consent can of course vary in practice, depending on the research protocol. The guarantee of prepublication information stems from an acknowledgment that it is both impractical and outside the occupational health remit to control media coverage of research, but nevertheless sought to accomplish the clear occupational health requirement to provide the results to the customer accurately and in a way which their questions could be answered and concerns could be addressed in a proper and appropriate manner. The requirement for consent, coupled with an acknowledgment of the duty of care, allows for detailed consideration of prepublication briefings at the protocol stage of research. The rules are therefore laid down before the results are known. This ensures worker knowledge irrespective of the nature of studies findings.

This ethical stance, particularly in relation to prepublication notification of results, was agreed with the editors of the principle health-based journals within the UK such as the British Medical Journal and the Lancet. While the policy was successfully implemented in a number of small radiation studies, questioners at various presentations expressed some skepticism in relation to the practicalities of dissemination of such information in major national studies. These concerns were essentially put to rest by the full implementation of the policy in relation to the publication of the UK's national study investigating Gardner's hypothesis by record linkage between the NRRW and childhood cancer registries. This piece of work involved data from 120,000 radiation workers with prepublication information available to workers in a number of different employment situations. Despite this scale of workers and companies throughout the UK, a media embargo requested by the British Medical Journal, as publishers of the article, remained intact (Draper, Little, Sorahan, & Kinlen, 1997). The success of the implementation of the standards was also published (Kalman, 1999). As a result of these experiences, the guidance on ethics for occupational physicians in the UK, at its 5th edition in 1999, for the first time, included specific guidance on occupational health research. This built upon the work undertaken by the occupational physicians in the nuclear industry, and has been maintained in subsequent editions (The Faculty of Occupational Medicine, in press). The principle ethical issues in research are identified as *worker recruitment*, *consent*, *confidentiality*, *data protection*, and *communication*.

8.4 RADIATION SCIENCE AND WORKERS

In addition to the research based on worker data, academic or indeed medical research has the potential to attract media interest in radiation health, and also therefore to affect workers and their families. Here, worker participation is not required, nor is there any contact with the employing organization and the academic researcher. Thus, any duty of care, and customer relationship, is much less clear. The last 20 years have, however, seen issues identical both in support for greater and lesser radiation controls. From radiation biology, the establishment of *genomic instability* (Kadhim et al., 1994) and a growing level of knowledge on the circumstances when it occurs, led to suggestions for much higher levels of radiation control. On the other hand, the clear demonstration of *adaptive responses* (Wolf, 1998) has led others to suggest radiation hormesis (low doses of radiation are good for you). In radiotherapy, there has been continued interest in tests for radiation sensitivity as part of treatment planning (Wolf, 1998). Here there are questions on the relevance of tests based around cell killing at doses of 10 s of Gray, when considering the occupational situation of risks of cancer at doses in the mili or micro sievert range. More recently, sensitivity is shown at lower diagnostic dose levels (Foray, Colin, & Bourguignon, 2012). In the absence of researcher information here, it falls on government, regulator or expert international body to provide advice in a timeous manner which others can use to advise workers. ICRP has taken on this role, and most recently, in 2005, confirming that the linear low threshold hypothesis remains the prudent base for radiation protection at low doses and low dose range (ICRP, 2004).

8.5 SUMMARY

In summary, radiation protection over the last 100 years has taken forward dramatic increases in radiation control, with resultant reduction in worker dose and risk, with increased worker safety. Even taking account of the expected healthy worker effects for large occupational epidemiological studies, radiation workers can be regarded as being involved in a safe industry.

It is important not to regard the words "nuclear" and "radiation" as synonymous. Despite the dramatic dose reductions in the nuclear industry, deterministic effects are now being seen in radiation workers in new areas of medical use. The consistent application of standards, together with the consistent provision of advice and information remain the keystone of effective worker safety from the hazards of ionizing radiation.

With radiation's huge media profile, it was not surprising that radiation workers were the test cases in terms of the development of ethics of worker involvement in occupational research. The identification that the researcher owes the worker a duty of care, and that the worker is a customer of the research, is achieving growing acknowledgment, not only confined to the radiation worker situation.

With the profile of ionizing radiation research, there is a need for the timeous update of workers and their advisors, as radiation knowledge develops. It is only by this mechanism that continued confidence in radiation protection by workers can be maintained.

REFERENCES

Baverstock, K., & Papworth, D. (1989). The UK radium luminiser survey. In D. Taylor (Ed.), *Risks from radium and thorotrast*. BIR Report 21, London: British Institute of Radiology.

Draper, G., Little, M., Sorahan, T., & Kinlen, L. (1997). Cancer in the offspring of radiation workers: a record linkage study. *BMJ*, *315*.

96/29/Euratom. (13 May, 1996). *Basic safety standards for the protection of the health of workers and the general public against the dangers arising from ionising radiation*. European Council Directive.

Foray, N., Colin, C., & Bourguignon, M. (2012). 100 years of individual radio sensitivity: how we have forgotten the evidence. *Radiology*, *264*, 627–631.

Gardner, M., Snee, M., Hall, A., Powell, Downes, S., & Terrell, J. (1990). Results of a case control study of leukaemia and lymphoma among young people near Sellafield nuclear plant in West Cumbria. *BMJ*, *300*, 423–429.

International Atomic Energy Agency (IAEA). (2001). *Investigation of an accidental exposure of radiotherapy patients in Panama: Report of team of experts, 26 May—1 June 2001*.

International Commission for Radiological Protection (ICRP). (1977). *Recommendation of the International Commission for Radiological Protection*. ICRP Publication 26, An ICRP. *1*(3).

International Commission for Radiological Protection (ICRP). (2004). *Development of the Draft 2005. Recommendations of ICRP*. A collection of papers, An ICRP. *34*(5).

International Commission for Radiological Protection (ICRP). (2013). *Radiological protection in cardiology*. ICRP Publication 120, An ICRP. *42*(1).

Ivanov, V. K., Gorski, A. I., Makutov, M. A., Tsyb, A. F., & Souchkevitch, G. N. (Nov 2001). Mortality among Chernobyl emergency workers: estimation of radiation risks. *Health Physics*, *51*(5), 514–521.

Kadhim, M. A., Morimone, S. A., Hepburn, M. D., Goodhind, D. T., Buckle, V. S., & Wright, E. G. (1994). Alpha particle induced chromosomal instability in human bone marrow cells. *The Lancet*, *344*, 987–988.

Kalman, C. J. (1999). Ethical requirements for occupational health research—compliance arrangement for a single company in relation to a recent major nuclear industry study. *Occupational Medicine*, *49*(4), 221–224.

Kendall, G. M., Muirhead, C. R., MacGibbon, B. H., O'Hagan, J. A., Conquest, A. J., Goodhill, A. A., et al. (Jan, 1992). *First analysis of the National Registry for Radiation Workers—Occupational exposure to ionising radiation and mortality*. NRPB—R251(ISBN 085951 3440).

Lloyd, D. C., Edwards, A. A., Fitzsimmons, E. J., Railton, R., Jeffery, P., Williams, T. G., et al. (Oct, 1994). Death of a classified worker probably caused by overexposure to gamma radiation. *Occupational and Environmental Medicine*, *S1*(10), 713–718.

Radiation (Emergency Preparedness and Public Protection). (2001). *Regulation UK statutory instruments*.

Rowland, R., & Lucas, H. (1984). Radium dial workers. In J. D. Boice & J. Fraumeni (Eds.), *Radiation carcinogens. Epidemiology and Biological Significance*. Raven Press.

Royal College of Physicians. (1996). *Guidelines in the practice of ethics. Committees in medical research involving human subjects*. London, UK: Royal College of Physicians.

Slovak, A. J. M., Kalman, C. J., Davies, N. F., & Pilling, K. (1994). The Gardner hypothesis—found wanting. *BMJ*, *308*.

Smith, P. G., & Doll, R. (1981). Mortality from cancer and all causes among British Radiologists. *British Journal of Radiology*, *54*, 187–194.

Smith, G. (1994). Increasing the accessibility of data. *BMJ*, 1519–1520.

Stoeckelhuber, B. M., Leibecke, T., Schulz, E., Melchert, U. H., Bergmann-Koester, C. U., & Helmberger, T., et al. (2005). Radiation dose to radiologist's hand during continuous CT fluoroscopy-guided interventions. *Cardiovascular and International Radiology*, *28*(5), 589–594.

The Equality Act. (2010). *UK legislation*.

The Faculty of Occupational Medicine. *Ethics Guidance for Occupational Health Practice*. Lichfield P and Brecker N 2012.

The Health and Safety at Work Act. (1974). *UK legislation*.

The Ionising Radiation Regulations. (1999). *UK statutory instrument*.

UK Department of Environment Food and Rural Affairs. (2002). *Consultation paper draft radioactive substances (natural gas), exception order*.

UK Strategy for Radioactive Discharges. (July, 2009).

Vano, E., Kleiman, N. H., Durma, A., Rehani, M. M., Echererri, D., & Cabrera, M. (Oct, 2010). Radiation cataract risk in interventional cardiology personnel. *Radiation Research*, *174*(4), 490–495.

Wolf, S. (1998). The adaptive response in radiobiology: evolving insights and implications. *Environmental Health Perspectives*, *106*(Suppl. 1), 277–283.

Chapter 9

ALARA: What is Reasonably Achievable?

Sven Ove Hansson
Royal Institute of Technology (KTH), Division of Philosophy, Stockholm, Sweden
E-mail: soh@kth.se

Chapter Outline

9.1 Introduction 143
9.2 ALARA and Optimization in the Radiation Protection System 145
9.3 A Three-Levelled Model 146
9.4 What is "Reasonably Achievable?" 148
9.5 The Scope of Economic Compromises 149
9.6 ALARA and Cost-Benefit Analysis 151
9.7 Related Principles of Protection 152
9.8 Conclusion 154

9.1 INTRODUCTION

Our knowledge about the dangers of ionizing radiation has passed through three stages. In the first few years after Röntgen's discovery of X-rays in 1895, the risks were unknown and researchers exposed themselves and others to high exposures that we now know to have lethal effects. In the second phase, beginning about 1900, it was recognized that acute effects such as skin burns can result from high short-term exposures. Efforts were made to keep doses well below the levels that had given rise to these effects. It was believed that below those levels there was no danger, so that exposure limits could provide complete protection against the negative health effects of radiation (Kathren & Ziemer, 1980).

The third phase has its origin in the discovery that accumulated exposure to ionizing radiation increases the risk of cancer, in particular leukemia. Beginning in the 1950s, radiation protectors extended their concerns from acute effects to long-term genetic and cancer effects. Such effects appeared to be stochastic, and there did not seem to be any threshold dose below which they disappear completely (Lindell, 1996). This was expressed as an uncertain hypothesis in a report by the International Commission on Radiological Protection (ICRP) in 1959:

Radioactivity in the Environment, Volume 19
ISSN 1569-4860, http://dx.doi.org/10.1016/B978-0-08-045015-5.00009-5

"The mechanism of leukemia induction by radiation is not known. It may be postulated that if the dose is lower than a certain threshold value no leukemia is produced. In this case it would be necessary to estimate the threshold dose and to make allowances for recovery, if any. There is not sufficient information to do this, but caution would suggest that an accumulated dose of 750 rem might exceed the threshold. The most conservative approach would be to assume that there is no threshold and no recovery, in which case even low accumulated doses would induce leukemia in some susceptible individuals, and the incidence might be proportional to the accumulated dose. The same situation exists with respect to the induction of bone tumors by bone-seeking radioactive substances."

(ICRP, 1959, p. 4)

Perhaps ironically, major advances in radiation protection were made in the Manhattan project, the same project that resulted in the worst radiation exposures that the world has seen. In order to protect their own personnel, the project assigned to Dr. Robert S. Stone in the Health Division of the Metallurgical Laboratory in Chicago the task of setting "tolerance levels" for radiation. He and his colleagues made it clear that since there is no absolutely safe level, radiation exposures should be kept at the lowest level that was practically possible to achieve. After the war, this principle was adopted by the country's National Committee on Radiation Protection (NCRT) that stated in 1954 that radiation exposures should "be kept at the lowest practical level" (Auxier & Dickson, 1983). The International Commission on Radiological Protection (ICRP) made a similar recommendation in 1959:

"[T]he Commission recommends that all doses be kept as low as practicable, and that any unnecessary exposure be avoided."

(ICRP, 1959, p. 11)

This recommendation has repeatedly been confirmed by the Commission. In the 1977 recommendations, it was rephrased as a requirement that "all exposures shall be kept as low as reasonably achievable, economic and social factors being taken into account" (ICRP, 1977, p. 3). Over the years, the principle has been known under several different names:

as low as practicable (ALAP)
as low as reasonably achievable (ALARA)
as low as reasonably attainable (ALARA)
as low as reasonably practicable (ALARP).
so far as is reasonably practicable (SFAIRP) (HSE, 2001a, p. 92).

ALAP appears to have been the first of these abbreviations. In the early 1970s, it was replaced by ALARA (mostly in the US) and ALARP (mostly in Europe) that were thought to be less imprecise (Wilson, 2002). Some authors have tried to make distinctions between these terms:

"What is achievable is a function of current science. But what is practicable is a function of economics and benefit–cost analysis… If ICRP were really serious about making its recommendations less tied to utilitarian and benefit–cost criteria, then it

would recommend exposures that were as low as possible/achievable, not those that are as low as practicable."

(Shrader-Frechette & Persson, 2002, p. 155)

In practice, however, ALARA and ALARP are taken as synonyms. The major difference is one of usage: In Britain where the principle is called ALARP it has a general use in workplace health and safety. In countries where it is called ALARA it tends to be restricted to radiation protection.

9.2 ALARA AND OPTIMIZATION IN THE RADIATION PROTECTION SYSTEM

In 1977 the ICRP introduced a system of three basic principles for radiation protection that are still in use. The three principles are:

"(a) no practice shall be adopted unless its introduction produces a positive net benefit;
(b) all exposures shall be kept as low as reasonably achievable, economic and social factors being taken into account; and
(c) the dose equivalent to individuals shall not exceed the limits recommended for the appropriate circumstances by the Commission."

(ICRP, 1977, p. 3)

These principles are known under the names justification, optimization, and (individual) dose limits. The combination of the three principles has been described as a "regulatory mix" that combines to provide better protection than what each of the three proposals could do alone (Lierman & Veuchelen, 2005). According to the justification principle, all radiation exposures have to be justified, which is usually interpreted as meaning that the benefits (advantages) should be shown to outweigh the risks (disadvantages). The third principle (dose limits) restricts the permitted exposure of each individual. But according to the second principle (ALARA), keeping the limits is not enough. Even if exposures are below the dose limits, efforts should be made to reduce them further. These further reductions in doses are usually conceived as referring to collective rather than individual doses, and therefore ALARA is often seen as a utilitarian principle (cf. chapter 3) (Lierman & Veuchelen, 2006).

The terminology "optimization" for ALARA indicates that some sort of compromise is aimed at. To optimize two or more factors (here: dose and costs) means to find the best balance between them. To optimize radiation doses while taking costs into account would therefore seem to mean that one finds a balance from which it would be undesirable to deviate in either direction: Higher doses should be avoided since they can cost-efficiently be reduced, and lower doses since they would involve indefensible costs. As we will see, such an "optimal balance" view is a common interpretation of ALARA. However, there is also another interpretation, namely that the principle requires lowering of doses as far as possible. According to the "as low as possible" view, further reduction of the doses can never conflict with the ALARA principle. The "as low as possible" view is arguably supported by the actual wordings of the acronym; it reads "as low as reasonably achievable", not "at the exact level that is reasonably achievable".

The legal status of the ALARA principle differs between jurisdictions. In Great Britain, an operator who has achieved what is considered to be an acceptable risk level may still be under an ALARA-based legal obligation to take further action to reduce exposures. In the US the ALARA principle does not have such legal force (French, Bedford, & Atherton, 2005; Lierman & Veuchelen, 2006).

9.3 A THREE-LEVELLED MODEL

According to some explications of ALARA, it is only intended to be applied to the middle part of the dose range. If the doses are sufficiently high, then they have to be reduced irrespective of the costs, and there is no need for considerations of what is "reasonable" or "practicable". On the other hand, if the doses are sufficiently low, then there is presumably no need for action. It is in the intermediate region, with doses that are neither negligible nor utterly unacceptable, that the ALARA principle will be applied. The upper limit of the ALARA region in this model has been called the "action level" and its lower limit the "inaction level" (Hendee & Edwards, 1986). Doses below the inaction level have also been referred to as "de minimis" doses. This term is derived from the legal phrase "de minimis non curat lex" (law does not concern itself with trifles). In the context of risk assessment, it expresses the view that some risks are so small that we have no reason to take action against them even if such action can be taken at no or negligible cost.

This three-levelled approach has been promoted in particular by the Health and Safety Executive (HSE) in Great Britain. In its influential policy document "Reducing risks, protecting people" from 2001, it divided situations involving risk into three categories: the unacceptable region, the tolerable region, and the broadly acceptable region (HSE, 2001a). An activity or practice falling in the unacceptable region should be "ruled out unless the activity or practice can be modified to reduce the degree of risk so that it falls in one of the regions below, or there are exceptional reasons for the activity or practice to be retained" (p. 42). The zone at the opposite end, the broadly acceptable region, contains risks that "are comparable to those that people regard as insignificant or trivial in their daily lives" and are therefore "generally regarded as insignificant and adequately controlled". The intermediate, "tolerable" region consists of risks that "people are prepared to tolerate in order to obtain benefits" (p. 43). It is in that intermediate region that the HSE recommends application of the ALARA principle (or in their own terminology: the ALARP principle). This approach is illustrated in Figure 9.1.

The HSE has also specified tentative limits between the three regions in terms of probabilities of fatalities. The limit between the broadly acceptable and the tolerable region has been set at an individual risk of death of one in a million per year (both for occupational and nonoccupational risk exposure). Hence, risks below that level would typically be taken to be negligible and therefore not in need of ALARA-based activities (p. 45). The limit between tolerable and

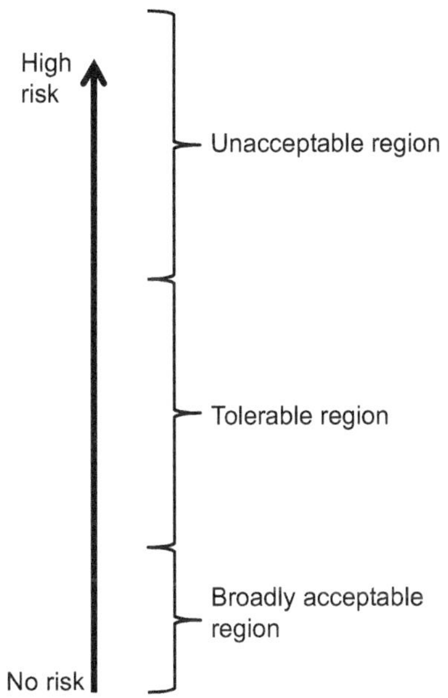

FIGURE 9.1 The three-leveled model of radiation protection.

unacceptable risks is tentatively set at one in 1000 per year for occupational risks and one in 10000 per year for nonoccupational risks (p. 46).

The idea that ALARA is only applicable above a certain "de minimis" dose is not in obvious conformity with the standard linear no-threshold assumption according to which there is no safe dose level above zero. If there is no safe level, then why should there be a level below which no efforts are made to reduce exposures? In the early 1980s this problem was discussed among three major contributors to the ICRP's work, namely the American Harald H. Rossi (1917–2000), the Swede Bo Lindell (b. 1922) and the Argentinian Dan J. Beninson (1931–1994). The debate was started by Rossi who defended "de minimis", although he conceded that there are strong reasons to believe that any dose of ionizing radiation can cause genetic damage. In defense of de minimis he said:

"One stops when one has reached a level where the radiation hazard has become de minimis... *It follows that a more accurate formulation of the ALARA philosophy is that: 'It is the obligation of those responsible for radiation protection to see to it that personal exposures are below maximum permissible limits and reduced as near as practicable to* de minimis *levels. Further reductions are not required'."*

(Rossi, 1980)

In a reply, Lindell and Beninson claimed that the level below which no further reductions are needed is not the same in all contexts but has to be determined from the specific circumstances:

"[I]n each situation, there is a level of dose below which it would not be reasonable to go because the cost of further dose reduction would not be justified by the additional

eliminated detriment. That level of dose, however, is not a de minimis level below which there is no need of concern, nor can it be determined once and for all for general application. It is the outcome of an optimization assessment which involves marginal cost-benefit considerations... It is not reasonable to pay more than a certain amount of money per unit of collective dose reduction, but if dose reduction can be achieved at a lesser cost even at very low individual doses, the reduction is, by definition, reasonable."

(Lindell & Beninson, 1981)

9.4 WHAT IS "REASONABLY ACHIEVABLE?"

The key terms ALAP, ALARA, and ALARP are all vague. When is a reduction of radiation doses "practicable", "achievable" or "attainable"? According to a strict interpretation these criteria are satisfied whenever a reduction is technically possible, irrespectively of what it would cost. According to a weak interpretation these acronyms refer to what can easily be afforded. In practice an interpretation that is intermediate between these two extremes seems to be aimed at. The additional term "reasonably" indicates that the strictest interpretation is not intended.

"Reasonable" is a common legal term that seems impossible to define in a precise manner. In a penetrating analysis, Corten (1999) drew attention to two major functions that this word has in legal discourse. First, it makes regulations adaptable, allowing them to be applied in different circumstances. The term is used to "introduce a degree of flexibility" in legal instruments in order to resolve "a contradiction between the essentially static character of legal texts and the dynamic character of the reality to which they apply" (p. 615). The term "reasonable" provides a legitimation for judges and administrators to interpret regulations flexibly, taking into account conditions and circumstances that the enactor of the regulations did not foresee. Secondly, the notion of reasonableness can provide legitimacy to a legal order "by presenting an image of a closed, coherent, and complete legal system". The notion "masks persistent contradictions regarding the meaning of a rule, behind a formula which leaves open the possibility of divergent interpretations" (p. 618).

The reasonableness incorporated in radiation protection (the "R" in ALARA) appears to have both these functions. First, it provides the flexibility needed to solve potential conflicts with economic and practical constraints and objectives. In this way "the degree of risk can be balanced against time, trouble, cost and physical difficulty of its risk reduction measures" (Melchers, 2001). Secondly, it creates the impression of a coherent and comprehensive system that is capable of solving all potential conflicts. Divergences between economic demands and the demands of health protection are presented as solvable within the system of radiation protection, rather than having to be treated as conflicts between radiation protection and external, economic interests. This "internalization" of economic compromises can of course be seen both as a way to resolve conflicts and as a way to hide them from view.

Even before the notion of "reasonable practicability" was used in radiation protection, it was established in legal language as a means to express that economic considerations must be taken into account. Its meaning was explained by the judge in an often quoted English court case from 1949 (Court of Appeal in Edwards vs National Coal Board):

"'Reasonably practicable' is a narrower term than 'physically possible' and seems to me to imply that a computation must be made by the owner, in which the quantum of risk is placed on one scale and the sacrifice involved in the measures necessary for averting the risk (whether in money, time or trouble) is placed on the other; and that if it be shown that there is a gross disproportion between them—the risk being insignificant in relation to the sacrifice—the Defendants discharge the onus on them."

(Jones-Lee & Aven, 2011)

Five years later the House of Lords confirmed this interpretation in another case (Marshall vs Gotham Co Ltd):

"The test of what is (reasonably practicable) is not simply what is practicable as a matter of engineering, but depends on the consideration, in the light of the whole circumstances at the time of the accident, whether the time, trouble and expense of the precautions suggested are or are not disproportionate to the risk involved, and also an assessment of the degree of security which the measures suggested may be expected to afford."

(Jones-Lee & Aven, 2011)

9.5 THE SCOPE OF ECONOMIC COMPROMISES

Since we do not have unlimited resources, a compromise has to be struck between radiation protection and economic considerations. It is less obvious what should be the scope (aggregation level) of such compromises (Hansson, 1998; pp. 106–109). As shown in Figure 9.2, there are at least four alternative such scopes that one could choose between in radiation protection. First, the compromise could be struck separately for each *individual workplace or company*. This would for instance mean that a financially strong company is required to provide its workers with better protection than a company with economic difficulties. Such an interpretation of ALARA would seem to be in accordance with the exact phrasing of the acronym; if a company can easily afford an improvement in protection then that improvement would seem to be reasonably achievable. The fact that there are other companies who cannot afford the same level of protection does not seem to be a decisive argument against it. But of course, such an interpretation will be more problematic from the viewpoint of workers in the economically less fortunate companies. They will receive less protection than their colleagues in the richer companies, and may have good reasons to call this difference unfair.

The second scope is that of the *industry or work process*. A choice of this scope would for instance imply that the same degree of protection is required

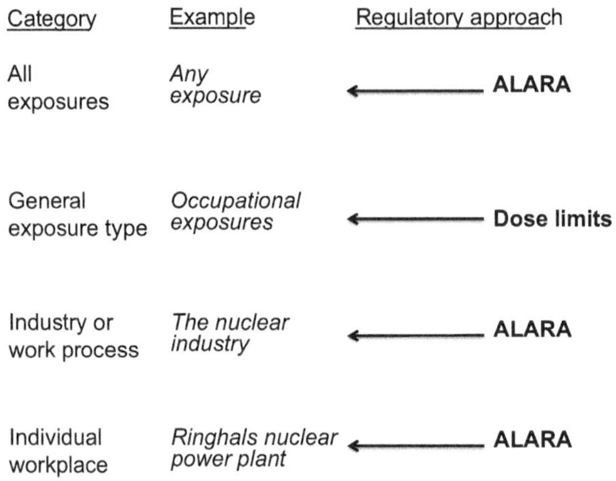

FIGURE 9.2 Different scopes (aggregation levels) for compromises between radiation protection and economic considerations.

of all nuclear power stations, and similarly of all industrial X-ray facilities, etc. However, the requirements need not be the same for the different types of industry. The requirements on the nuclear industry could presumably be based on what a representative such plant can afford, and similarly for other types of plants.

The third scope is *general exposure type*. We can divide radiation exposures into a small set of major types: occupational exposure, exposure through food, medical exposure of patients, etc. These types can be used to delineate scopes for radiation protection. This would mean for instance that the same requirements are applied to all occupational exposures, and similarly the same requirements for all exposures through ambient air, etc.

Finally, a standard for radiation protection can have *all exposures* as its scope. This would mean that the same standards are applied to all types of exposures, be they occupational or ambient.

In practice, individual dose limits are applied to general exposure types. Both the ICRP and national standard-setting bodies have separate exposure limits for occupational and nonoccupational exposures, the latter typically being about one order of magnitude lower. In contrast, the ALARA principle is applied to the other three scopes, but usually not to that of general exposure types. On some occasions, a company that can afford to decrease its exposures is required to do so with reference to ALARA, even though other similar companies with the same problem cannot afford to do so. On other occasions what is "reasonably practicable" is judged on the level of a whole industry or work process. This means for instance that the protection requirements in dentists' offices are decided uniformly for all such offices, presumably based on what they can typically afford, and similarly for other types of workplaces. Finally, attempts have been made to operationalize ALARA in a uniform way for all exposures by means of cost-benefit analyses.

9.6 ALARA AND COST-BENEFIT ANALYSIS

By cost-benefit analysis (CBA) is meant a decision-aiding technique that weighs advantages against disadvantages in numerical terms (Hansson, 2007). In a typical CBA, the various advantages and disadvantages associated with an option are all assigned a monetary value. Usually, a fixed value is assigned to the loss of a human life, and risks of losses in life are valued as fractions of that value. Hence, if the loss of a life is assigned the value $7.000.000, then the value of a risk of 1 in 1000 of an accident in which 8 people would die is

$$0.001 \times 8 \times 7.000.000 \text{ dollars} = 56.000 \text{ dollars}$$

Cost-benefit analysis has largely been used as a means to avoid excessive regulatory demands. Government agencies entrusted with the protection of humans and the environment are required to show that the monetary costs of their regulations are lower than the value of the resulting gains in risk reduction. This way of thinking may seem to be quite different from that of ALARA, but in particular in the USA the ALARA principle has often been interpreted as a requirement that dose reductions be optimized with the help of cost-benefit analysis.

According to Kathren, Munson, and Higby (1984), in the tolerable region ("ALARA region") as depicted in Figure 9.1, "the value of a person-rem is constant". Below that region, the value of a person-rem "tends to zero", whereas above it that value "increases sharply". This means that in its area of application the ALARA principle is "treated as being simply a restatement of the standard cost-benefit criterion that from the point of view of social welfare maximization a safety project should be undertaken only if its costs do not exceed its benefits" (Jones-Lee & Aven, 2011).

In Britain the Health and Safety Executive (HSE) has chosen to explicate ALARA in terms of cost-benefit analysis, but with another method that puts higher demands on safety. If cost-benefit analysis is applied in the standard way, then safety obligations are satisfied when a level has been reached where the costs of further improvements would be larger than the calculated gains. But instead of this standard criterion, the HSE applies what they call a "test of gross disproportion". According to that test, safety obligations are only satisfied when the costs of further improvements would be *much* larger than the calculated gains (HSE, 2001a, p. 67). In the agency's own words:

"[A] computation must be made in which the quantum of risk is placed on one scale and the sacrifice, whether in money, time or trouble, involved in the measures necessary to avert the risk is placed in the other; and that, if it be shown that there is a gross disproportion between them, the risk being insignificant in relation to the sacrifice, the person upon whom the duty is laid discharges the burden of proving that compliance was not reasonably practicable." (p. 62)

According to the HSE, the extent of this "bias" toward safety differs between different contexts. In each particular case it "has to be argued in the light of all the circumstances applying to the case", including the needs that there may be to apply a precautionary approach (p. 67). In another document, the HSE introduced a numerical specification of this type of bias toward safety:

"In any assessment as to whether risks have been reduced ALARP, measures to reduce risk can be ruled out only if the sacrifice involved in taking them would be grossly disproportionate to the benefits of the risk reduction…[W]e believe that the greater the risk, the more that should be spent in reducing it, and the greater the bias on the side of safety. This can be represented by a 'proportion factor', indicating the maximum level of sacrifice that can be borne without it being judged 'grossly disproportionate'."

(HSE, 2001b)

In 2006 the British railway inspectorate specified such proportion factors (also called "disproportion factors") in numerical terms. These factors are intended to be applied to the "value for preventing a fatality" (VPF), which is essentially a more palatable term for what cost-benefit analysts usually call a "value of life". The inspectorate noted that the disproportion factors applied in practice by the HSE were in general around 1–2 in cases with a very low baseline risk (i.e. close to the "broadly acceptable" region in Figure 9.1) and around 10 at very high baseline risk (i.e. close to "unacceptable" in the same diagram). Based on this they proposed the following rules of thumb:

low baseline individual risk and no societal risk: disproportion factor 1–2
low baseline individual risk and societal risk: disproportion factor 3
high baseline individual risk: disproportion factor 3–10 depending on the level of individual risk (HMRI, 2006).

The recommendation is to multiply the VPF with the appropriate disproportion factor and then use this augmented VPF in comparisons with the costs. This use of disproportion factors will have essentially the same effect as using higher VPFs (or "values of life") than what are commonly used in cost-benefit analysis. It is an open question whether this calculative approach corresponds to the meaning of the acronyms ALARA and ALARP.

9.7 RELATED PRINCIPLES OF PROTECTION

Several principles that are used in other areas are quite close in meaning to ALARA. One of the more important of these is *best available technology* (BAT) that is used in several jurisdictions to promote efficient pollution-limiting technologies. Just like ALARA, BAT is usually used in combination with a numerical limit, e.g. an effluent limit. It has the function of requiring further reductions below that limit to the extent that this is practically possible. Just like ALAP, the earliest version of ALARA, BAT has a wording that seems to require reductions at any price, but in practice it has been applied in ways that take economic limitations into account. Just like ALAP was replaced by ALARA and ALARP, BAT has been replaced in several contexts by acronyms that explicitly leave room for practical and economic considerations, such as:

best available technology not entailing excessive costs (BATNEEC);
best practicable means (BPM); and
best practicable environmental option (BPEO).

Similar phrases used primarily in the United States include the following that can all be described as environmental counterparts of ALARA:

best available control technology (BACT);
lowest achievable emissions rate (LAER); and
reasonably available control technology (RACT).

The *substitution principle* (also called the product choice principle) requires that hazardous (or potentially hazardous) chemical substances be replaced by less hazardous alternatives. This principle has usually been defined as requiring that dangerous substances or technologies be replaced by less dangerous ones, while at the same time retaining the functionality of the original substance as far as possible and keeping costs as low as possible (Hansson, Molander, & Rudén, 2011).

ALARA has often been compared to the *precautionary principle*, and it has been claimed that the two principles "have a common denominator: they are general rules for decision makers on risk assessment and risk management under scientific uncertainty" (Lierman & Veuchelen, 2005). However, the similarity between the two principles should not be exaggerated. Although they are both intended to ensure that environmental and health concerns are taken seriously, they do this in different ways and at different stages in the decision-making process.

There is no consensus on what the precautionary principle means, but most versions of the principle refer to situations in which there is a valid (scientifically based) suspicion of a potential danger, but not full evidence of its existence. The principle requires or allows that in such a situation, measures are taken to protect the environment and/or human health. There are two major types of definitions of the principle: argumentative and prescriptive definitions. An argumentative version of the precautionary principle is found in Principle 15 of the Rio Declaration (UNCED, 1993). It requires that "lack of full scientific certainty shall not be used as a reason for postponing cost-effective measures to prevent environmental degradation". Prescriptive versions of the principle prescribe actions. Perhaps the most famous of these is the so-called Wingspread Statement: "When an activity raises threats to the environment or human health, precautionary measures should be taken, even if some cause-and-effect relationships are not fully established scientifically" (Raffensperger & Tickner, 1999; pp. 354–355).

Hence, the precautionary principle deals with how we should react to uncertain information. If there is no uncertainty, then there is no need to apply the precautionary principle, although there may be reasons to be cautious in other ways. (It is important not to confuse the precautionary principle with cautious decision-making in general.). In contrast, the ALARA principle does not primarily relate to uncertainty. If we knew for certain that the linear no-threshold model is accurate, then we would still have good reasons (in fact better reasons) to apply ALARA in radiation protection, in spite of there being no uncertainty. In other words, the two principles deal with different phases in the decision-making process. The precautionary principle concerns what factual assumptions we should base our decisions on, whereas ALARA pertains to what decisions we should make given these assumptions.

9.8 CONCLUSION

Like most safety principles, ALARA is fairly vague and in need of interpretation (Möller & Hansson, 2008). Its interpretations in practical applications have sometimes diverged in different directions. We have identified four major interpretative issues that would need to be settled in order to make the principle fully precise:

1. Does ALARA require an "optimal balance" that it is undesirable to deviate from in either direction? Or does it require that exposures be "as low as possible", so that a reduction of exposures can never conflict with the ALARA principle?
2. Does ALARA require that even very low doses be reduced when this can be done at sufficiently small costs? Or is there a dose level below which ALARA should not be applied?
3. Should ALARA-based compromises between radiation protection and economic considerations be made separately for each company, in a unified manner for a whole branch of industry, or even in a unified manner for society as a whole?
4. Can the demands of ALARA be operationalized with cost-benefit calculations? If they can, should standard monetary VPF be used, or should higher values be applied in order guarantee a bias toward safety?

Each of these questions is connected with the underlying ethical issue how to find the best compromise between the demands of safety and those of economy and production. Therefore, the task of clarifying what ALARA means, and in particular, what it should mean is basically an ethical task.

REFERENCES

Auxier, J. A., & Dickson, H. W. (1983). Guest editorial: concern over recent use of the ALARA philosophy. *Health Physics*, *44*, 595–600.

Corten, O. (1999). The notion of 'reasonable' in international law: legal discourse, reason and contradictions. *International and Comparative Law Quarterly*, *48*, 613–625.

French, S., Bedford, T., & Atherton, E. (2005). Supporting ALARP decision making by cost benefit analysis and multiattribute utility theory. *Journal of Risk Research*, *8*, 207–223.

Hansson, S. O. (1998). *Setting the limit. Occupational health standards and the limits of science*: Oxford University Press.

Hansson, S. O. (2007). Philosophical problems in cost-benefit analysis. *Economics and Philosophy, 23*, 163–183.

Hansson, S. O., Molander, L., & Rudén, C. (2011). The substitution principle. *Regulatory Toxicology and Pharmacology, 59*, 454–460.

Health and Safety Executive. (2001a). *Reducing risks, protecting people. HSE's decision-making process*. Norwich: Her Majesty's Stationery Office. http://www.hse.gov.uk/risk/theory/r2p.2.pdf.

Health and Safety Executive. (2001b). *Principles and guidelines to assist HSE in its judgements that duty-holders have reduced risk as low as reasonably practicable*. http://www.hse.gov.uk/risk/theory/alarp1.htm.

Hendee, W. R., & Marc Edwards, F. (1986). ALARA and an integrated approach to radiation protection. *Seminars in Nuclear Medicine, 16*, 142–150.

Her Majesty's Railway Inspectorate. (2006). *HMRI ALARP guidance and general principles*. London. http://www.rail-reg.gov.uk/upload/pdf/risk-alarpguidance.pdf.

International Commission on Radiological Protection. (1959). *Recommendations of the ICRP: ICRP Publication No. 1*. New York: Pergamon.

International Commission on Radiological Protection. (1977). Recommendations of the ICRP: ICRP Publication No. 26. *Annals of the ICRP, 1*(3), 1–53. (Oxford: Pergamon).

Jones-Lee, M., & Aven, T. (2011). ALARP—what does it really mean? *Reliability Engineering and System Safety, 96*, 877–882.

Kathren, R. L., & Ziemer, P. L. (1980). The first fifty years of radiation protection—a brief sketch. In Ronald L. Kathren & Paul L. Ziemer (Eds.), *Health physics: A backward glance* (pp. 1–9). Elmsford, N.Y.: Pergamon Press.

Kathren, R. L., Munson, L. H., & Higby, D. P. (1984). Application of risk-cost benefit techniques to ALARA and de-minimis. *Health Physics, 47*, 195.

Lierman, S., & Veuchelen, L. (2005). The optimisation approach of ALARA in nuclear practice: an early application of the precautionary principle. Scientific uncertainty versus legal uncertainty. *Water Science and Technology, 52*(6), 81–86.

Lierman, S., & Veuchelen, L. (2006). The optimisation approach of ALARA in nuclear practice: an early application of the precautionary principle? Scientific uncertainty versus legal uncertainty and its role in tort law. *European Environmental Law Review, 15*(4), 98–107.

Lindell, B. (1996). The history of radiation protection. *Radiation Protection Dosimetry, 68*, 83–95.

Lindell, B., & Beninson, D. J. (1981). ALARA defines its own limit. *Health Physics, 41*, 684–685.

Melchers, R. E. (2001). On the ALARP approach to risk management. *Reliability Engineering and System Safety, 71*, 201–208.

Möller, N., & Hansson, S. O. (2008). Principles of engineering safety: risk and uncertainty reduction. *Reliability Engineering and System Safety, 93*, 776–783.

Raffensperger, Carolyn, & Tickner, Joel (Eds.), (1999). *Protecting public health and the environment: Implementing the precautionary principle*. Washington, D.C.: Island Press.

Rossi, H. H. (1980). What are the limits of ALARA? *Health Physics, 39*, 370–371.

Shrader-Frechette, K., & Persson, L. (2002). Ethical, logical and scientific problems with the new ICRP proposals. *Journal of Radiological Protection, 22*, 149–161.

United Nations Conference on Environment and Development. (1993). *The Earth summit: The United Nations Conference on Environment and Development (UNCED), Rio De Janeiro 1992*. Introduction and commentary by Stanley P. Johnson. London: Graham & Trotman.

Wilson, R. (2002). Precautionary principles and risk analysis. *Technology and Society Magazine, IEEE, 21*(4), 40–44.

Part III

Lessons Learned from Nuclear Accidents

Chapter 10

Lessons Learned from the Chernobyl Accident in Norway

Astrid Liland* and Lavrans Skuterud
Norwegian Radiation Protection Authority, Østerås, Norway
**Corresponding author: E-mail: astrid.liland@nrpa.no*

Chapter Outline

10.1 Introduction **159**
10.2 Radioactive Contamination—A Societal Challenge **161**
10.3 Early Lessons from the Chernobyl Accident **162**
10.3.1 Information Needs 162
10.3.2 The Importance of Local Monitoring Stations 163
10.3.3 Setting Permissible Levels for Radioactive Substances in Foodstuffs 163
10.3.4 Consequences and Countermeasures in Food Production 164
10.3.5 Monitoring of Affected People 165
10.3.6 Psychosocial Effects 166
10.4 Long-Term Challenges and Lessons Learned **167**
10.4.1 Information Needs 167
10.4.2 Zoning and Compensation 167
10.4.3 Technical and Social Countermeasures 168
10.4.4 Involvement of Affected Groups 171
10.4.5 Psychosocial Effects 171
10.5 Health Concerns **172**

10.1 INTRODUCTION

Norway was the country outside the former Soviet Union that experienced the largest consequences of the fallout from the Chernobyl accident (Liland, Lochard, & Skuterud, 2009; Tveten, Brynildsen, Amundsen, & Bergan, 1998). As a non-nuclear country Norway had limited emergency plans for nuclear accidents and

Radioactivity in the Environment, Volume 19
ISSN 1569-4860, http://dx.doi.org/10.1016/B978-0-08-045015-5.00010-1

the authorities faced an immense challenge in handling the situation. In the fall-out affected regions natural pastures are important in milk and meat production for sheep, cows, goats, and reindeer (see Figure 10.1) and the contamination of foodstuffs was significant, see Table 10.1. The contamination greatly exceeded the permissible levels in foodstuffs for radiocesium, which in June 1986 were set at 600 Bq/kg for basic foodstuffs and 370 Bq/kg for milk and infant food. In July 1986, the authorities declared that all producers should be financially compensated

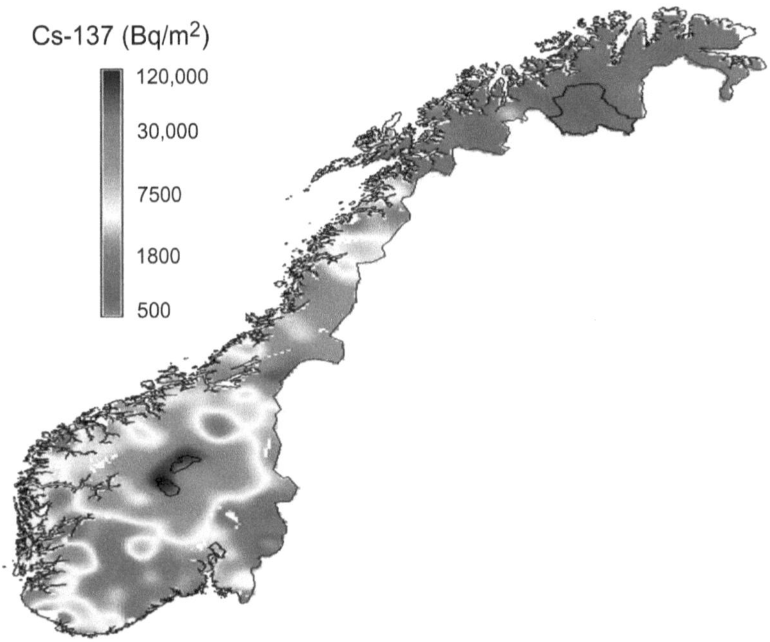

FIGURE 10.1 Map of Chernobyl Cs-137 deposition in Norway. Interpolated map based on average municipality values from Backe, Bjerke, Rudjord, and Ugletveit (1986). (For color version of this figure, the reader is referred to the online version of this book.)

TABLE 10.1 Some Maximum Values Measured in Foodstuffs in Norway in 1986

Produce	Activity Concentration
Cow's milk	1200 Bq/l
Goat's milk	2900 Bq/l
Lamb meat	40 000 Bq/kg
Reindeer meat	150 000 Bq/kg
Freshwater fish	30 000 Bq/kg

for their economic losses due to the consequences of the accident. The affected regions were sparsely populated by humans but the problems they faced in dealing with the situation were significant. Authorities, experts, producers and lay people worked together to develop management strategies that proved successful for handling the situation. Norway is still performing mitigating actions and extensive monitoring due to radioactive contamination 27 years after the Chernobyl accident.

10.2 RADIOACTIVE CONTAMINATION—A SOCIETAL CHALLENGE

When a large territory is contaminated by radioactive substances, every part of society is affected in the contaminated regions—environment, economy, production, living conditions, and health. It is thus a societal problem, not a radiation protection or health issue alone.

In addition, people generally have little knowledge about radioactivity and radiation and are not prepared to tackle a contamination event. The fear of something that is invisible and at the same time ubiquitous and invasive in people's life can be very difficult for people to handle. This has also been seen for other contaminants like pesticides (Baudé, Levinson, Ollagnon, & Hériard Dubreuil, 2012). From our Chernobyl experience in Norway and Belarus and Fukushima experience in Japan, we know that people have general worries such as:

- How dangerous is the radiation and the radioactive contamination?
- Is the food safe? What should I feed my children?
- Will this deteriorate the health of me and my family, and what about future generations?
- Can we continue to live here?
- Can we continue to be farmers, fishermen, and reindeer herders?
- Can the community continue the production of food, feed, and goods?

In such situations, experience shows that it may be important for authorities to provide information and knowledge to the affected population so that they will be able to understand the challenge they are facing. It has proven useful for authorities to engage in dialogues with the people in contaminated regions to find possible solutions to their problems. Local residents have the best expertise on relevant local conditions and local resources that could be used in a rehabilitation strategy with the aim of normalizing the living conditions, but in most cases they need assistance from various experts to find the sound solutions to the challenges.

Situations where large areas within a country would be contaminated for a decade or more would represent a huge challenge to any country. There is a range of possible countermeasures that could be implemented after a nuclear or radiological accident, directed both at the population and at the production of food, feed, and goods. However, actions in the recovery phase need to be chosen with care, taking account of the wider societal aspects and preferably in elaboration with people from local, regional, and national levels. The goal of the remediation strategies would be a return to a normal situation which is not a preaccident situation, but a situation where people can live and produce in a contaminated area with acceptable risk and living conditions due to the implementation of mitigating actions. Most probably, the management strategy would also need to be revised with time to accommodate changes in contamination levels, environmental factors, technical solutions, and societal development.

10.3 EARLY LESSONS FROM THE CHERNOBYL ACCIDENT

10.3.1 Information Needs

In the early period after the Chernobyl accident in 1986, the information needs in Norway were immense. Many of the questions were related to food safety and health concerns. People wanted to know what was dangerous and what was safe. Could they become seriously ill because they had stayed outdoors for a long time during the days of fallout? Could they eat fish or drink the milk? Some lost the joy of gardening or outdoor activities because of the uncertain risk from touching the grass and ground. It seemed difficult for the public to compare the health risks from this new danger to other health risks (Daugstad, 1996).

Different authorities gave conflicting messages in the early period (April/May) after the accident adding to the general confusion. Sometimes, the statements were categorical even if the facts were insecure or missing and they were later perceived as too reassuring and unreasonably trying to minimize the risks. The publics' information needs were immense and the situation turned into an information crisis as health authorities did not manage to cover the needs (Norges offentlige utredninger, 1986). Further, the credibility of the health authorities were weakened when it was known that they had deliberately held back some information that the public and media felt they were entitled to know (Reistad & Larsen, 1996).

People also want to know the real facts even if the truth is frightening: "We can handle the truth" (Bay-Larsen, Oughton, Liland, Eikelmann, & Hansen, 2009). Experience has shown that the public responds better to possible worst case scenarios than to attempts from authorities to minimize the dangers. Sandman (2001) expresses this as "Tell people how scary the situation is (…) and watch them get calmer" in his guidelines for risk communication.

A public opinion poll performed shortly after the accident showed that 32% of the public understood most of the information provided, 50% understood some while 8% did not understand anything of the information provided. When asked if they trusted the information given, only 2% stated they had high confidence and 32% reasonable confidence, while 38 and 22% had little or very little confidence, respectively, in the information given (Weisæth, 1989).

10.3.2 The Importance of Local Monitoring Stations

During the summer of 1986, the Norwegian authorities initiated the creation of a network of local monitoring stations for radiocesium in selected local food control laboratories across the country (Strand, Strand, & Baarli, 1987). Local monitoring stations were important for providing information to people on the contamination in local foodstuffs and environmental samples. But the distribution of such instruments was scarce and people in many districts had to wait several weeks to get the results. In one district, the community started a fund-raising so that the local food control office could buy an instrument of their own. For an efficient transmission of the measurement results, the results for local samples were broadcasted daily on the local radio rather than spending time writing reports for every sample since the number of samples was overwhelming. "Everyone" in the community listened to this broadcasting. After some time, the national authorities suddenly decided that this local food-control office was not allowed to publish the measurement results before they had been approved by regional or national authorities. This created strong reactions at the local level. Why were they not allowed to know their situation straight away? After strong pressure, this practice was abolished (Røine, 1996) and local food control offices around the country played an important role in building public trust in a very difficult situation (Daugstad, 1996).

10.3.3 Setting Permissible Levels for Radioactive Substances in Foodstuffs

Permissible levels is a challenging topic as the public in general interpret these levels as directly related to toxicity, while in practice other concerns like the ALARA principle (As Low As Reasonably Achieveable), consumer's confidence and trade are also important in the decision-making process. Therefore, countries may adopt different strategies to permissible levels, with Norway's strategy (described below) being an example differing significantly from Japan's recent decision on reducing the permissible level for radiocesium in foods to 100 Bq/kg.

At the time of the Chernobyl accident, Norway had no permissible levels for radionuclides in foods. In May, values of 1000 Bq/kg and 300 Bq/kg were chosen as temporary values for I-131 and Cs-137, respectively. During May and June, the information on the extent of the fallout and the results of the food sampling provided a better basis for assessing the contribution by various foodstuffs to the public's radionuclide intake. This was used to assess more permanent permissible levels based on radiation protection principles. However, the levels adopted in Norway in June 1986 were the results of consultations among Nordic countries and were also in accordance with European Economic Community (EEC) levels: 370 Bq/kg for radiocesium ($^{134}Cs + ^{137}Cs$) in milk and infant food and 600 Bq/kg for all other foodstuffs. Mitigating actions, mainly food bans, were imposed in the food production systems to comply with the new limits in 1986, while better countermeasures were developed the following years.

The consequences for reindeer husbandry became evident during late summer and early autumn, with average radiocesium concentrations in reindeer meat in some areas approaching 100 times the permissible level. Furthermore, the experience from the nuclear bomb fallout in northern Norway showed effective half-times for ^{137}Cs in reindeer of about 7 years (Westerlund et al., 1987), indicating that reindeer husbandry would be affected for more than a generation. Reindeer herders therefore felt their whole existence was endangered (Stephens, 1987). These semidomesticated animals are free-ranging in herds of hundreds and thousands giving limited possibilities for mitigating actions like fencing and clean feeding. Therefore, in November 1986 the government decided to increase the permissible level for radiocesium in reindeer meat to 6000 Bq/kg in an attempt to maintain a meaningful business base for reindeer herders—and also protecting the associated Sámi culture and life-style. The decision followed a radiation protection assessment justifying the elevated level by the low average consumption of reindeer meat in the general population (~0.5 kg per person per year). The elevated level applied to traded reindeer meat, while high consumers like reindeer herders were recommended special actions to reduce the contamination in reindeer meat consumed in the Sámi household (e.g. clean feeding, dietary advice, specific preparation of meat etc.). From 1987 the level of 6000 Bq/kg also applied for wild freshwater fish and game meat.

Although it was not explicitly stated at the time the permissible level was raised to 6000 Bq/kg, it was the intention to reduce the level as the contamination in reindeer declined and it was justified on a radiation protection basis (i.e. ALARA). In 1994, the level was reduced to 3000 Bq/kg, while later assessments (in 2001 and 2009) concluded that reductions to 1500 Bq/kg will not be cost-efficient. Justification on a radiation protection basis has been the most important criteria in these assessments, but other aspects also need to be taken into account. These include e.g. consumer needs and practical consequences for producers. For the producers, a lowering of levels will impose more application of remedial actions, and may be perceived as reintroducing the accident and its consequences in areas that have been declared "clean" for many years. The pragmatic approach with higher permissible levels for traded reindeer meat could also be viewed as an expression of sympathy with the affected population.

10.3.4 Consequences and Countermeasures in Food Production

Meat and milk were produced as normal in the contaminated regions in 1986, but much of the food could not be used for human consumption. The milk was collected at the farms and the animals slaughtered as usual, but the milk and meat was either used as feed for fur farming or discarded and disposed of as waste. It was a heavy burden for the farmers to produce food that would only be condemned, even if the authorities would give them a financial compensation for their loss. They could not use the products from the farm in their own household either and were obliged to buy food from the stores instead (Ekerbakke, 1996).

Clean feeding and live monitoring of animals were introduced as countermeasures in meat production from 1987. Although this reduced the need for condemnation, it brought about several practical problems for farmers. The sheep farming in Norway is based on a limited stock housed during wintertime. The lambs are born in spring and spend the whole summer season outdoor on the pastures. In autumn, the lambs are slaughtered before the cold weather necessitates housing again. Due to the very high contamination in animals (several 1000 Bq/kg of radiocesium), the clean feeding times could extend for several months. The barns were too small to appropriately accommodate so many sheep and the animal welfare was questionable (Røine, 1996).

From 1989 Prussian blue (a cesium binder) was added to feed concentrates, salt licks, and rumen boli to reduce the uptake of radiocesium in animals grazing contaminated pasture. This reduced the length of the clean feeding periods, but entailed other extra work for the farmers. Although the Norwegian authorities had declared on 31 July 1986 that the producers should be compensated for any financial loss, the farmers and reindeer herders in affected regions clearly express that the compensation received is not sufficient to cover all their real costs of implementing countermeasures. A lot of their extra labor time is not covered, even if the direct costs are covered (Bay-Larsen, Oughton, Liland, Eikelmann, Hansen, 2009; Daugstad, 1996).

Nonetheless, the implementation of countermeasures ensured that it was possible to continue farming in the contaminated regions and produce milk and meat that complied with the permissible levels in foodstuffs. Very few farmers gave up their farms due to the accident, even if the work became more laborsome.

For the reindeer herders, the situation was even more difficult. In some areas, all meat produced the first three years were discarded from human consumption. This was a depressing time for the herders. The first positive sign came in 1989 when the use of boli with Prussian blue reduced the radiocesium levels in meat considerably so that it could again be used for humans. But they realized that the contamination would persist for a long time and had to adapt the husbandry practice for good. The use of Prussian blue, altered slaughtering season, live monitoring, and clean feeding were all new practices that had to be implemented. The traditional successful breeding based on weight and health was no longer of value since the animals were now selected for slaughtering based on their individual Bq level (Håkenstad, 1996).

10.3.5 Monitoring of Affected People

Monitoring of persons was used in northern Norway from 1965 as a method to survey internal doses for people exposed to fallout from the nuclear weapons testing. After 1986, the program was enlarged to include people affected by the Chernobyl accident. It gave the authorities information on the contamination levels caused by the accident and the successfulness of the remediation strategies. The first years after Chernobyl, the focus was on reindeer herders, hunters/gatherers and general rural population groups in the most affected areas in Norway (Strand, Selnæs, Bøe, Harbitz, & Andersson-Sørlie, 1992). The studies showed that reindeer herders were by far the most exposed group and later monitoring was limited to this group. The elevated dietary intake of radiocesium by reindeer herders were expected based on over 20 years' experience on radiocesium in the lichen—reindeer—humans food-chain in northern Norway (Westerlund et al., 1987). Therefore, special arrangements were put in place to reduce the dietary intake. These included:

- replacing the reindeer meat from their own herds by meat from less contaminated areas;
- special advices on how to prepare and reduce radiocaesium levels in contaminated meat; and
- recommendations on how much meat could be consumed depending on contamination levels.

Dietary surveys were also performed to provide more detailed information on intake of foods and radiocaesium. Combining diet information and knowledge of contamination levels in major foodstuffs with the results of the repeated monitoring of people proved very useful in the dialogue with the affected persons in Norway: In most cases reasonable explanations for observed internal contamination could be elaborated, and for the affected people it has been reassuring to study the success of their personal efforts to reduce contamination intake. The monitoring thus turned out to be an important tool in helping the population in coping with the situation (Tønnessen et al., 1996). Similar experiences were made in Belarus, and this initiated the development of a prototype software tool for this purpose (Levy, Bataille, Croüail, & Skuterud, 2008).

10.3.6 Psychosocial Effects

Shortly after the accident, a survey was performed among almost 1000 Norwegians on their reactions to the accident. 25% said they kept their children more indoors, 20% perceived the situation as unpleasantly stressing, and women were more worried than men. It was clear that people had little prior knowledge on radiation risks that most probably increased the perceived risk compared to the real risk (Weisæth, 1989).

As described above, the producers in the affected regions encountered many challenges that influenced their life. The uncertainty linked to the duration of the problem and the associated risks added psychosocial burdens to the affected people. For instance, the early practice of producing food that was discarded for human consumption was ethically unacceptable for many producers.

For people in animal husbandry, the consumer trust in their products is of utmost importance and an invisible enemy that reduced the quality of their produce was difficult to handle. The initiation of mitigating actions directly involving the producers helped relieve the psychosocial effects since it enabled producers to do something to improve their situation. Some farmers have also pointed to the handling of the Chernobyl accident as a community voluntary work that had a positive spin-off for the engagement and social behavior in the affected territory (Bay-Larsen et al., 2009). Nonetheless, a nuclear accident with serious contamination of large areas will entail a number of psychosocial effects that should be properly addressed by the health authorities.

10.4 LONG-TERM CHALLENGES AND LESSONS LEARNED

10.4.1 Information Needs

It is clear that information needs are large both in early and late phases after an accident. In the affected territories in Norway local measurement stations were essential, both for food safety, health, industry, and trade. It was a means to increase public trust, and gave them a feeling of self-control that reduced the psychosocial stress reactions. Authorities at both national and regional/local level should continue a good information and communication strategy over the years. To improve understanding and acceptance, the information should be tailored to meet the needs of the public, include the new generations as they grow up and include newcomers to an affected production system.

Today, the speed of information flow, the wealth of information channels, and the use of social media pose both a challenge and a possibility for informing and communicating with the public. For instance, it can be difficult for the public and the media to know which experts to trust when they give conflicting messages. After a nuclear emergency, journalists may also want to seek alternative voices because they do not know which information sources they can rely on in this radically new situation. Internet provides a wealth of sites that claim they have the right knowledge and solution to consumer's challenges and it can be difficult for people to know what is trustworthy or not. On the other hand, authorities have a large set of information and communication tools available that could be an asset to them if used wisely.

10.4.2 Zoning and Compensation

The use of geographical zoning for managing the situation may be necessary in the emergency phase. In areas like Belarus, geographical zones defined by the ground contamination levels have been used to regulate economic compensation schemes for the population. In Norway, no compensation was paid to others than the producers having direct economic losses or expenses because of the contamination. Although the Norwegian system may not have been perfect (see above), it is our impression that compensation schemes based on expenses/losses are better than those based only on contamination levels. In Belarus the latter have been shown to result in stigma and resignation. The affected inhabitants in Belarus were relying on the aid from the government or from foreign countries instead of enabling them to improve their own situation. In Norway, zoning was based on the levels of radiocesium in live animals, and just used to indicate if and for how long countermeasures were needed. Compensation was paid to farmers and reindeer herders for their extra labor and economic losses incurred due to countermeasures application. Reindeer herders were also paid "compensation for alternative diets", enabling them to replace their own highly contaminated reindeer meat with less contaminated products. The affected people in Norway have not been victims of stigma or passivity to our knowledge. It is thus reasonable to ask whether zoning and compensation of people are appropriate tools taking into account the risk of inducing a passive attitude among the population and consequently favoring a possible recession of the living conditions in the contaminated territories. Compensations to support mitigating actions performed by the affected individuals seem to be more positive than compensations based solely on living in a contaminated territory (Liland et al., 2009).

10.4.3 Technical and Social Countermeasures

Section 3.4 discussed some of the consequences of the Chernobyl fallout for food production in Norway. In the long-term management, the following measures have been or are still being applied to reduce contamination levels and consequences:

- Monitoring of radiocesium in animals before slaughter ("live monitoring"; to avoid condemnation of meat above the permissible levels);
- Clean feeding of animals before slaughter (following monitoring);
- Caesium binder (Prussian blue) in concentrates (pelleted feed), salt licks, and rumen bolus to prevent absorption of ingested radiocesium in the animals;
- Change of slaughter time (in reindeer husbandry);
- Elevated permissible levels for reindeer, game, and freshwater fish; and
- Dietary advice and monitoring of internal contamination.

Although most of the measures were initiated by national/central authorities or institutes, the practical implementation was elaborated in close cooperation with representatives of farmers and herders in the affected areas. The extensive use of monitoring of animals before slaughter is one of the success stories of Norwegian Chernobyl management: In addition to reducing public expenses on compensation for condemned meat, clean feeding of contaminated animals bypasses the psychological burden that production for condemnation can be for farmers and herders. Another success was the implementation of Prussian blue in various forms. Bentonite and zeolite were initially tried as cesium binders. However, the high doses needed of these clay minerals involved also animal welfare issues (e.g. appetite and drinking water requirements). More efficient and large-scale application of cesium binders was therefore obtained after developing methods of administering Prussian blue, which gave larger reduction at much smaller quantities (Hove, Staaland, & Pedersen, 1991). Norwegian methods and experiences on the use of Prussian blue were also made available to the contaminated areas of Russia, Ukraine, and Belarus (IAEA, 1997). More details on the Norwegian remediation strategies and their high cost-effectiveness is given in Brynildsen, Selnæs, Strand, and Hove (1996) and Tveten et al. (1998).

Countermeasures directed at the public included dietary advice and monitoring of affected groups. The identified vulnerable groups were people harvesting from nature, self-sufficient farmers, and reindeer herders. Only the reindeer herders are still being followed up today. Figure 10.2 illustrates contamination levels in two groups of Sámi reindeer herders in central Norway following the Chernobyl fallout. While average values currently are 40–80 Bq/kg in women and 60–140 Bq/kg in men, individual values range from 10 to 300 Bq/kg. The maximum radiocesium concentration observed in any reindeer herder in Norway after the Chernobyl fallout was about 4200 Bq/kg. The suite of measures introduced in reindeer herding and reindeer herding families reduced the ingestion doses to reindeer herders 80–90% the first year and

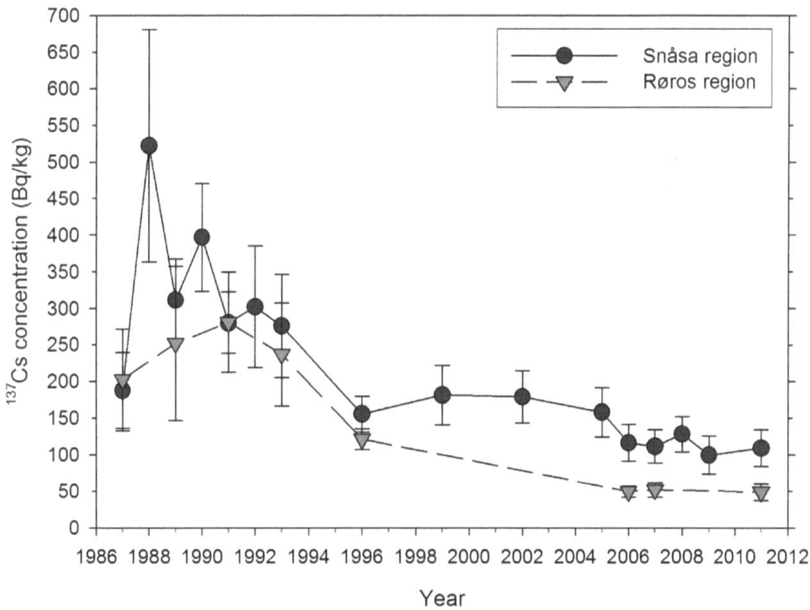

FIGURE 10.2 Average concentrations of ^{137}Cs in reindeer herders in two areas of central Norway, Snåsa (Nord-Trøndelag county) and Røros (Sør-Trøndelag county). Bars indicate standard error. See text for information on range in individual values. (For color version of this figure, the reader is referred to the online version of this book.)

totally about 73% during the period 1986–2009 (Skuterud & Thørring, 2012). Due to the serious consequences of the contamination of the reindeer herder's households, the reindeer herders also received "compensation for alternative diets" so that economic constraints should not be a reason for consumption of contaminated meat.

When people are monitored, the measurement time (15–20 min) allows for discussion with an expert on individual diets, health risks, concerns, etc. (see Figure 10.3). This gives an opportunity to communicate one-to-one on personal concerns. The experience in Norway shows that this is very valuable since we all perceive risks differently and have different information needs.

Due to changing contamination levels, a long-term countermeasure strategy will probably not be static. Permissible levels may also differ between the early and long-term phase of an accident, as was the case in Norway. The management strategy is therefore likely to change with time. So does the rest of the society and stakeholder views may also change accordingly. Long-term contamination therefore necessitates long-term management and commitment (e.g. Skuterud, 2012). These long-term aspects are therefore another reason why early involvement with affected populations will be recommended (see below).

FIGURE 10.3 Monitoring of a reindeer herder (left) in the mobile laboratory of the Norwegian radiation protection authority. The monitoring gives the opportunity to communicate one-to-one between reindeer herders and scientists, addressing the concerns of the former. The man to the right is Lavrans Skuterud. *Source: photo—Geir Tønset, Adresseavisen. (*For color version of this figure, the reader is referred to the online version of this book.*)*

10.4.4 Involvement of Affected Groups

Experience from both Norway and Belarus has shown that the involvement of affected groups in a long-term rehabilitation management is very important for improving living conditions and favoring a positive and sustainable development of long-term contaminated territories (Lochard, 2004). One reason for this is that the *reality* for people is what is going on in their local community (Liland et al., 2009). Central accident management and information on country average consequences, although important, is not enough to address the needs at the local level. The establishment of local monitoring and information centers has proven to be very successful in this respect (Bataille, Crouail, & Lochard, 2008; Liland et al. 2001, 2010). Furthermore, the rehabilitation strategies should be elaborated with participants from local, regional, and national levels as part of a participatory process. There are several positive aspects of this. First of all, people at local/regional level would have the best knowledge to find solutions adapted to their local community with help from authorities and experts. It also gives people the chance to self-help, which avoids resignation and increases the self-confidence that they can improve their own situation. Last but not least, a rehabilitation strategy elaborated *with* the people is more likely to gain wide acceptance and successful implementation than central directives elaborated solely by authorities *for* the people (Liland et al., 2009).

This participatory process should be followed by a long-term health follow-up of the affected groups, including psychosocial effects. This follow-up is important for reassurance of the people, particularly due to the uncertainties in the Linear-No-Threshold model, which means that experts, at present, cannot give people a precise answer to the risk from low doses. Including monitoring of people in such a program may make the contamination situation more tangible and controllable for the population (Mehli, Skuterud, Mosdøl, & Tønnesen, 2000). Monitoring would also give people their individual exposure values that could greatly differ from average values. The actual results of the monitoring are important to show people if the self-help measures are effective and sufficient enough to reduce their intake of radionuclides. The follow-up could last as long as countermeasures are still needed or as long as the affected groups deem it necessary.

10.4.5 Psychosocial Effects

The publication of the Chernobyl Forum (WHO, 2006) confirmed that the psychosocial effects of the accident were much higher than the radiation effects. The term health also includes psychosocial aspects: "Health is a state of complete physical, mental and social well being and not merely the absence of disease or infirmity" (WHO Constitution). In Norway, the production for condemnation or giving up the family farm due to the accident are examples of strong psychosocial burdens for individuals in affected territories (Bay-Larsen et al., 2009).

For people in animal husbandry, producing quality food is of utmost importance. The extensive use of countermeasures to reduce the radiocesium levels in milk and meat enabled the producers to continue their livelihood and produce food for human consumption. In the long run, we must assume that this improved the physical and psychosocial health of the populations in the affected areas (Strand & Reitan, 1996). The use of countermeasures in food production is more ethically and socially acceptable than condemnation and much more rewarding, in a broader sense, for the farmers.

The strong focus on countermeasures in food production systems resulted in little focus on other aspects of the contamination situation in the long term. The health authorities were only engaged to a limited extent after the first couple of years. No health follow-up was proposed for the affected people, a part from monitoring of internal contamination in selected groups. It took time before the authorities understood that the social and personal consequences of the accident persisted in the affected populations. The problem appeared to be solved at the national authority level; however, this view was not shared by people in the contaminated regions.

In Norway, local participants at a postaccident management seminar in 2008 (summarized in Bay-Larsen et al., 2009; Liland et al., 2010) expressed that they still experienced psychosocial stress due to the Chernobyl accident. They were still affected by the uncertainty and worry over the possible health consequences. Many would have wanted more information and assistance from the local doctor or health service. A Sámi participant at the seminar expressed her views on the consequences of the Chernobyl accident in the following way:

"Reindeer herding is not an occupation—it is a way of life that involves the whole family and where there is a strong link to the nature and the animals. Since it is based on nature it is particularly vulnerable to environmental contamination. With a catastrophe like the Chernobyl accident, the reindeer breeders lost control of the situation. Since they lacked knowledge on how to handle radioactive contamination, they were at the mercy of the authorities to inform them on what to do. Their prior experience and knowledge were no longer sufficient. The implementation of mitigating actions entailed negative consequences for economic, social, and cultural aspects of their life. Examples are more laborsome husbandry practice, changed diet and changed use of animal parts for traditional crafts which also influences on the knowledge of the Sámi language of future generations since traditions are lost. The monetary compensation received cannot mitigate all these impacts. Today, the consequences of the Chernobyl accident are born by the individual reindeer herder and his/her family" (Bay-Larsen et al., 2009).

10.5 HEALTH CONCERNS

It is well known that uncertainty may breed fear—particularly in connection with exposure to radioactive contamination (e.g. Gardner, 2008). The main concerns of an affected population is likely to be those about health effects, but the need for information does not only relate to health risks—it relates to all aspect of the contamination and exposure situation—for inhabitants, consumers, and producers. The public often discusses radioactive contamination and radiation with the "mysterious" properties "can't hear, see or smell"—discussions that easily induce unnecessary fear in addition to potentially leading to undesirable (i.e. more risky) behavior. Inhabitants in a contaminated area will inevitably start making their own interpretation of the information they receive, and of the contamination situation around them. Therefore, the inhabitants will benefit from information on radiation, properties of the contamination and expected behavior in the environment and in humans, which will in turn also benefit the authorities. It is too easy to underestimate the range of information the public is able to utilize. Another challenge is the public's confidence in the information providers: Even in non-nuclear Norway, it may be difficult for representatives of authorities to reach a level of trust where the information on radiation and radioactivity is accepted as unbiased by the local population.

Knowledge of radiation doses—external as well as internal (e.g. obtained from measurements of persons)—is a premise for assessing risks. And it may be essential in helping inhabitants of a contaminated area in managing and coping with the situation. Nevertheless, the dose information may not be enough to provide answers to the population's concerns. The monitoring of the reindeer herders in Norway is a typical example. Many years of study have given a good understanding of the doses they have received (Skuterud & Thørring, 2012), but as the doses are low compared to levels currently known to result in somatic health effects, the results apparently does not give satisfactory answers to their primary questions:

- What is the actual risk that my kids and I are exposed to?
- Did I, my wife, father, aunt,...develop cancer because of the Chernobyl fallout?

The continuous monitoring of the population may also induce worries due to the following dilemma: If the risks are so low, why do the authorities bother to monitor us at all? Furthermore, if the risks are so marginal even for reindeer herders with their high consumption of reindeer meat, why do the producers have to take actions before they can trade their products to the general public, which consumes just small amounts of the product?

Many Norwegians in the contaminated areas still wonder if the increasing number of cancer incidence or other diseases, like diabetes in adolescents, during the last decades is related to the Chernobyl accident (Bay-Larsen et al., 2009). The same worries of increased disease rates are also shared by neighbors of the IFE (Institute for Energy Technology) research reactor at Kjeller, Norway (Dalbak, Clausen, Reitan, & Tønnessen, 2002). Such concerns need to be addressed appropriately by the authorities and possible investigations or surveys undertaken to try to answer their questions.

Assessing health consequences at the individual level is one challenge—consequences at a population level another. It is challenging to conduct health surveys of relatively small population groups living in contaminated areas with a design expected to give unambiguous results. Even in cases of existing preaccident health registers, the risk of false answers increases when the population is broken up into subpopulations. There can be false positives as well as false negatives. This also applies to registers of affected population created after an accident. In Norway, there has therefore been some hesitation about conducting health surveys among the Chernobyl-affected Sámi reindeer herders in central Norway because they are a separate ethnic group, they have a special diet and lifestyle, and they are relatively few in numbers (about 1000 individuals). Also, there is little (if any) relevant health statistics of pre-Chernobyl. The above reasons are not good explanations for the affected persons, and in the long term (related to the expected latency period of cancers) there is an increasing need for health surveys. Furthermore, due to possible improvements in analytical procedures etc. in the future, it might for scientific purposes be useful to establish registers and biobanks with the population's consent. Preferably, such registers should be linked to other contaminated populations and international studies for the possible improvement and confidence of the results.

Having discussed some pros and cons about health surveys among the Chernobyl-affected Sámis in central Norway, it should be mentioned that cancer incidence and mortality in the general Sámi population in the Nordic countries has been studied (summarized in Hassler, Soininen, Sjölander, & Pukkala, 2008). The studies were performed because reindeer herding Sámis were also exposed to radioactive fallout as a consequence of the nuclear weapons testing in the 1950–1960s, with average levels of internal contamination similar to those following the Chernobyl accident. The studies show that the risk of developing and dying from cancer is low among the Sámi, both in comparison with general populations and regionally matched reference populations of non-Sámi (Hassler et al., 2008). The reasons for the low incidence are assumed to be a lifestyle that traditionally included high physical activity, and genetic factors (Haldorsen & Tynes, 2005).

REFERENCES

Backe, S., Bjerke, H., Rudjord, A. L., & Ugletveit, F. (1986). *Nedfallet av cesium i Norge etter Tsjernobylulykken. (Cesium fallout in Norway after the Chernobyl accident)*. Østerås, Norway: National Institute of Radiation Hygiene. SIS Rapport 1986:5 Language: Norwegian.

Bataille, C., Crouail, P., & Lochard, J. (2008). Rehabilitation of living conditions in the post-Chernobyl context: implementation of an inclusive radiation monitoring system in the Bragin district in Belarus. In P. Strand, J. Brown & T. Jølle (Eds.), *International conference on radioecology and environmental radioactivity* (pp. 129–132). Østerås, Norway: Norwegian Radiation Protection Authority. 15–20 June 2008, Bergen, Norway. Oral and oral poster presentations proceedings, part 2.

Baudé, S., Levinson, E., Ollagnon, H., Hériard Dubreuil, G. (2012). *Action n°36 du Plan d'action contre la pollution par la chlordécone en Guadeloupe et en Martinique 2011-2013: Facilitation de l'élaboration d'une stratégie territoriale de développement durable de la qualité de vie dans le contexte de la pollution à la chlordécone en Martinique et en Guadeloupe.* Audit patrimonial – Rapport de synthèse de l'ensemble de la démarche menée en Martinique. (Language: French).

Bay-Larsen, I., Oughton, D., Liland, A., Eikelmann, I. M., & Hansen, H. S. (2009). *Erfaringsbasert kunnskap i norsk atomberedskap – medvirkning fra berørte parter. Rapport fra EURANOS-prosjektet. (Experience-based knowledge in nuclear and radiological emergency preparedness – involvement of national stakeholders. Report from the EURANOS project).* StrålevernRapport 2009:8, Østerås: Norwegian Radiation Protection Authority. (Language: Norwegian).

Brynildsen, L. I., Selnæs, T. D., Strand, P., & Hove, K. (1996). Countermeasures for radiocesium in animal products in Norway after the Chernobyl accident—techniques, effectiveness, and costs. *Health Physics, 70*(5), 665–672.

Dalbak, L. G., Clausen, F., Reitan, J. B., & Tønnessen, A. (2002). *Helsebekymringer relatert til virksomheten på IFA/IFE Kjeller. En studie blant lokalbefolkningen. (Health concerns related to the operations at IFA/IFE, Kjeller. A study among the neighbouring population).* StrålevernRapport 2002:2, Østerås: Norwegian Radiation Protection Authority. (Language: Norwegian).

Daugstad, K. (Ed.), (1996). *Ti år etter Tsjernobyl. (Ten years after Chernobyl).* Oslo: Landbruksforlaget. (Language: Norwegian).

Ekerbakke, S. (1996). Ytring fra grasrota. In *Ti år etter Tsjernobyl. (Ten years after Chernobyl).* (p. 15). Oslo: Landbruksforlaget. (Language: Norwegian).

Gardner, D. (2008). *Risk: The science and politics of fear.* McClelland & Stewart.

Håkenstad, S. (1996). Ytring fra grasrota. In *Ti år etter Tsjernobyl (Ten years after Chernobyl)* (pp. 48–49). Oslo: Landbruksforlaget. (Language: Norwegian).

Haldorsen, T., & Tynes, T. (2005). Cancer in the Sami population of North Norway, 1970–1997. *European Journal of Cancer Prevention, 14*, 63–68.

Hassler, S., Soininen, L., Sjölander, P., & Pukkala, E. (2008). Cancer among the Sami—a review on the Norwegian, Swedish and Finnish Sami populations. *International Journal of Circumpolar Health, 67*(5), 421–432.

Hove, K., Staaland, H., & Pedersen, Ø. (1991). Hexacyanoferrates and bentonite as binders of radiocaesium in reindeer. *Rangifer, 11*(2), 43–48.

International Atomic Energy Agency. (1997). *The use of Prussian blue to reduce radiocaesium contamination of milk and meat produced on territories affected by the Chernobyl accident.* Report of United Nations Project E 11. IAEA-TECDOC-926, Vienna: International Atomic Energy Agency.

Levy, F., Bataille, C., Croüail, P., & Skuterud, L. (2008). A tool for interpretation of whole body monitoring results in a long-term contaminated environment: the CORPORE application. In P. Strand, J. Brown & T. Jølle (Eds.), *International conference on radioecology and environmental radioactivity, 15–20 June 2008, Bergen, Norway. Posters proceedings, part 2.* (pp. 196–199). Østerås, Norway: Norwegian Radiation Protection Authority.

Liland, A., Lochard, J., & Skuterud, L. (2009). How long is long term? reflections based on over 20 years of post-Chernobyl management in Norway. *Journal of Environmental Radioactivity, 100*, 581–584. http://dx.doi.org/10.1016/j.jenvrad.2009.04.006.

Liland, A., Oughton, D., Bay-Larsen, I., Eikelmann, I. M., Hansen, H. S., & Skuterud, L. (2010). Norwegian testing of the EURANOS framework for post-accident rehabilitation preparedness. *Radioprotection, 45*(5), S215–S224.

Liland, A., Skuterud, L., Bergan, T., Forseth, T., Gaare, E., & Hellstrøm, T. (2001). *Overvåkning av radioaktiv forurensning i næringsmidler og det terrestre miljø 1986–1998. (National monitoring of radioactive contamination in food and the environment 1986–1998)*. StrålevernRapport 2001:1, Østerås: Norwegian Radiation Protection Authority. (Language: Norwegian).

Lochard, J. (2004). Living in contaminated territories: a lesson in stakeholder involvement. In *Current trends in radiation protection* (pp. 211–220). EDP Sciences.

Mehli, H., Skuterud, L., Mosdøl, A., & Tønnesen, A. (2000). The impact of Chernobyl fallout on the southern Saami reindeer herders of Norway in 1996. *Health Physics*, *79*(6), 682–690.

Norges offentlige utredninger. (1986). *Informasjonskriser. (Information crisis'). Norges offentlige utredninger* (Vol. 1986, pp. 19). Oslo: Universitetsforlaget. (Language: Norwegian).

Reistad, O., & Larsen, A. K. (1996). Norsk atomulykkesberedskap. In *Ti år etter Tsjernobyl. (Ten years after Chernobyl).* (pp. 103–120). Oslo: Landbruksforlaget. (Language: Norwegian).

Røine, T. (1996). Ytring fra grasrota. In *Ti år etter Tsjernobyl. (Ten years after Chernobyl).* (pp. 24–25). Oslo: Landbruksforlaget. (Language: Norwegian).

Sandmann, P. (*Anthrax, bioterrorism, and risk communication: guidelines for action.* The Peter M. Sandman Risk Communication Website: http://www.psandman.com/col/part1.htm#finding12001 (last accessed 08.03.13).

Skuterud, L. (2012). Living with an existing exposure situation due to accidental contamination: The need for long-term management and involvement - for how long? *Proceedings of the IRPA13 congress Living with Radiation - Engaging with Society, Glasgow, 13–18 May 2012*. International Radiation Protection Association (IRPA). Paper no. TS10b.1 (http://www.irpa-13glasgow.com/wp-content/uploads/2012/05/Full-Papers-TS10.zip. ; last accessed 05.06.13).

Skuterud, L., & Thørring, H. (2012). Averted doses to Norwegian Sámi reindeer herders after the Chernobyl accident. *Health Physics*, *102*(2), 208–216.

Stephens, S. (1987). Chernobyl fallout: a hard rain for the Sami. *Cultural Survival Quarterly*, *11*(2), 66–71.

Strand, P., & Reitan, J. B. (1996). Stråledoser og mulige konsekvenser for miljø og helse. In *Ti år etter Tsjernobyl. (Ten years after Chernobyl).* (pp. 77–98). Oslo: Landbruksforlaget. (Language: Norwegian).

Strand, P., Selnæs, T. D., Bøe, E., Harbitz, O., & Andersson-Sørlie, A. (1992). Chernobyl fallout: internal doses to the Norwegian population and the effect of dietary advice. *Health Physics*, *63*, 385–392.

Strand, T., Strand, P., & Baarli, J. (1987). Radioactivity in foodstuffs and doses to the Norwegian population from the Chernobyl fall-out. *Radiation Protection Dosimetry*, *20*(4), 211–220.

Tønnessen, A., Skuterud, L., Panova, J., Travnikova, I. G., Strand, P., & Balonov, M. I. (1996). Personal use of countermeasures seen in a coping perspective. Could the development of expedient countermeasures as a repertoire in the population, optimise coping and promote positive outcome expectancies, when exposed to a contamination threat? *Radiation Protection Dosimetry*, *68*, 261–266.

Tveten, U., Brynildsen, L. I., Amundsen, I., & Bergan, T. D. S. (1998). Economic consequences of the Chernobyl accident in Norway in the decade 1986–1995. *Journal of Environmental Radioactivity*, *41*(3), 233–255.

Weisæth, L. (1989). Reactions in Norway to fall-out from the Chernobyl accident. In T. Brustad, F. Landmark & J. B. Reitan (Eds.), *Radiation and cancer risk*. New York: Hemisphere.

Westerlund, E. A., Berthelsen, T., & Berteig, L. (1987). Cesium-137 body burdens in Norwegian Lapps, 1965–1983. *Health Physics*, *52*(2), 171–177.

WHO. (2006). *Health effects of the Chernobyl accident and special health care programmes*. Report of the UN Chernobyl Forum Expert Group "Health", Geneva: World Health Organization.

WHO Constitution. http://apps.who.int/gb/bd/PDF/bd47/EN/constitution-en.pdf (last accessed 08.03.13).

Chapter 11

Lessons from the Fukushima Daiichi Nuclear Disaster

Michio Miyasaka
School of Health Sciences, Niigata University, Nishi-ku, Niigata, Japan
E-mail: miyasaka@clg.niigata-u.ac.jp

Chapter Outline

11.1 What Happened at Fukushima 177
11.1.1 Accident Causes 178
11.1.2 Human Costs 180
11.1.3 Information Disclosure and Evacuation 181
11.1.4 A Built-in Moral Hazard 183
11.2 "Safety Culture" as a Monoculture 184
11.2.1 The Safety Culture Ideology 184
11.2.2 Applied Ethics as a Multiculture 186
11.3 Applied Ethics for Nuclear Science 188
11.3.1 The Utilitarian (Risk-Based) Approach 188
11.3.2 The Deontological (Rights-based) Approach 191

11.1 WHAT HAPPENED AT FUKUSHIMA

On March 11, 2011, the Fukushima Daiichi Nuclear Power Station (FDNPS) was hit by an earthquake and subsequent tsunami that would lead to the meltdown of multiple reactors, three hydrogen gas explosions, and a massive release of radioactive material into the land, sea, and air. Radioactive levels remain extremely high in and around the FDNPS, making detailed investigations impossible to this day. This means that we still know very little about what caused the Fukushima disaster, a level 7 (severe) accident on the International Nuclear and Radiological Event Scale. Investigation committees on

Radioactivity in the Environment, Volume 19
ISSN 1569-4860, http://dx.doi.org/10.1016/B978-0-08-045015-5.00011-3

the accident were separately established by the government (Investigation Committee on the Accident at the Fukushima Nuclear Power Stations of Tokyo Electric Power Company, 2012), the Diet (The National Diet of Japan Fukushima Nuclear Accident Independent Investigation Commission, 2012), and Tokyo Electric Power Company (TEPCO, 2012) and their findings are hereafter referred to as the Government Report, Diet Report, and TEPCO Report, respectively. Private interests have likewise established two investigatory committees of their own, namely Fukushima Genpatsu Dokuritsu Jiko Chosa Iinkai (2012) and FUKUSHIMA Project Iinkai (2012). The five investigation reports issued by these committees are, as can be expected, similar on some points and dissimilar on others. A clear description of the cause of the accident and the damage it wrought is of course necessary for any logically sound analysis of it. Note, however, that such information would also reveal the magnitude of the blame accruing to the electric power utility that operated the facility and to the government that regulated the facility. Thus, asking one of those parties for even a simple description of what happened can run afoul of conflicting interests.

11.1.1 Accident Causes

Which was the primary, direct cause of the accident: seismic motion or a tsunami wave? It matters, because the answer reveals whether government regulations, based on hypothetical scenarios for seismic events, were indeed sufficient in the face of hard reality and whether the electric power utility faithfully observed those regulations. Also, the assignment of cause can have a significant impact on the economics of nuclear power generation. Conventionally, cost calculations are cited to support the contention that nuclear power is considerably less expensive than other energy sources. However, this could be because the calculations presume a level of safety that, in actual practice, has been shown to be insufficient to prevent a serious accident. Here too, asking for a clear identification of the cause of the accident also risks an entanglement in conflicting interests.

The TEPCO Report (TEPCO, 2012) claimed that a maximum acceleration of 550 gals was observed at 1st floor basement of the Unit 2 reactor building, and thus "[i]t can be said that the seismic ground motion of the recent earthquake was roughly on par with the assumptions that were made for the seismic safety assessment for this facility." As described in the report, "the tsunami run-up reached the ground level of major buildings"; that is, 10 m above sea level at Units 1 to 4, and 13 m at Units 5 and 6. TEPCO emphasized that in preparing countermeasures, it relied on *Tsunami Assessment Methodology for Nuclear Power Plants in Japan* by the Japan Society of Civil Engineers (JSCE, 2002), in which the estimations given for FDNPS were 5.4–6.1 m; "[h]owever, the March 11 tsunami greatly exceeded those estimations."

In short, TEPCO concluded that the seismic motion was within the expected range covered by countermeasures, while the tsunami was not. The company's claim raises the question of fairness over the estimation, which could be regarded as a conflict of interest, because TEPCO could reduce its responsibility for the accident at FDNPS by finding that the shaking from the earthquake was within the expected range and was not the cause of the damage to the nuclear reactors, and rather that the subsequent tsunami, which was beyond expectations, was the cause. Such an approach is in part enabled by the government's guidelines for the design of nuclear reactors. According to the Nuclear Safety Commission (NSC), which was in charge of the government's nuclear regulations until September 2012, nuclear reactor facilities "must be designed so that their safety functions would withstand forces produced by seismic motion that can be expected to occur, albeit extremely rarely, and that would significantly impact them while in operation"(NSC, 2006). In addition, the NSC defined an "active fault to be taken into consideration" as a fault for which activity after the late Pleistocene cannot be denied, and stated that such a fault could be identified by whether or not displacement or deformation it caused could be observed in the strata from the last interglacial period or on the relevant geomorphic surface. As these points suggest, the NSC's regulatory rules regarding earthquakes were concrete and detailed. In contrast, the NSC's regulatory rules regarding tsunami were relatively abstract and open to interpretation. The rules stated that nuclear reactor facilities had to be designed so that their safety functions would not be significantly impacted by a tsunami that could be expected to occur, albeit extremely rarely, while in operation; however, the NSC did not set criteria for evaluating a "significant impact". As a result, attributing the FDNPS accident to the tsunami, rather than to the earthquake, made it easier for TEPCO to claim its compliance with government regulations and reduce its responsibility.

Japan, one of the most earthquake-prone countries in the world, ranks third in terms of number of installed reactors (WNA, 2013). Nuclear power plant sites have been investigated by geologists, who mapped what could be active faults in and around the site. Yet even such scientific endeavors can be rife with conflicts of interest, particularly should they have some bearing on a determination of plant safety. Faults thought to be potentially active have been discovered near several proposed or actual plant sites, but electric power utilities tend to be extremely reluctant to accept such findings, for they mean that a plant cannot be built on that site or that any plant existing on the site must be decommissioned. Tsunamis, on the other hand, are rare occurrences, at least relative to earthquakes, and there is not much evidence on which to base predictions. This said, it is interesting to note that every commercial nuclear power plant in Japan is positioned on a coastline. Naturally, there are safety standards intended to guard against earthquake-induced tsunamis, and they inevitably reflect some presumptions of scale (wave height and run-up height). However, as mentioned above, whereas very specific standards have been developed for resistance to seismic shocks and stresses, there is little in the way of concrete standards for resistance to tsunami surges.

This is not to say that scientists have overlooked the threat presented by tsunamis; indeed, in Japan, there have been substantial debates on the subject. One concerns the lessons to be drawn from the Jogan Earthquake, which historians tell us struck the area around Sendai in the northern part of Honshu in 869. This quake, too, generated a strong tsunami. Satake and colleagues published a paper in 2008 in which they concluded that "the tsunami deposits extend more than 3 km from the estimated coast line" (Satake, Namegaya, & Yamaki, 2008). However, TEPCO (2012, pp. 26–33), citing limitations with research methodology, did not accept these findings or at least the gravity of the lesson to be drawn from them. According to the Investigation Committee on the Accident at the Fukushima Nuclear Power Stations of Tokyo Electric Power Company (Government Report, TEPCO, 2012), planners did consider adopting more stringent tsunami safety standards back around 2008. They entrusted entities such as the Japan Society of Civil Engineers to examine this issue in more detail and, at the time, stated that a final conclusion will be forthcoming in 2012 or later. Neither the NSC nor the Nuclear and Industrial Safety Agency of the Ministry of Economy, Trade and Industry (METI), however, has ordered TEPCO to take any action on that front, citing a lack of legal authority to do so.

11.1.2 Human Costs

According to the Nuclear and Industrial Safety Agency, a nuclear regulatory branch of METI, the total amount of radioactive materials discharged from the FDNPS into the air was estimated at approximately 1.6×10^{17} Bq for iodine 131 and at approximately 1.5×10^{16} Bq for cesium 137 (Nuclear Emergency Response Headquarters Government of Japan, 2011). Data show that the radiation spiked within 6 days of the quake. Since then, radiation levels have gradually declined as short-lived radioisotopes have decayed (e.g. iodine-131 with a half-life of 8 days). However, low-dose radiation remains; the pattern is asymptotic, reflecting the presence of long-standing radionuclides (e.g. cesium-134 and cesium-137, with half-lives of 2 and 30 years, respectively). The monitoring data at the main gate and west gate of FDNPS, as stated in the TEPCO Report, show that the highest dose was in the range of more than 1–10 mSv/h between March 12–16, simultaneous with the hydrogen gas explosions at Units 1, 3, and 4 and venting operations implemented at Units 1 and 3. It also shows that asymptotic lines were approaching 10 μSv/h (TEPCO Report, 2012; pp.354–370). According to Ministry of Education, Culture, Sports, Science and Technology (2011), the highest dose was from 0.5 μSv/h to more than 1.5 μSv/h in prefectures to the south of Fukushima (e.g. 0.496 μSv/h in Tokyo, 1.318 μSv/h in Utsunomiya, 1.504 μSv/h in Mito).

It remains unclear just what degree of risk is presented by such exposure. On the one hand, the total amount of radioisotopes released was huge and it was dispersed widely in the air, soil, and water. On the other hand, the risk presented by a long-term exposure (especially internal exposure) to low-dose radiation remains a contentious issue in the radiation community. As mentioned in the Government Report (2012, p.332), the International Commission on Radiological Protection (ICRP) recommendations classify the harmful effects of radiation exposure into two categories: "deterministic effects" where death or cell malfunction deterministically occurs with high radiation dose and "stochastic effects" where malignant disease or hereditary effects are stochastically caused by relatively low-dose radiation (ICRP, 2007). No deterministic effects have been confirmed. Instead, stochastic effects have been the focus of scientific debate. The National Diet of Japan Fukushima Nuclear Accident Independent Investigation Commission (The Diet Report, (2012)), cited the estimated cumulative effective dose of external exposure between March 11 and July 11, 2011 as being relatively high among 14,412 residents of three regions in Fukushima Prefecture: 6092 (42.3%) people had been exposed to a dose of 1–10 mSv and 99 (0.7%) people to more than 10 mSv.

Scientists tend to limit their consideration of human costs to identifiable health-related issues. But what about residents who can no longer work because radioactive contamination has made their professions unviable? Should this sort of thing not be included within our list of risks related to nuclear power plants? A farmer with produce he cannot sell; a fisherman with seafood nobody buys. What about people forced to abandon the land on which they have lived all their lives? What about children, who are particularly susceptible to radiation, many of whom were sent to distant locales by worried parents? These are all clear, indisputable consequences of the Fukushima accident. What degree of responsibility is borne by the electric power utilities and the government to take steps to avoid such costs? Here, before even getting into the issue of what caused the accident, we once again run into the problem of conflicts of interest—the debate over what caused the accident is being advanced by two parties, the electric power utilities and the government, each of which has its own interests in the outcome of the debate. This said, there is a point at which the interests of the government and electric power utilities coincide, as they both stand to gain by playing down the issue. The electric power utility wants to minimize its costs, including the compensation to be paid to aggrieved parties, while the government wants to minimize its responsibility for not having adequately regulated this and other electric utilities.

11.1.3 Information Disclosure and Evacuation

It is difficult to give a fair, uncontested account of the measures taken to protect residents from radiation following the onset of the accident. There were of course many uncertainties at the time; and, within this admittedly murky environment, the government steadily escalated its evacuation orders and recommendations. At 9 p.m. on the evening of the earthquake and tsunami (March 11), Prime Minister Naoto Kan issued the first evacuation order, directing residents within a radius of 3 km from the Fukushima plant to leave the area. Subsequently, after efforts to vent the plant (i.e. release pressure from the reactors into the atmosphere) did not go as hoped and a first hydrogen explosion wracked the Unit 1 reactor building, the evacuation radius was extended to 10 km and then to 20 km. Likewise, residents beyond the 20 km evacuation zone but within a 30 km radius were told to stay indoors under a shelter-in-place order. Sometime later, on March 25, the central government, recognizing the serious deterioration in living conditions and inability to bring relief supplies into the area, directed local governments to oversee the "voluntary evacuation" of residents then hunkering down under the shelter-in-place order.

Such government evacuation orders and recommendations have since been subject to extensive debate. Here, much of the criticism is directed toward the System for Prediction of Environmental Emergency Dose Information (SPEEDI), which was not employed well in its intended function of helping to formulate evacuation plans. SPEEDI is a system that calculates the dispersal of radioactive substances in and around a nuclear power plant in the event of an accident. All Japanese nuclear power plants, including the Fukushima facility, are provided with this system. However, as the Diet Report details (2012, Chapter 4, pp.55–66), at Fukushima, the Emergency Response Support System, which collects and processes data relating to reactor condition, went down as a result of the accident, leaving SPEEDI to perform its calculations with default values. Starting on March 16, the NSC began to infer what it could about conditions within the reactor by making use of available data, most notably concentrations of radioactive substances measured in the surrounding atmosphere. The NSC completed these calculations on March 23 and finally released some estimates of the dispersal of radioactive substances. It was found from inference and actual measurements; however, that levels of radioactivity could be relatively high even at locations outside of the 30 km radius—given this data, some residents were quick to notice that they had actually fled to areas having a level of higher radioactivity. The problem, of course, is that radioactive substances are not evenly dispersed over nice, clean, concentric circles, but travel wherever the wind takes them. So, for a number of reasons—most notably, the government's decision to draw evacuation zones as concentric circles, failure of the data collection system necessary for accurate forecasting, and delays in switching over from one evacuation category to another—the end result of such problems is that a good number of residents were subject to radiation that they should have been able to avoid.

Much criticism has also been directed at another issue here, the concept of "voluntary evacuation". The Government Report, Diet Report, and investigation reports by the two private entities are all notable in this regard. For instance, the Diet Report (2012, p.22) has this to say:

It is the natural right of citizens to decide to evacuate from locations that are possibly contaminated with radioactive substances in order to safeguard their own health, so leaving the evacuation decision to the citizens might seem like a decision that respects their liberty. We must conclude, however, that doing so was inappropriate. It is the endowed duty of democratic states to protect the lives and safety of citizens, as part of the social contract between citizens and the state.

What is particularly noteworthy regarding this point is that the Commission takes issue not with the failure of the government to provide its citizens with enough information to decide whether to evacuate on their own, but rather with the failure of the government to protect those citizens.

On a more theoretical level, we can say that there are two main approaches to information disclosure and the associated evacuation response. The first we will call the "paternalistic" option. It takes a high degree of specialist knowledge to ascertain the risk of a serious accident at a nuclear power plant and to predict the likely extent of any associated harm. Accordingly, under the paternalistic approach, specialists assess the information, arrive at some decision among themselves, and finally tell local residents what they need to know (and, of course, it is the specialists who decide just what the residents need to know). The second we will call the "autonomous" option. This entails disclosing all information, including that which is scientifically ambiguous and open to interpretation, and letting the residents decide on their own. As clearly evident in our analysis, the government clearly followed the first approach, the paternalistic option, in this case. Indeed, the Government Report, Diet Report, and two private entity investigation reports either explicitly or implicitly went with this approach as well. That is, they all presume a disaster-response model under which the government makes some decision and then issue directives to the citizenry in accordance.

11.1.4 A Built-in Moral Hazard

These four reports follow this paternalistic perspective in their criticism of what the government and the electric power utility did not do (i.e. their omissions) with regard to the Fukushima accident. For instance, the utility, despite having been told of the risk of a large tsunami, did not do much to guard against one. Likewise, the government's supervision of that utility could hardly be called strict. Also, the parties concerned were slow to disclose important information to local residents, information necessary for such people to decide for themselves whether or not to flee.

We can point to several systemic factors behind these omissions. One factor is that electric power utilities are granted a monopoly to operate within their respective areas. That is, in Japan, a single large entity is, in principle, entrusted with all power generation, transmission, and marketing (sales) functions within a particular region. Such utilities are conceptually placed as "nurturers" of industrial activity within their specific areas. In addition to supplying local industry with electricity (under a monopolistic arrangement that allows them to set prices free of competitive pressure), they tend to hold substantial equity stakes (shareholdings) in various manufacturers of industrial equipment. Indeed, it is difficult for a company to conduct its business within such an area without maintaining good relations with the local electric power utility. As Samuels (1987) points out, Japan has the most fully private electric power sector in the world, where state initiatives succeed only when bolstered by considerable private support. This holds true for mass media outlets as well, which are eager to secure advertising revenue for the utilities themselves and from companies under their influence. Indeed, in the aftermath of the Fukushima accident, the major media companies are said to have intentionally avoided the use of the word "meltdown" until the Japanese government officially made it acceptable to do so (Diet Report, 2012, Chapter 3, pp.80–81).

From the point of view of TEPCO, the company had built its plant in full compliance with government safety directives and, if some situation were to occur beyond what is envisioned by such directives, then it should be the responsibility of the government to take care of it. In short, this called into question the very concept of liability on the part of a for-profit company. Ramseyer calls this a "moral hazard inherent in private ownership" and notes that Japanese electric companies did not have to pay the full cost of a melt-down. They bear the costs of an accident "only up to the fire-sale value of their net assets. Beyond that, they pay nothing—and the damages from a nuclear disaster easily soar past that point" (Ramseyer, 2012).

11.2 "SAFETY CULTURE" AS A MONOCULTURE

Above we have seen how, either explicitly or implicitly, the powers that be have come to adopt a paternalistic perspective. That is, government-appointed experts assess information pertaining to nuclear science or nuclear power generation, the government assigns safety standards to electric power utilities and other related companies as it sees fit, and the government tells local residents nothing more than what the government decides they need to know. Furthermore, we have seen how this perspective harbors conflicts of interest, how electric power companies have regional monopolies, and how the government has been lax in regulating those companies. We next turn our attention to the relation between this paternalistic perspective and what has come to be called "safety culture". We criticize various monocultural aspects of this culture and then apply ethics of consideration, particularly applied ethics.

11.2.1 The Safety Culture Ideology

According to International Atomic Energy Agency (IAEA 1991, p.1), the term "safety culture" was first introduced in its summary report on the postaccident review meeting on the Chernobyl accident (International Nuclear Safety Advisory Group, 1986), and further expanded on in Basic Safety Principles for Nuclear Power Plants (International Nuclear Safety Advisory Group, 1999). IAEA defines safety culture as the "assembly of characteristics and attitudes in organizations and individuals which establishes that, as an overriding priority, nuclear plant safety issues receive the attention warranted by their significance". Safety culture "refers to the personal dedication and accountability of all individuals engaged in any activity that has a bearing on the safety of nuclear power plants," which can be achieved only through their "good practices". However, good practices are "not sufficient" if applied mechanically. There is a requirement to go beyond the strict implementation of good practices so that all duties important to safety are carried out correctly, with alertness, due thought and full knowledge, sound judgment and a proper sense of accountability." In other words, safety culture can be considered as an idealized representation of professionalism. Its ethical perspective is one of education and training, marked by a reliance on the discretion of scientists, engineers, and corporate leaders. In an organizational context, it is supported by "leadership and management", it demands the firm commitment of top management, and it entails periodic safety assessments and lessons learned from actual experience.

Indeed, whenever some accident or incident occurs at a Japanese nuclear power plant, the investigation reports that follow almost invariably stress a need to reaffirm this safety culture. The reports following the Fukushima accident are certainly no exception. For example, the Diet Report (2012, Chapter 5, pp.30–69) presents as one of several "institutional issues at TEPCO" a conflict between managerial (business) issues and the need to maintain a "safety-first attitude". Here, the report points to a downplaying of "repeated remonstrations regarding safety culture". Also, under the category of "organizational issues concerning regulatory bodies," it contains a mention of "structural problems," issues similar to what we discussed earlier in this paper. These too, we are told, harbor conflicts with safety culture. Other examples can be found in the Government Report (2012, p. 476). Here it states that the Fukushima accident "showed quite a number of problems with TEPCO such as insufficient capability in organizational crisis management; hierarchical organization structure being problematic in emergency responses; insufficient education and training assuming severe accident situations; and apparently no great enthusiasm for identifying accident causes. TEPCO should receive with sincerity the problems the Investigation Committee raised and should make further efforts to solve these problems and build a higher level safety culture on a corporate-wide basis." Yet, particularly notable about these reports is the manner in which they all, while pointing out various organizational or structural problems, position the underlying issue as an incomplete, imperfect, or somehow deficient safety culture, and they all uniformly direct electric power utilities to "make further efforts" in this regard. In other words, neither the Diet Report nor the Government Report contains any criticism of safety culture or its ideology.

What is this safety culture? As stated above, it is an idealized representation of professionalism embraced by those in the broadly defined field of nuclear science. It calls on government officials, electric power utility managers, nuclear sciences, power engineers, and other concerned parties to devote themselves to "safety," the one and only absolute truth, the shared value at the pinnacle of their belief system. In this sense, safety culture is a monoculture. As above, it has "safety" at its pinnacle. Below that other values are arranged within a complex hierarchical structure, with the position of each essentially defining its importance. When some accident or incident occurs, investigators turn their eye to these values, examining which were respected and which weren't, which were implemented and which weren't, where control was sufficient and where it wasn't, where training was appropriate and where it wasn't. But with regard to the ultimate value of "safety," there must be a shared agreement, an overriding belief, some initial assumption they can all accept. Disagreement is not permitted, at least not for long—that is, if there is any divergence of opinion, it is to be taken as temporary, a transient step toward a new agreement. For example, with regard to the Fukushima accident, there is the issue of the degree of risk presented by long-term exposure to relatively low-dose radiation. And, at this very moment, scientists are supposed to reach some conclusion, thereby concluding their debate on the issue, hopefully as soon as possible, and presenting a set of unified assessment standards for all to accept.

Applied ethics, a concept that gained sway in the latter half of the twentieth century, offers another approach. Applied ethics accepts the inevitability of conflicts among multiple incompatible values and provides a methodology for addressing such conflicts (Beauchamp & Childress, 1979). It is not unusual for even experts to arrive at differing assessments of the risk presented by a certain phenomenon. Under applied ethics, attention focuses not as much on the hierarchy of values as on their selection. We can get a better grasp of this admittedly abstract concept by examining the debate over the medical sciences in Japan. This is a relatively recent issue, at least in Japan, and it provides an apt illustration of a paradigm shift from a monoculture to a multiculture.

11.2.2 Applied Ethics as a Multiculture

In Japan, medical ethics has gradually transformed from its tradition of monoculture to a particular discipline of multiculture under the rubric of "bioethics" (Iwashita, 1994). Bioethics did not take firm root in Japan as a formal discipline until the 1980s or even 1990s, but it had been imported from the United States as a set of concepts and methodology, which first arrived in Japan as a field of study approximately 20 years earlier, preceding even the related field of engineering ethics. Bioethics can itself be traced back to the Nuremberg trials, in which a number of German doctors were prosecuted for crimes against humanity and war crimes, and back also to the World Medical Association's Declaration of Helsinki (1964).

The Helsinki Declaration held that tests on human subjects can be justified only if conducted under the free consent of those subjects (World Medical Association, 1964). Into the 1970s, this "human subject" designation was extended to include patients; in other words, patients were deemed to have their own rights. In Japan's case, the introduction of medical ethics is a particularly interesting case of transformation from a monoculture to a multiculture. From the 1970s, translations of various American textbooks on bioethics came to be published, and the concept of "informed consent" gradually began to take root in Japanese hospitals, universities, and research institutes.

The advance of bioethics is also notable for bringing about another change to Japanese professional society, specifically a new emphasis on pluralistic decision-making. Here, too, credit goes to American bioethics, which also led to change that could not be ignored by Japanese medical practitioners. Pluralistic decision-making is notable for its principle-based approach. That is, a number of basic principles are assumed, including respect for autonomy, nonmaleficence, beneficence, and justice, which provide multiple options for application to specific issues so as to arrive at some morally appropriate solution (Beauchamp & Childress, 1979). An advantage to this approach is that it can be applied to issues over a broad range of areas, and indeed it has been useful in assessing possible courses of action in such areas as reproductive medicine, organ transplantation, regenerative medicine, and terminal care.

Another feature of pluralistic decision-making methodology is that it allows the end user (in the case of bioethics, the patient or, sometimes, even a healthy individual) to decide on a course of action from among several options presented. This is a marked change from the conventional paternalistic approach to medicine; this time around, it is the patient who makes the decision. Medical expertise is still valuable and relevant, of course. The doctor explains the risks and benefits of each option to the patient; and, once the patient makes a decision, the doctor goes from there. On the other hand, the patient does have a right to accept or reject the doctor's explanation; or, if not quite convinced either way, to seek further explanation, perhaps from another doctor in the form of a second opinion.

We have not yet seen a similar paradigm shift in ethics in the field of nuclear science, however. What we have seen is a continued reliance on experts to assess scientific risks and benefits, especially risks. When experts disagree, divergent opinions seem to carry little weight, at least relative to what we now see in the field of medicine. With regard to Japanese nuclear power plants, what we see instead are a broad discretionary scope on the part of electric power utilities, weak regulations on the part of the government, and minimal participation on the part of end users (local residents). The methodology is not one under which specialists prepare multiple options for assessment by the general public. Quite the opposite—under a safety culture, the public is presented with only what are essentially the unanimous results of experts' risk assessments. Furthermore, should there be some divergence of opinion among those experts, the public will hear very little about it and the grounds for it.

11.3 APPLIED ETHICS FOR NUCLEAR SCIENCE

Does the Fukushima accident provide an opportunity to induce a paradigm shift from a safety culture to applied ethics? How can pluralistic applied ethics be established in nuclear science or nuclear power generation? We need to examine these questions from a methodological perspective. The application of applied ethics to nuclear science is more complex and multifaceted than its application to the medical sciences. In medicine, risks and benefits occur between individuals (i.e. the physician and patient). In the case of a nuclear power plant, risks and benefits occur in a more complex context encompassing a company and a group of people (i.e. the electric power utility and residents of areas around the plant and the end users of electricity). Each and every engineer at a nuclear power plant must consider what kinds of risks and benefits his actions may pose for a large number of residents. Moreover, since nuclear power generation is related to national infrastructure, consideration must be given to the relationship between nuclear power generation and national interests or political strategies in energy. In discussing such multifaceted issues, ethicists have often used two approaches: a utilitarian or risk-based approach and a deontological or rights-based approach. The advocates of the former would argue that it is possible to rank various options by appropriately evaluating the risks and benefits of NPPs; the proponents of the latter would raise the question of who is qualified to participate in decision making in such evaluations.

11.3.1 The Utilitarian (Risk-Based) Approach

We may be able to assume that a loose government-regulation is partly attributed to the existence of a national interest in raising national wealth by supporting economic activities. Does this assumption explain many of the problems associated with the Fukushima accident? Placing overriding priority on safety, the concept behind the prevailing safety culture is similar to the classical principle of nonmaleficence in medicine—*primum non nocere* (first, do no harm). However, even those who believed in the safety culture of the nuclear power industry did not seem to place the highest priority on the safety of local nuclear power plant residents as doctors would have done. Analysis of the issue of risk avoidance from an ethical perspective reveals conflicting views about its thoroughness. Needless to say, the view supporting the most thorough risk avoidance is based on the precautionary principle. In this view, the party intending to use certain substances for commercial purposes has the burden of proof regarding their safety. Since the precautionary principle is adopted in medicine, no pharmaceutical company can commercialize new drugs unless its safety is proven. If this example were applied straightforwardly to the electric power industry, the electric power companies would not be allowed to operate nuclear reactors unless they had proven foreseeable minimal risks had been eliminated. In reality, as mentioned above, standards that are far looser than this have been adopted, either explicitly or implicitly. TEPCO did not need to verify the risk of low-dose radiation exposure or the risk of a large tsunami based on past tsunamis. It was also not necessary to create a list of opposing views on safety or risks. Some commentators have even ranked the policy importance of nuclear power risks in the following (decreasing) order: proliferation, theft, sabotage, accident, and routine emissions (Nelkin, 1977).

The fact that those who have commercially used nuclear reactors have been able to avoid a strict obligation to prove safety like that met in the field of medicine could be attributed to expanded application of the utilitarian approach. For example, the postaccident responses made by the electric power utility and the government (including information disclosure and evacuation advisories) can be closely examined from the standpoint of the utilitarian approach. As mentioned above, the health risk for local residents caused by radiation exposure resulting from the accident remains uncertain scientifically. At the time of and immediately after the accident, it was impossible to predict the seriousness of the risk. In such a situation, what were the options in terms of information disclosure that would maximize the sum of social utility? Should information on the worst risk be disclosed, or should only optimistic information be disclosed? The answers to these questions when the values of things relevant only to local residents are considered in the calculation of social utility would be different from when the values of other things are also considered.

11.3.1.1 Considered Values of Things: Local Residents' Benefits

In terms of the values of things relevant to local residents, limited disclosure of information has to provide them with certain benefits. What kinds of benefits could the limited release of information on the accident offer to them? In the situation immediately after the accident, if only the health risks to local residents had been considered, it would have been better to adopt a pessimistic prediction or the worst-case scenario, evacuate them, and then modify the scenario to a more optimistic one. So why was not this process used? The reasons discussed in Japan include concern over health risks that are scientifically unfounded. Those who have this concern argue that since health damage by radiation exposure is proven only in a limited manner (as mentioned above), the government should limit the information disclosed on it and pursue policies that reflect scientifically founded judgment. The prevention of financial damage caused by rumors is also being discussed. This is because disclosing the possibility of a negligible level of radioactive pollution in some areas may cause sales of agricultural or marine products from there to drop and incur economic losses for local residents.

People later became more concerned because of the control over information disclosure held by TEPCO, which held most of the relevant primary information, and by the government agencies that had regulated the industry (although this control was insufficient). Financial damage caused by rumors actually increased even though the central and local governments measured the level of radioactive contamination in food, building materials, fertilizers, and other things. The reason for this is that people suspected the government of controlling the information disclosed and of providing only optimistic evaluation results. But didn't people expect this given the controlled information disclosure that occurred at the time of the accident? As formal networks stopped functioning, people exchanged information using social network systems. They also obtained information published by the military, government agencies, and scientists from the United States and Germany. For instance, people were surprised to know that the U.S. Defense Department ordered military-related personnel to evacuate from the area within 50 miles (about 80 km) of FDNPS. With today's information infrastructure, it is impossible in reality to justify nondisclosure of information to people based on utility.

11.3.1.2 Conflicts between Benefits for Local Residents and Benefits for the Nation

As for the values of things that are not relevant to local residents, we need to examine benefits for companies and benefits for the nation. It is allowed that electric power utilities, which play a part in a nation's basic infrastructure, are strongly supervised by the national government. Particularly, with regard to nuclear power generation as an industry, the national government can impose strong restrictions on the companies because the scale of effect of an accident and the responses to it can exceed an individual company's ability to handle them. However, as mentioned above, the Japanese government's regulation of electric power companies was lax. Did the government expect growth in national wealth (the fruits of prosperity) by giving the electric power utilities greater discretion? Economically, benefits to the nation are, for example, increased production and creation of employment opportunities through industrial development. There are cases where such benefits are obtained at the cost of risks posed to local residents. This has been fiercely discussed in connection with repeated incidents of pollution in Japan. Harm to local residents' health is weighed in view of the benefits products bring to many people, industrial development, and creation of employment opportunities, and criticism was leveled at the government for putting priority on the latter. In addition, what has been explicitly claimed, especially after the Fukushima accident, is the benefit of nuclear power plants to the nation in a military context; that is, the plants are seen to offer the potential for producing nuclear weapons. Furthermore, some argue that even if a country has no intention of possessing nuclear weapons, it is in its indirect interests to maintain its people's receptive attitude toward nuclear weapons in order to help its allies conduct their own nuclear weapon strategies.

In cases where local residents' interests clash with national interests, risk assessments by the national government and disclosure of relevant information are extremely important ethical issues. Since the government has authority to regulate nuclear technologies and holds relevant information, dictatorship can result if the government makes decisions solely based on national interests. It is essential, therefore, to have effective measures that restrict the government's authority over risk assessment and information disclosure regarding nuclear power generation and that secure transparency. The Fukushima accident made it clear that a low level of transparency is disadvantageous the public. However, no matter which view is adopted, there remains the question of fairness; in other words, the question of who should be involved in decision making.

11.3.2 The Deontological (Rights-based) Approach

11.3.2.1 Right to Participation in Building a Nuclear Power Plant (Spatial Justice)

With regard to nuclear power plants and their nuclear waste, ethicists discuss intergenerational ethics (Shrader-Frechette, 2000, 2002). They question the justifiability of making future generations bear the risks associated with managing radioactive substances with half-lives of more than tens of thousands of years for the sake of the current generation's interests. At the same time, there is the argument that, in order to meet increased demand for energy and prevent global warming, "we must 'bite the bullet' and accept nuclear as a viable energy source" (Hale, 2011). The Fukushima accident, however, raised questions not just about this notion of temporal justice, but about spatial justice too. In other words, it is necessary to seriously consider the rights of not only future generations, but also people living in remote places because the risks associated with running a nuclear power plant are not something that only a small number of neighboring residents should bear. The common thread among the nuclear power plants in Japan is that they bring benefits to urban residents while putting risks on the shoulders of residents of remote places suffering a population decline. The nuclear power plant that stands on the underpopulated coastal area in Fukushima Prefecture produced electricity for the residents of Tokyo. This view was largely ignored by Japanese scientists, scholars, journalists, and politicians.

In the judicial context, the Japanese courts have repeatedly declared that only those who lived in areas close to a nuclear power plant were entitled to sue and challenge the electric power utilities. In 1973, in the earliest lawsuit, neighboring residents of the Ikata Nuclear Power Plant demanded cancellation of a permit to construct Reactor 1. Subsequently, more than 10 lawsuits demanding the cancellation of construction permits or the end of operation were filed against nuclear power plants and related facilities (a uranium enrichment facility and a nuclear waste disposal facility). These lawsuits are time-consuming. Under the Japanese system of conducting three trials, a decision has been made by the Supreme Court for only two of the six lawsuits that began in the 1970s; the other lawsuits are still continuing today. In the meantime, the construction and operation of the plants subject to these lawsuits has continued. In 1992, in its decision in a lawsuit demanding the cancellation of a construction permit for the Fukushima Daini Nuclear Power Station (which is located about 11 km from FDNPS), the Supreme Court denied the plaintiff's claim while admitting the defendant's burden of proof regarding safety, making the following statement:

> *This decision only recognizes safety in basic design. However, the safety of [a nuclear power plant] requires detailed design based on basic design, construction, and operation as well as the best efforts of the people involved in each stage. Given global environmental pollution due to coal-fired power generation, there is no other way but to promote nuclear power generation while increasing its safety*
>
> ([No name given] vs. Tanaka, 1992).

The Supreme Court limited the burden of proof regarding safety to basic design only. By not making design details or actual operation the subject of its judgment, it gave substantial discretion to the power utilities engaging in commercial operation and other operators. Also, surprisingly, the political opinion that nuclear power generation should be promoted was clearly stated. At the same time, the Supreme Court said that the power utilities bore the burden of proof regarding safety (though limited to the basic design of the nuclear reactor) and, at least, allowed residents in an area within 50 km of the plant to file a lawsuit (Sato, 1993). In reality, however, the area destroyed by the accident was not subject to the judgment, and the area polluted by dispersed radioactive substances was far larger than the area that qualified for lawsuit filing. As a result of the Fukushima accident, the public lost significant trust in the government, the electric power utilities, nuclear scientists, and courts. For each discovery of radioactive contamination after the accident, the government repeatedly commented that it would not cause immediate health damage, but many people doubted such statements. Is the government implying the possibility of future health damage? Even if some health damage occurs due to prolonged low-dose radiation exposure, isn't it difficult to scientifically prove the causal relationship between the radiation exposure and the health damage? Isn't it the victims that bear the burden of proof?

11.3.2.2 Safety Discourses and Engineers' Obligations

Lastly, we analyze, in a historical context, the reason why ethics in nuclear science has continued to be regarded within a monoculture. The paradigm shift that occurred in medical ethics was caused by revelations of inappropriate judgments by doctors and scientists. These cases overturned the assumption that experts were benevolent practitioners who would not make mistakes, and established a new assumption that experts could make ethically inappropriate judgments depending on the situation.

In nuclear science, it is necessary to reexamine ethical concepts and methodologies based on lessons from historical events including accidents and experts' improper actions. Nuclear science evokes complex feelings in the minds of the Japanese because of the memory of the atomic bombs dropped on Hiroshima and Nagasaki. In order to separate nuclear power generation from this memory, its peaceful use was iconized. The Nobel Prize awarded in 1949 to Hideki Yukawa, who is known for his theory on particle physics, is said to have inspired postwar Japan. In popular culture, cartoons about the robots Astro Boy and Doraemon (both of which have a nuclear reactor inside their body) became icons of technology acquiring a sense of humanity. The slogan "a technology-based country" (a resource-scarce country achieving development through science and technology) became the ideal political image of Japan in the last half of the twentieth century.

Strongly optimistic statements about nuclear power generation were made in the context of the tight postwar relationship between Japan and the United States and with belief in Japan's high-level achievements in the nuclear science field. The incident in which Daigo Fukuryu Maru, a Japanese tuna fishing boat, was exposed to and contaminated with fallout from a hydrogen bomb experiment conducted by the United States in 1954 strengthened the public's opposition to the military use of nuclear power. But the incident barely raised any concern about the peaceful use of nuclear power. When the Three Mile Island accident occurred in 1979, it was argued that mishandling by the employees, which caused the accident, could not happen in Japan. The cause of the Chernobyl disaster in 1986 was attributed to the quality of management techniques in place, which were far lower than those observed in Japan. More than 10 accidents at domestic nuclear power plants and related facilities, ranging from minor to relatively serious ones, were attributed to lapses in the safety culture. Once people started to talk about the fact that the safety culture had lost its effectiveness or had been mythicized, some conservatives began to openly discuss the need for nuclear power generation in a military sense. Shigeru Ishiba, an influential Liberal Democratic member of the Diet, stated that although he did not think that Japan should have nuclear weapons, a signal indicating its ability to produce them if so desired would act as a deterrent to neighboring countries (Anonymous, 2011, p. 85). The Yomiuri Shimbun, which has the largest circulation among Japanese newspapers, pointed out in its editorial on September 7, 2011 that Japan was allowed to use plutonium, which could be used as a material for nuclear weapons, and that this situation was actually functioning as potential nuclear deterrence in diplomacy (Yomiuri Shimbun Editorial, 2011). The newspaper then argued that nuclear power generation should not be abolished so as not to abandon the country's potential ability to possess nuclear weapons.

If a problem is placed in a political or military context, there is the risk that scientists will be pushed away from the values that they deem fundamentally important. There are risks that a scientifically unfounded option will be selected as the separation between facts and an evaluation will become vague. This will cause scientists to lose their identity as such as well as the respect that they once commanded from the public. Scientists can contribute to a culture in which no one has the courage to call unclear matters unclear. What the public expects of scientists is the ability to distinguish things that are scientifically certain from those that are not and the ability to show facts as facts. Aspects of ethics in the technological field that are added in the post-Fukushima era will grant people the right of choice and contribute to improving their independence so that they can obtain sufficient information in order to choose the energy policies they want.

ACKNOWLEDGMENTS

I am deeply grateful to Noriko Miyasaka for her support and warm encouragement. I also thank the Japan Society for the Promotion of Science for a grant (KAKENHI Grant Number 22242001) that made it possible to complete this study.

REFERENCES

Anonymous. (5 Oct 2011). *Ronkyaku chokugeki (An interview with Shigeru Ishiba)*. : SAPIO. pp. 85–87.

Beauchamp, T. L., & Childress, J. (1979). *Principles of biomedical ethics*. New York: Oxford University Press.

Fukushima Genpatsu Dokuritsu Jiko Chosa Iinkai. (2012). *Fukushima genpatsu jiko dokuritsu chosa iinkai chosa kensho hokokusho*. Tokyo: Discover21. (in Japanese).

Fukushima Project Iinkai. (2012). *Report: Genpatsu jiko no honshitsu*. Tokyo: Nikkei BP Consulting. (in Japanese).

Hale, B. (2011). Fukushima daiichi, normal accidents, and moral responsibility: ethical questions about nuclear energy. *Ethics, Policy and Environment*, *14*(3), 263–265.

International Atomic Energy Agency. (1991). *Safety culture: A report by the international nuclear safety advisory group*. (pdf). Vienna: International Atomic Energy Agency. Available at: <http://www-pub.iaea.org/MTCD/Publications/PDF/Pub882_web.pdf>. (Accessed 15.01.13).

International Commission on Radiological Protection. (2007). The 2007 recommendations of the International Commission on Radiological Protection (ICRP Publication 103). *Annals of the ICRP*, *37*, 2–4.

International Nuclear Safety Advisory Group (1986). *Summary report on the post-accident review meeting on the Chernobyl accident, safety series No.75-INSAG-l*.

International Nuclear Safety Advisory Group (1999). *Basic safety principles for nuclear power plants, safety series No.75-INSAG-3*.

Investigation Committee on the Accident at the Fukushima Nuclear Power Stations of Tokyo Electric Power Company. (2012). *Final report*. (online) Available at http://icanps.go.jp/eng/. (Accessed 15.01.13).

Iwashita, R., & Sakai, A. (1991). The physician–patient relationship and medical ethics in Japan. *Cambridge Quarterly of Healthcare*, *3*, 60–66.

JSCE (Japan Society of Civil Engineers): The Tsunami Evaluation Subcommittee, The Nuclear Civil Engineering Committee. (2002). *Tsunami assessment methodology for nuclear power plants in Japan*. (pdf). Tokyo: JSCE (Japan Society of Civil Engineers). Available at: <http://committees.jsce.or.jp/ceofnp/system/files/JSCE_Tsunami_060519.pdf>. (Accessed 15.01.13).

Ministry of Education, Culture, Sports, Science and Technology. (2011). *Monitoring information of environmental radioactivity level*. Available at: <http://radioactivity.mext.go.jp/old/en/>. (Accessed 15 January 2013).

Nelkin, D. (1977). Thoughts on the proposed science court. *Technology and Human Values*, *18*, 20–31.

[No name given] vs Tanaka. (1992). Saikou Saibansho Minji Hanreishu (Supreme Court reports: civil cases). *166*, 509. (in Japanese).

Nuclear Emergency Response Headquarters Government of Japan. (*Report of Japanese government to the IAEA ministerial conference on nuclear safety—the accident at TEPCO's Fukushima Nuclear Power Stations*. (online) Available at: <http://www.kantei.go.jp/foreign/kan/topics/201106/iaea_houkokusho_e.html>2011 (Accessed 15.01.13).

Nuclear Safety Commission. (2006). Hatsuden-yo genshiro shisetsu ni kansuru taishin sekkei shinsa shishin (The guideline of examination for earthquake-resistant design of power-generation reactors, in Japanese).

Ramseyer, J. M. (2012). Why power companies build nuclear reactors on fault lines: the case of Japan. *Theoretical Inquiries in Law*, *13*(2), 457–486.

Samuels, R. J. (1987). *The business of the Japanese state: Energy markets in comparative and historical perspective*. New York: Cornell University Press.

Satake, K., Namegaya, Y., & Yamaki, S. (2008). Numerical simulation of the AD 869 Jogan tsunami in Ishinomaki and Sendai plains. *Katsudanso/Kojishin Kenkyu Hokoku*, *8*, 71–89. (in Japanese).

Sato, H. (1993). Ikata/Fukushima daini genpatsu sosho saikosai hanketsu no ronten. *Jurist*, *1017*, 36–42. (in Japanese).

Shrader-Frechette, K. (2000). Duties to future generations, proxy consent, intra- and intergenerational equity: the case of nuclear waste. *Risk Analysis*, *20*(6), 771–778.

Shrader-Frechette, K. (2002). *Environmental justice: Creating equality, reclaiming democracy*. Oxford: Oxford University Press.

The National Diet of Japan Fukushima Nuclear Accident Independent Investigation Commission. (2012). *Main report*. (online) Available at: <http://warp.da.ndl.go.jp/info:ndljp/pid/3856371/naiic.go.jp/en/index.html>. (Accessed 15.01.13).

Tokyo Electric Power Company. (2012). *Fukushima nuclear accident analysis report*. (online) Available at: http://www.tepco.co.jp/en/index-e.html. (Accessed 15.01.2013).

World Medical Association. (1964). *WMA declaration of Helsinki—Ethical principles for medical research involving human subjects*. (online) Available at: http://www.wma.net/en/30publications/10policies/b3/. (Accessed 15.01.13).

World Nuclear Association. (2013). *World nuclear power reactors & uranium requirements*. (online) Available at: http://www.world-nuclear.org/info/reactors.html. (Accessed 15.01.13).

Yomiuri Shimbun Editorial. (2011). Shasetsu. Yomiuri Shimbun 7 Sep.

Chapter 12

Environmental Injustice Inherent in Radiation Dose Standards

Kristin Shrader-Frechette
Department of Philosophy and Department of Biological Sciences, University of Notre Dame, Notre Dame, IN, USA
E-mail: kshrader@nd.edu

Chapter Outline

12.1	Introduction	197
12.2	The Fukushima Accident	198
12.3	Fukushima Environmental-Injustice Victims	199
12.4	Poor People: EIJ Victims	200
12.5	Poor People and Nuclear/Radiation Regulations	201
12.6	Children: FD EIJ Victims	203
12.7	Children and Nuclear-Radiation Regulations	204
12.8	Nuclear Workers: FD EIJ Victims	206
12.9	Workers and Nuclear-Radiation Regulations	207
12.10	Future Generations, FD, and Nuclear-Radiation Regulations	208
12.11	Conclusions	210

12.1 INTRODUCTION

Consider the following statements.

(A) "The only possible physical/health consequences of the Fukushima disaster would be the occurrence of cancer."

(B) "Besides thyroid cancer, no other cancers have been recognized to have increased after...[the] Chernobyl [nuclear accident]."

(C) "Instruction to evacuate [Fukushima] has...been issued...where you might receive physical effects from radiation...If you do not receive... instruction to evacuate, you can stay."

Radioactivity in the Environment, Volume 19
ISSN 1569-4860, http://dx.doi.org/10.1016/B978-0-08-045015-5.00012-5

(D) "Below the 20 mSv dose, it is OK to lead a normal life...you do not need to worry."

(E) "If you live in an area outside of the evacuation area, you do not need to worry about giving or receiving any radiation at all" (MEXT, 2011).

Statements (A)–(E) above are obviously false, yet the Japanese government posts them on an official government website for Fukushima-Daiichi (FD) information. Even if one looks at a scientifically "old" source, a classic US National Academy of Sciences report, published 6 years before the Japanese postings, it is clear that (A) is false because radiation effects also include genetic damage, transmissible mutations, autoimmune disease, cardiovascular disease, etc. (NAS, 2006). (B) is false because the Chernobyl accident is documented as having caused many cancer and non-cancer effects, including increased autoimmune disease, genetic mutations, chromosomal aberrations, etc. (NAS, 2006). (C)–(E) are false because the evacuation occurred only for those receiving>20 mSv doses, yet doses<20 mSv are risky, as there is no safe dose of ionizing radiation; radiation effects are linear, with no threshold (LNT) for increased risk/harm (NAS, 2006; Walker 2000). Because the Japanese government misrepresented the FD accident, both in ways (A)–(E) and by underestimating FD radiation releases (Shrader-Frechette, 2011, 2012), this chapter shows that FD is a classic situation of environmental injustice—disproportionate/inequitable pollution affecting the most vulnerable people.

12.2 THE FUKUSHIMA ACCIDENT

What happened at FD? There were three nuclear meltdowns, releases of "extremely intense radioactivity" (Kerr, 2011; Tabuchi & Pollack, 2011), four explosions spewing highly radioactive debris for miles, roofs/walls being blown off several reactors, and gaping holes ripped in nuclear containment. Reported nuclear-plant-radiation doses were at least as high as 500 milliSieverts (mSv)/hour (Bradford, Lyman, & Kanter, 2011; Friedman, 2011; PSR, 2011; Suzuki, 2011). For this dose, according to the classic (International Agency for Research on Cancer) dose–response curve for ionizing radiation, after just 2 h, all fatal cancers—among exposed-workers—could be attributable merely to FD radiation (Cardis et al., 2005, 2007).

FD radiation levels were so high that, for weeks, workers could not enter the reactor building, even for minutes; hence it was impossible to stop the 6 runaway reactors and 6 uncontrolled spent-nuclear-fuel-storage pools. Instead, radioactive fuel melted through the thick steel/concrete of several reactor bottoms. Tons of irradiated water also leaked from FD into the Pacific, at doses of 1000 mSv/hour; because of no room to store tons of other contaminated water, Japanese officials deliberately dumped it directly into the Pacific, without warning nearby countries (Kerr, 2011; Tabuchi & Pollack, 2011). Yet a 1000 mSv radiation dose is enough to cause nausea/vomiting; triple this dose is lethal, and people could die after only 3 h of exposure. Even 19 miles from the crippled plant, the UN's International Atomic Energy Agency said waterborne-radiation doses were thousands of times above normal (Belson & Tabuchi, 2011a, 2011b; Suzuki, 2011). As a result, the international physicians' group, Physicians for Social Responsibility (PSR), winner of the 1985 Nobel Peace Prize, estimated that FD radiation releases could be "several times that of the Chernobyl [nuclear-accident] release." The FD cesium release, alone, was equivalent to that from 168 Hiroshima bombs (Bradford et al., 2011).

Yet, if German physicians and scientists, as well as the World Health Organization are correct, Chernobyl caused as many as 38,000 to 60,000 premature cancer deaths—something that suggests FD could cause even more fatalities. The Reactor Safety Team of the US Nuclear Regulatory Commission (NRC) also warned that the runaway FD reactors and fuel-storage pools posed "threats that could persist indefinitely" (Belson & Tabuchi, 2011a, 2011b; Suzuki, 2011)—that there could be additional "catastrophic explosions" (Glanz & Broad, 2011). Former US energy secretary Robert Alvarez said the FD disaster could cause "a catastrophic release of radiation that will not necessarily happen all at once—that could make contaminated areas of land "uninhabitable for decades" (Brumfiel, 2011). Consequently, as the Japanese Chief Cabinet Secretary Yukio Edano warned, it is "unlikely that evacuations will be resolved soon" (Fountain, 2011a); as the Japanese Nuclear Safety Commission warned, "radiation levels are unlikely to decline for some time" (Broad & Jolly, 2011; Shrader-Frechette, 2011, 2012).

12.3 FUKUSHIMA ENVIRONMENTAL-INJUSTICE VICTIMS

Who are the main environmental-injustice (EIJ) victims of this devastating nuclear accident? The most obvious ones are uneducated and uninformed people who have been misled about government claims (A)–(E), and therefore are EIJ victims insofar as they cannot give free, informed consent to risks that have not been adequately disclosed. These uneducated and uninformed citizens also are EIJ victims insofar as they have FD-related anxiety/stress, because in addition to misleading them, the government blames these victims, not itself and its poor nuclear management for FD-related stress. Recall that, after the Three Mile Island and Chernobyl accidents, once the US and USSR governments misled citizens about these accidents and their effects, they also blamed the victims for being "radiophobic" and nuclear hypochondriacs. (Shrader-Frechette, 2011). Similarly, after FD, the Japanese government blamed the victims, not itself. It claims on its official FD website that

(F) "People's psychological stress from having anxiety about being exposed to [FD] radiation is more serious than the actual effect of radiation."

(G) "Overreaction [to FD radiation] causes disruption to normal life and prejudice in society. Everything requires balance."

(H) "Some people have misunderstandings [about FD]. The government will work hard to help them understand correctly" (MEXT, 2011).

Besides the uneducated and uninformed, other key FD EIJ victims include poor people, children, and nuclear workers. Subsequent sections of the chapter consider them, in order. However, Japanese EIJ victims have only recently been recognized. Scholars say that, despite the well-known Minimata tragedy, only after 2006 was there any Japanese recognition of both many "environmental-pollution diseases in Japan" and the "state of social exclusion" of EIJ victims—mainly because Japan appears to be a relatively egalitarian society (Mori, 2008).

12.4 POOR PEOPLE: EIJ VICTIMS

Although Japanese social inequalities within the population tend to be less expressed than in other nations (Kagamimori, Gaina, & Nasermoaddeli, 2009), poor people in Japan are EIJ victims insofar as poor people anywhere always bear the brunt of environmental hazards. This is probably both because they tend to be less educated and informed than other citizens, and thus cannot protect themselves from pollution as well as others can; also, they often cannot afford to move away from dangerous situations. In Japan, EIJ among the poor may be increasing because Japan's economic inequality has been increasing for decades; Japanese economic inequality "is now higher than the OECD [Organization for Economic Cooperation and Development] average, and the ratio of people with income below the poverty line…ranks in the highest group among OECD member countries" (Ishisa, 2010; Oshio & Kobayashi, 2009). Indeed, Japanese economic inequality, as measured by income/assets/spending, is worse than in the US—long considered the most economically unequal society in the developed world (Ohtake, 2008). Because of this increasing economic inequality, and because social stratification in Japan is quite rigid, the Japanese middle class is also smaller than that in the US and much smaller than that in the western Europe. Moreover, Japanese citizens tend to assume that Japanese poor are poor "due to their individual moral failings…a situation that still affects Japanese social-welfare policies" (Tipton, 2008). Consequently, "the [Japanese] government has been reluctant to acknowledge poverty…and has not put any effort into measuring poverty"; as a result, both poverty and EIJ have worsened, so that Japanese childhood poverty predicts adult poverty, and so that poor people are subject to "social exclusion" (Abe, 2010; Shrader-Frechette, 2012).

Given such circumstances, it is unsurprising that, after the FD accident, poor people—especially the sick and elderly poor who lived near the FD plant—became EIJ victims. They could not afford to pay anyone to take them away. In fact, even before the FD tragedy, the 1995 Kobe earthquake clearly illustrated the plight of elderly poor and sick poor affected by EIJ. The earthquake "disproportionately affected parts of the city where there were more elderly, people with disabilities," partly because low-income housing is closer to the shore, "built on silt, or in reclaimed areas…[having] weak soil or landfill that liquefied during the quake." However, high-income housing was "located in the mountains facing the bay," where "roads were wider" and "access was better." Thus, after the 1995 Kobe earthquake, just as after 2005 US Hurricane

Katrina, the national government dallied "for days" while people were dying in the streets. In both cases, the government response was "too little, too late." Japanese citizens, like those of New Orleans, "were outraged that news media were able to get helicopters in for live coverage of the ongoing tragedy, while the national government was seemingly still in denial" (Nakamura, 2009).

After the FD disaster, many Japanese poor were abandoned by government and "marooned" for weeks without roads, electricity, or water; those who were able tried to "forage for firewood...lug water from the marsh," and later, to wait for relief workers giving them "their one meal a day"—a piece of bread, a can of tuna, and a pack of instant noodles; for weeks, many poor people had no medicine and no way to get to the hospital to receive either medicine or medical care (Fackler, 2011a). Weeks after the accident began, there was still a "fuel shortage in the disaster zone, hampering relief efforts and leaving [poor] evacuees—even the sick and the elderly—without transportation or heat," despite the frigid, snowy conditions (Broad & Jolly, 2011). Why were the poor so hurt by FD?

One reason is that the poor tend to live near noxious facilities, such as the FD reactors. Within the first month after the accident began, many nearby residents—too poor to go anywhere—already had been subjected to major quantities of radiation. Why? Despite outcries from other nations, the Japanese government insisted that it had plans neither "to expand the [12-mile] zone" of evacuation, nor to begin soil clean-up of cesium. Yet, to avoid poisoning the food chain for thousands of years, soil "decontamination has to be done very quickly," or it cannot be done at all. Not until a month after the disaster began did the government begin considering expanding the evacuation zone that the US and international nuclear agencies had long ago recommended (Broad & Jolly, 2011; Reich, 2011; Shrader-Frechette, 2011, 2012; Kaiser 2011; see Fountain 2011b, 2011c; Jolly 2011b). The result of this delay? Poor people faced EIJ.

12.5 POOR PEOPLE AND NUCLEAR/RADIATION REGULATIONS

Poor people living near FD also face EIJ because of lenient radiation-dose standards and poor nuclear-siting regulations. At least in the US, those living near nuclear facilities tend to be poor, and the same is likely true in other nations (Alldred & Shrader-Frechette, 2009; Shrader-Frechette, 2011). For poor people living near nuclear plants, even without any accidents, their radiation risks are higher than those faced by others. Why? Nuclear facilities are allowed to release radiation (up to 25 mrem/year/facility). The result? Statistically-significant increases have been documented in infant/fetal mortality near the US reactors (Fountain, 2011a; Normille, 2011), in childhood leukemia near German reactors (Mangano, 2008), and in cancer near UK reactors (Spix, 2008). These data suggest that (even without any accidents) those living near reactors—who also tend to be poor—face higher health risks (Mangano, 2002; Shrader-Frechette,

2002). As already mentioned, both the US National Academy of Sciences and scientific consensus are that the only safe dose of radiation is zero, and that the damaging effects of ionizing radiation are cumulative and LNT. This means that, all other things being equal, higher ionizing-radiation doses cause greater molecular damage (thus more cancer, genetic defects, immune deficiencies, etc.). This is why, in 1974, the US Environmental Protection Agency (EPA) said that normal US radiation releases, between 1970 and 2020, could cause as many as 24,000 additional US deaths, even without any accidents (Mangano, 2000).

If the preceding data are correct, nuclear regulations contribute to poor people's being EIJ victims in several ways. First, radiation-dose standards allow predictable, near-reactor increases in radiation-related harms. Second, nuclear-siting regulations do not prevent the facilities from having a greater probability of being built in poor neighborhoods; instead government could make meeting environmental-justice (EJ) criteria necessary for nuclear siting. Third, regulations do not ensure that poor people near nuclear facilities are able to exercise free informed consent to them. In Fukushima, for instance, local residents are EIJ victims because their poverty forced them to accept the risky FD nuclear reactors in the first place. It is well known that poor/minority residents are often forced to accept very risky facilities, merely because they are poor. They need either jobs or help with the local tax base, if they are to receive even the normal government services for citizens. In addition, poor residents in the Fukushima Prefecture live in an economically depressed area that, because of its poverty, is a target of those wishing to site nuclear facilities. Thus, although the Tokyo Electric company, owners of the 6 FD reactors and storage pools, has long been plagued with "cover-up scandals" about safety, Fukushima residents agreed to provide the home for the FD reactors because of the much-needed monies they would bring. By the end of the 1970s, the plant employed thousands of workers, and local people "received substantial subsidies from the national government as well as property tax receipts from Tokyo Electric [the owner of FD]. By 2008, the subsidies alone added 13 billion yen ($157 million)" to town finances. In 2007, only when the town accumulated a debt of $121 million, did it approve two new reactors at FD. In 2010, the town "received about $45 million from the government, or 60 percent of its total revenue" in subsidies because it had accepted these new reactors. Obviously, however, poor people living near FD did not genuinely consent to these reactors if they needed them, to obtain even basic services that citizens elsewhere already enjoyed (Shrader-Frechette, 2011, 2012; Takahashi, 2011).

Of course, the Japanese government is not completely to blame for FD EIJ. After FD, it was busy dealing with the tsunami. As the Japanese Prime Minister put it, "The first priority is the [nuclear] accident response [preventing worse explosions, core melts, and contamination]. Then it [government] needs to help those who have been affected" (NRC/NAS, 2006). The Japanese poor also were abandoned after FD because, at least in the first several weeks of the disaster, potential rescuers would not come into the heavily irradiated disaster zone; weeks after the accident had begun, victims faced "starvation...trapped in their homes or refugee shelters" because "delivery trucks refused" to come within tens of miles of FD (Jolly, 2011a; The Economist, 2011). Radiation fears also contributed to FD EIJ. As a result, for weeks the sick and elderly were abandoned; bodies of hundreds of people...lay unburied near the plant because they were contaminated by radiation...The police and morgue workers were unable to handle them safely" (Fackler, 2011b; Schenkman, 2011). Of course, Japanese-government authorities did not do adequate planning for how to deal with the FD disaster. As a result, the poor were hurt most.

Even if the Japanese government had planned adequately, however, this alone would not have kept Fukushima poor people from being EIJ victims. As already noted, they still faced weak radiation-protection standards, siting regulations that allow EIJ, and lack of citizen consent to nuclear siting. Thus, FD shows that EIJ can occur even before disaster strikes, if legal/ethical strategies for stopping/preventing EIJ are inadequate. Under current law, one can sue for redress only retroactively, only after demonstrable harm has occurred. Yet as the FD case shows, after demonstrable EIJ and radiological harm has occurred, it is too late to protect people. The plight of Fukushima poor people after FD likewise suggests that because catastrophes can cripple disaster-related government actions and agencies, avoiding EIJ requires avoiding situations that could lead to such crippling catastrophes (Shrader-Frechette, 2011, 2012).

12.6 CHILDREN: FD EIJ VICTIMS

Among poor people living near FD, of course children are some of the most troubling EIJ victims. Why? Most adults have defenses against premature disease and death caused by pollution, but children often do not (UNICEF, 2005). Because their organ/detoxification systems are still developing, they are less able to withstand most toxins. Per unit of body mass, children also take in more air/water/food/pollutants than adults do. For these reason, quantitative risk assessors use the rule of thumb that because children are about 10 times more susceptible to pollutants than adults receiving the same dose, children have "particular need for special protection from pollutants" (APHA, 2000).

Not surprisingly, children have been FD EIJ victims mainly because, despite their heightened sensitivity/vulnerability, radiation-dose regulations provide them no additional protection, beyond what adults receive. After the FD disaster, children and adults living near the disaster area were allowed to receive 2000 mrem/year, or 20 mSv/year radiation. Thus PSR says the FD "impact on the health of Japanese children is being glossed over." They say there are about 350,000 children < age 18 living in Fukushima Prefecture and that, if they received only two years of allowable RD-radiation exposure, at least 2500 of them would die prematurely of cancer. After four years of exposure, 5000 of these children would die prematurely. After 8 years of this exposure, 10,000 of the children would die prematurely of cancer—and yet the nuclear industry and

Japanese government are covering up these health consequences. Instead, the Japanese government and nuclear industry are claiming the children's doses are safe, as the beginning quotes from the Japanese-government website illustrate. Yet, as already mentioned, no dose of ionizing radiation is safe—especially not a dose (20 mSv/year) that is about 10 times higher than natural background and about 25 times higher than each nuclear plant is allowed to release in an entire year. Thus, PSR says that the allowed Japanese exposure of the children at doses up to 20 mSv/year—mainly to save money and avoid more cleanup—is "medically unacceptable." Already by June 30, 2011, newspapers around the world confirmed that Fukushima children tested positive for internal radiation exposure—through food/soil/water. Not only can the children not consent to this risk, but also the Japanese government is continuing to cover up the risk because it "is not adequately monitoring radiation contamination of soil, food, water, and air and is not providing the parents with sufficient information to protect their children." Virtually the same analysis of "unconscionable" medical risks to children was provided by Toshiso Kosako—a scientific advisor who resigned from the Japanese government to protest the FD radiological harms imposed on Japanese children (Bradford et al., 2011; von Hippel, 2011; Shrader-Frechette, 2011, 2012).

The continuing failure of the Japanese government both to advise parents how to protect their children, and to collect crucial radiation-exposure data is important. Without data, the FD crisis cannot be fully understood, and people—especially children—cannot be protected. Also, unless the data are collected immediately, they disappear, the crisis cannot be examined accurately, and parents/physicians will be less able to consent to government actions. Without continuing-monitoring data, government/industry can continue to deny the problem, deny medical/health coverage to victims, and deny their rights to protection/informed consent. After all, one cannot consent to risks that are misrepresented. Although the Japanese government still has not released full reports about FD radiation releases, Hirosaki University scientists themselves did measurements. They warned that many FD victims have received four times the allowed 20 mSv dose, although the Japanese government said all doses were "far below levels that warrant concerns about human health." The government also admits that the most damaging radiation is from internal doses, from food and water. Yet, its 20 mSv standard includes only external radiation and ignores internal contamination—resulting from isotopes such as cesium-134 and cesium-137, with a half-life of more than 30 years—that will continue for decades (Butler, 2011; Cyranoski, Brumfiel, & Nature Magazine, 2011; Shrader-Frechette, 2011, 2012).

12.7 CHILDREN AND NUCLEAR-RADIATION REGULATIONS

Given that children are both more sensitive to ionizing radiation and less able to consent to it than adults are, one would think that regulations would better protect children. However, just as the FD accident imposed heavy EIJ radiation-dose burdens on children, so do normal radiation-dose regulations. In at least two ways, current radiation-dose regulations force children to be EIJ victims.

First, given children's much higher radiosensitivity, international/national radiation standards do not provide more-protective radiation regulations for them. Instead, children have the same radiation-dose limits as members of the public. Why? Regulators assess hazards of ionizing radiation in terms of the reference-man model. As the International Commission on Radiation Protection (ICRP) put it: "Reference man is defined as being between 20 and 30 years of age, weighing 70 kg [154 pounds], is 170 cm [5 feet 7 inches] in height, and lives in a climate with an average temperature of from 10° to 20 °C. He is a Caucasian and is a Western European or North American in habitat and custom" (Makhijani, Smith, & Thorne, 2006). Reference-man models for radiation risks focus only on adult male cancer risks, although children's cancer risk can be up to about 40 times higher, and women's cancer risk is roughly 50-percent higher, when all receive the same radiation dose (Makhijani et al., 2006; NRC/NAS, 2006). Moreover, because the reference-man model addresses only cancer, it ignores noncancer radiation harms such as increased genetic defects, immune-system damage, blood diseases, spontaneous abortion, birth defects, and neonatal mortality. (Makhijani et al., 2006). The reference-man model likewise ignores the permanent 25-point drop in drop for every sievert (100 rem) of in-utero ionizing-radiation exposure during critical weeks (8–26) of brain formation (ICRP, 2005; NRC/NAS, 1990). Obviously, current radiation standards do not adequately protect children.

Second, although regulations can provide the fetus with special protections, international standards protect them less well than adults. According to regulations of the US and those of some other nations, female nuclear workers need not declare their pregnancy unless they choose. If they do, they and their fetus are allowed to receive radiation doses of 500 mrem/year (not 5000 rem/year that other US radiation workers may receive); however, this fetal dose is 5 times higher than 100 mrem total/year, what members of the public may receive, and 5 times higher than the fetus may receive in much of Europe, e.g. Germany (Makhijani et al., 2006). The obvious question is why US/Japanese fetuses are less well protected than European, and why US/Japanese children are less well protected than adults, when infants can be up to 38 times more radiosensitive than adults (Makhijani et al., 2006). Indeed, if fetuses/infants were protected at the same level as US members of the public, their annual radiation dose will be a maximum of (1/38)(100 mrem), or 2.6 rem. Because the US/Japanese fetal-radiation standard = 500 rem/year, the fetus is 500/2.6 or 192 times less protected than adults. Yet if children are the most vulnerable members of society, if some current radiation regulations protect them less well than adults, this suggests that radiation regulations violate EJ. They follow the preferences of the powerful, not the needs of the vulnerable (Shrader-Frechette, 2011, 2012).

12.8 NUCLEAR WORKERS: FD EIJ VICTIMS

Another major group of EIJ victims are nuclear workers. They are allowed to receive much higher doses of ionizing radiation than members of the public. Yet, typically, only low socioeconomic-status people tend to take such high risks (Shrader-Frechette, 2011, 2012).

Although many FD nuclear workers are heroes, having sacrificed to save others, many also are EIJ victims. They are likely unable to genuinely consent to higher radiation doses that can harm them and their families. Why not? Post-accident FD workers are mainly unskilled, temporary workers who are often forced, by economic necessity, to accept whatever jobs they can get. Indeed, even without an accident, 80–89 percent of all 83,000 Japanese nuclear workers are temporary contract workers who receive about 16 times more radiation than normal radiation workers—whose doses are recorded. They often are offered $350 to $1000 per day, for work that lasts only a few minutes because the radiation doses are so high (Kakuchi, 2011; Shrader-Frechette, 2007).

Several additional problems suggest that, after the FD disaster, nuclear-cleanup workers have been EIJ victims. One reason is that, only weeks after the accident began, the Japanese government raised the allowable, post-FD radiation doses, for plant workers, to 250 mSv/year (Tabuchi, 2011)–250 times what members of the public can receive. Because FD onsite doses were at least as high as 500 mSv/hour, and even higher inside the plant, this means some FD workers received their exposure limit of 250 mSv/year within seconds/minutes. According to the classic IARC radiation dose–response curve (Cardis et al., 2005, 2007), 25 percent of the fatal cancers among FD workers, receiving 250 mSv/year, would be caused by this FD exposure. Yet because of the cumulative nature of radiation risk, some FD workers must have had cumulative doses higher than 250 mSv. For instance, for 30-year-old FD workers, normal-background radiation alone—without any X-rays, and so on—already has given them 3 mSv/year, or 90 mSv cumulative background dose. Even without any medical exposures, IARC indicates that these 30-year-olds who received 250 mSv from FD could attribute 34 percent of their fatal cancers to FD radiation.

If FD-accident workers were never told that their exposure would likely kill many of them prematurely, they obviously did not give genuine consent to FD risks and are EIJ victims. If they were told, this also suggests EIJ. Why? It is improbable that hundreds of mentally healthy persons could agree to roughly a one-third chance of death, in exchange for temporary work—for seconds or minutes—that pays only $170–$350/hour. Rational workers might risk death for a huge sum, to feed their families, but risking death for a one-time, $170–$350 payment seems less likely. Under such circumstances, it again seems probable that many FD-accident workers did not genuinely consent and are EIJ victims. In the poor Japanese economy, they simply may have been desperate for work (Beauchamp & Childress, 1989; Faden & Beauchamp, 1986; Kakuchi, 2011; Tabuchi, 2011).

Another FD-accident-worker problem is that they may have received much higher doses than they were told. Neither the utility nor Japanese regulators have released full statistics on radiation levels inside the containment buildings where these men have worked. Nor have they provided full information on total FD radiation releases (Bradsher & Tabuchi, 2011). "The company refused to say how many contract workers had been exposed to radiation" and what protective clothing and respirators they had. Even so, radiation-protective gear protects workers only from skin/lung contamination. Even in the best gear, gamma rays irradiate their entire body and pass through it (Fountain, 2011a; Tabuchi, 2011). Such partial protection and failure to disclose radiation levels again suggest nuclear-cleanup workers face EIJ, disproportionate radiation burdens.

Moreover, FD workers' EIJ and consent problems are exacerbated by the fact that officials allow reporters no access to them. Hence, it is difficult to know how they fare as they clean up FD. Local health experts say that "many former workers are experiencing health problems," and that "most day workers were evacuated" from FD after the earthquake. Since then, however, temporary FD workers have been "off limits to reporters." Although virtually all workers injured by high radiation during the FD accident "were subcontractor employees," the company has "refused to say how many contract workers had been exposed to radiation" (Kakuchi, 2011; Tabuchi, 2011), again suggesting cover-up and EIJ (Shrader-Frechette, 2011, 2012).

12.9 WORKERS AND NUCLEAR-RADIATION REGULATIONS

Because of several regulatory shortcomings, however, nuclear workers—just as with poor people near noxious facilities—are not merely EIJ victims after an accident. Rather, even without accidents, they are often EIJ victims, perhaps more so in the US/Japan. There regulatory standards allow nuclear workers to be exposed to roughly 50 times more ionizing radiation—50 mSv/year—than that to which members of the public can be exposed—1 mSv annually. This double standard is ethically questionable for at least 5 reasons. First, as already mentioned, there is no safe dose of ionizing radiation except zero (NRC/NAS, 2006), and these higher allowable worker exposures mean higher numbers of job-related cancers/deaths. Second, it is not obvious that workers should have to trade their lives and health for work. Third, most European nations do not allow this double standard, and they have far stricter worker-radiation standards than do workers in the US/Japan, precisely because the Europeans believe that workers ought not have to trade their lives and health for employment.

A fourth reason to suspect EIJ, because of workers' radiation-dose double standard—used in Japan and most English-speaking nations— is that because radiation-dose regulations cover only single sources (e.g. a nuclear plant) or exposure classes (workplace, medical, or public) and are defined for average occupational exposure, workers typically do not know their precise cumulative/individual/relative risks from radiation. Because cumulative/individual/relative risk information is necessary to calculate radiation dose, therefore damaging effects, this information is necessary for risk disclosure, consent, and EJ. However, many temporary nuclear workers—without full-time employment, adequate compensation, union representation, or health benefits—typically receive the highest workplace radiation risks, and yet they are unlikely even to be told their doses. Why not? The nuclear industry is not required to "count" exposures of temporary workers when it reveals average worker doses to regulators (Shrader-Frechette, 2011, 2012).

A fifth reason nuclear workers likely face regulation-related EIJ is that regulations do not prohibit "jumpers," temporary workers who jump from nuclear plant to nuclear plant, performing high-risk work and receiving the maximum annual dose at each plant. Globally, "thousands of untrained, itinerant, temporary laborers…handle the bulk of the dangerous work at nuclear power plants". In Japan, "these contractors were exposed to levels of radiation about 16 times as high as the levels faced by full-time Tokyo Electric employees[during the] last year" before the FD accident. Yet these temporary workers are part of "Japan's two-tiered work force, with an elite class of highly paid employees at top companies and a subclass of laborers who work for less pay, have less job security, and receive fewer benefits. Critics charge that such two-tiered regulatory practices both endanger worker health and undermine safety at nuclear reactors. "This is the hidden world of nuclear power," says a former university physics professor in Tokyo. In 2010, in Japan 88 percent of nuclear workers "were contract workers"–without employee benefits and without the protection of workplace radiation standards. At FD in 2010, 89 percent were temporary-contract workers. In Japan, the elite 10 percent of workers are nuclear-plant operators. Under these elite are the 89 percent of workers who are "contractors, subcontractors and subsubcontractors—with wages, benefits and protection against radiation dwindling with each step down the ladder…Some workers are hired from construction sites, and some are local farmers…Others are hired by local gangsters." They have a "constant fear of getting fired," and they try to hide injuries and doses to avoid being fired. Unions urge "operators not to forgo radiation exposure records and not to force workers to lie to government inspectors about safety procedures." Yet, Japanese workers say that union leaders and their families have been physically threatened by "thugs who kicked down their doors." They say that, "once you enter a nuclear-power plant, everything's a secret." Obviously secrecy, nondisclosure of risks, and a double standard for dose limits, make it difficult for many nuclear workers not to be EIJ victims (Kakuchi, 2011; Tabuchi, 2011).

12.10 FUTURE GENERATIONS, FD, AND NUCLEAR-RADIATION REGULATIONS

However, perhaps the most hidden of nuclear EIJ victims are members of future generations. They have high potential for being EIJ victims because, in principle, they cannot give free informed consent to radiation risks, for instance, from nuclear waste. They are out of sight, out of mind. Yet, because regulations tend to protect future generations less well than current generations, obviously future generations would not consent to unequal treatment, even if they had a voice. Two key ways in which, even without nuclear accidents, future generations are disproportionately harmed, include genomic instability and less-protective nuclear-waste standards.

Because even low doses of ionizing-radiation exposure can cause genomic instability, inheritable damage that weakens the gene pool (Barber, Hickenbotham, Hatch, et al, 2006; Dubrova, 2003; Dubrova, Hickenbotham, Glen, et al, 2008; Hatch, Derijck, Black, et al, 2007; Huumonen 2012; Kadhim, Salomaa, Wright, et al, 2012; Karotki & Baverstock, 2012; Morgan, 2012), genomic instability amounts to an EIJ imposed by current generations on future generations. Moreover, the EIJ of genomic instability increases, as radiation exposures increase—causing greater rates of future disease/death. Radiation's delayed, transgenerational effects thus explain why cancer rates, not merely cancer incidences, continue to increase in Japan, even among those born several generations after Hiroshima/Nagasaki exposures. Such transgenerational effects of genomic instability also explain harms to children who were born many years after parental radiation exposure, as in the large cluster of leukemia cases in children living near the Sellafield, UK, nuclear facility (Shrader-Frechette, 2012, 2011).

Besides genomic instability, future generations face EIJ because of inequitable regulations that allow later generations to receive much higher doses of ionizing radiation than current generations. In the US, for instance, government-proposed regulations (not adopted yet) for buried, high-level nuclear waste from reactors require that no member of the public receive more than 0.25 mSv/year from any nuclear facility (USA CFR, 2002). However, for 10,000–1,000,000 years after closure, the standard allows average public exposures of 1 mSv/year (US EPA, 2005). One problem with this standard is that average exposure for future generations could be four times higher than for current generations. Another problem is that the standard protects people only from average doses, not from extraordinarily high doses to some individuals.

While the pragmatic reasons for imposing greater nuclear-waste risks on future generations are clear, given the difficulty of preventing future waste leaks, imposing dangerous, higher risks on future generations amounts to EIJ. One reason is that there is no defensible ethical reason (based on merit or benefits to future generations) that could justify forcing future people to bear higher risks, especially higher risks that benefit only present generations. Instead, the main reason for the different standards appears to be that current generations wish to save money by providing less secure radiation protection for the future. Obviously no one would accept a four-times-higher risk than that received by other people, especially if there were no moral reasons to justify it. Instead, present generations could set aside resources, in trust, so that future generations will have necessary funds/materials to protect themselves when the waste leaks, as everyone admits it will.

Another EIJ difficulty with the proposed waste regulations' protecting only against average, not high, individual risks, is that virtually all environmental/health regulations guarantee individual protection against specific pollutant levels. However, the proposed US radwaste standard requires only that average/mean doses meet the 0.25 mSv/year standard. This means people receiving higher-than-average exposures would get no protection. That is, EPA adopted "the arithmetic mean as the statistical measure to be applied at all times" (US EPA, 2005). Thus, if 6000 nearby residents annually received waste-related-radiation doses of 0.01 mSv/year, but 1 person received a fatal dose of 600 rem (6 Sv), this death would be legal/allowable, despite its EIJ. Using average doses as the criterion for acceptable pollution levels is thus like accepting the claim of a city police chief who says he will prosecute no murderers unless the city's number of murders exceeds the per-capita national average.

However, each individual arguably has rights to equal protection, not merely to protection that achieves some overall-average dose for all. This "average" regulatory-dose scheme thus guarantees future EIJ victims—those receiving high doses whose protection is excluded, provided "average" dose-limits are met. Given the way nuclear-waste doses are calculated (US EPA, 2005), the proposed regulation also excludes all persons whose radwaste doses are higher than those to the "Reasonably Maximally Exposed Individual" (RMEI). Yet because the RMEI is defined as neither drinking water nor eating food grown near the site, this RMEI would not have the highest radwaste exposures—as pointed out in a US National Academy of Sciences report (NRC, 1995), and those with higher doses would be EIJ victims (Shrader-Frechette, 2012).

According to the proposed regulations, future people also are likely to be EIJ victims because they would not even be protected against actual harm (specific doses), but only against projected doses. That is, the US Nuclear Regulatory Commission will use "projected doses presented by DOE [Department of Energy]" to determine regulatory compliance (US EPA, 2005). Such projections make EIJ against future persons more likely because, instead of guaranteeing justice and equal treatment, based on actual, individual doses, DOE could manipulatively lower average doses by expanding dose distributions—and then claiming that high doses to some were acceptable. Obviously dose projections could be manipulated more easily than dose measurement, especially by those who might profit by not controlling nuclear pollution. If so, future victims of manipulated dose projections also would be EIJ victims.

12.11 CONCLUSIONS

The FD accident disproportionately harmed at least three groups of EIJ victims—poor people, children, and nuclear workers. However, EIJ during normal nuclear operation affects at least the previous three groups plus future generations. EIJ affects these four group because nuclear-related regulations treat people inequitably for apparently pragmatic, rather than ethical, reasons—such as saving money, rather than protecting those most needing protection. Although society may never overcome EIJ originating from individual biases, the moral of this chapter is that, at least, society ought to be able to overcome EIJ originating from poor regulations.

REFERENCES

Abe, A. K. (2010). Social exclusion and earlier disadvantages: an empirical study of poverty and social exclusion in Japan. *Social Science Japan Journal*, *13*(1), 5–30.

Alldred, M., & Shrader-Frechette, K. (2009). Environmental injustice in nuclear-plant siting. *Environmental Justice*, *2*(2), 85–96.

APHA. (2000). The precautionary principle and children's health. *Policy Statements*. http://www.apha.org/legislative/policy/Pols2000_rev.pdf. Last accessed 18.02.06.

Barber, R. C., Hickenbotham, P., Hatch, T., et al. (2006). Radiation-induced transgenerational alterations in genome stability and DNA damage. *Oncogene*, *25*, 7336–7342.

Beauchamp, T. L., & Childress, J. F. (1989). *Principles of biomedical ethics*. New York: Oxford University Press.

Belson, K., & Tabuchi, H. (2011a). Japan finds contaminated food up to 90 miles from nuclear sites. *New York Times*, *20*, 80A.

Belson, K., & Tabuchi, H. (2011b). In setback, radioactive water is leaking directly into sea, Japan says. *New York Times*, *3*, 11A.

Bradford, P., Lyman, E., and Kanter, A. (2011). Remembering 3–11. Washington DC: Physicians for social Responsibility.

Bradsher, K., & Tabuchi, H. (Mar. 2011). Last defense at troubled reactors. *New York Times*, *14*. www.nytimes.com/2011/03/16/world/asia/16workers.html?pagewanted=all. Last accessed 20.09.11.

Broad, W. J., & Jolly, D. (Mar. 2011). UN's nuclear chief says Japan is 'Far From the End of the Accident'. *New York Times*, *27*, 10A.

Brumfiel, G. (2011). Water leak stops at Fukushima, but big problems remain. *Nature News Blogs*. http://blogs.nature.com/news/thegreatbeyond/2011/04/water_stops_atfukushima_but_b.html. Last accessed on 06.04.11.

Butler, D. (2011). Fukushima health risks scrutinized. *Nature*, *472*, 13–14.

Cardis, E., et al. (2005). Risk of cancer after low doses of ionizing radiation. *British Medical Journal*, *331*, 77–80.

Cardis, E., et al. (2007). The 15-country collaborative study of cancer risk among radiation workers in the nuclear industry. *Radiation Research*, *167*, 396–416.

Cyranoski, D., Brumfiel, G., & Nature Magazine. (2011). Fukushima crisis is still hazy. *Scientific American*. http://www.scientificamerican.com/article.cfm?id=fukushima-crisis-is-still-hazy. Last accessed 07.09.11.

Dubrova, Y. E. (2003). Radiation-induced transgenerational instability. *Oncogene*, *22*, 7087–7093.

Dubrova, Y. E., Hickenbotham, P., Glen, C. D., et al. (2008). Paternal exposure to ethylnitrosourea results in transgenerational genomic instability in mice. *Environmental and Molecular Mutagenesis*, *49*, 308–311.

Fackler, M. (2011a). As food is rationed, resolve is plentiful. *New York Times*, *29*, 10A.

Fackler, M. (2011b). Japanese city's desperate cry resonates around the world. *New York Times*, *7*, 1A.

Faden, R., & Beauchamp, T. L. (1986). *A history and theory of informed consent*. New York: Oxford University Press.

Fountain, H. (2011a). Radiation exposure could curtail workers' efforts. *New York Times*, *14*. www.nytimes.com/2011/03/15/science/15workers.html. Last accessed 20.09.11.

Fountain, H. (2011b). Wind and rain steer radiation's reach. *New York Times*, *20*, 8A.

Fountain, H. (2011c). Cleanup questions as radiation spreads. *New York Times*, *1*, 10A.

Friedman, S. M. (2011). Three mile island, Chernobyl, and Fukushima. *Bulletin of the Atomic Scientists*, *67*(5), 55–76.

Glanz, J., & Broad, J. G. (Apr. 2011). "US sees new threats at Japan's nuclear plant. No. 55367, *New York Times CLX*, *6*, 1A. 12A.

Hatch, T., Derijck, P. D., Black, G., et al. (2007). Maternal effects of the scid mutation on radiation-induced transgenerational instability in mice. *Oncogene*, *26*, 4720–4724.

von Hippel, F. (2011). The radiological and psychological consequences of the Fukushima Daiichi accident. *Bulletin of the Atomic Scientists*, *67*(5), 27–47.

Huumonen, K., Immonen, H.-K., Baverstock, K., et al. (2012). Radiation-induced genomic instability in *Caenorhabditis elegans*. *Mutation Research/Genetic Toxicology and Environmental Mutagenesis*, *748*(1–2), 36–41.

International Commission on Radiological Protection (ICRP). (2005). *Recommendations*. Stockholm: ICRP.

Ishida, A. (2010). Trend analysis of anticipated lifetime income inequality among post-war Japanese youth Atsushi Ishida anticipated lifetime income inequality in Japan. *International Journal of Japanese Sociology*, *19*(1), 64–81.

Japanese Ministry of Education, Culture, Sports, Science, and Technology (MEXT). (2011). *To educators, for correctly understanding radioactivity.* Translated by Setsuko Shiga, University of Notre Dame, Notre Dame, Indiana, Tokyo: MEXT and Japan Pediatrics Association.

Jolly, D. (2011a). Japan Weighs nationalizing a stricken utility. *New York Times*, *30*, 8A.

Jolly, D. (2011b). Radioactive iodine detected in ocean. *New York Times*, *1*, 10A–11.

Kadhim, M., Salomaa, S., Wright, E., et al. (2012). *Non-targeted effects of ionising radiation—implications of low dose risk*. Mutation Research/Reviews in Mutation Research.

Kagamimori, S., Gaina, A., & Nasermoaddeli, A. (2009). Socioeconomic status and health in the Japanese population. *Social Science & Medicine*, *68*(12), 2151–2160.

Kaiser, J. (2011). Japan soil measurements surprisingly high. *Science*. http://news.sciencemag.org/scienceinsider/japan_quakes. Last accessed 10.04.11.

Kakuchi, S. (2011). *Fukushima blows lid off exploited labour*. Rome: Inter Press News Agency http://ipsnews.net/news.asp?idnews=104978. Last accessed 27.09.11.

Karotki, A. V., & Baverstock, K. (2012). What mechanisms/processes underlie radiation-induced genomic instability? *Cellular and Molecular Life Sciences*, *69*(20), 3351–3360.

Kerr, R. (2011). Fukushima radiation. *Science*. http://news.sciencemag.org/scienceinsider/2011/04/fukushima-radiation-modeling-shows.html. Last accessed on 10.03.11.

Makhijani, A., Smith, B., & Thorne, M. (2006). *Science for the vulnerable: Setting radiation and multiple exposure environmental health standards to protect those most at risk*. Takoma Park, MD: IEER Press.

Mangano, J. J. (2000). Improvements in local infant health after nuclear power reactor closing. *Environmental Epidemiology and Toxicology*, *2*, 32–36.

Mangano, J. J. (2002). Infant death and childhood cancer reductions after nuclear plant closings in the United States. *Archives of Environmental Health*, *57*, 23–31.

Mangano, J. (2008). Excess infant mortality after nuclear plant startup in rural Mississippi. *International Journal of Health Services*, *38*, 277–291.

Mori, M. (2008). Environmental pollution and bio-polictics: the epistemological constitution in Japan's 1960s. *Geoforum*, *39*, 1466–1479.

Nakamura, K. (2009). Disability, destitution, and disaster: surviving the 1995 Great Hanshin earthquake in Japan. *Human Organization*, *68*(1), 82–88.

National Research Council (NRC). (1995). *Radiation dose reconstruction for epidemiologic uses*. Washington, DC: National Academy Press.

National Research Council/National Academy of Sciences (NRC/NAS). (1990). *Health effects of exposure to low levels of ionizing radiation: Beir V*. Washington, DC: National Academy Press.

National Research Council/National Academy of Sciences (NRC/NAS). (2006). *Health risks from exposure to low levels of ionizing radiation: Beir VII, Phase 2*. Washington, DC: National Academy Press.

Normile, D. (2011). Conflicting U.S. and Japan evacuation policies sow confusion. *Science*. http://news.sciencemag.org/scienceinsider/japan_quakes. Last accessed 10.04.11.

Ohtake, F. (2008). Inequality in Japan. *Asian Economic Review*, *3*, 87–109.

Oshio, T., & Kobayashi, M. (2009). Income inequality, area-level poverty, perceived aversion to inequality, and self-rated health in Japan. *Social Science and Medicine*, *69*(3), 317–326.

Physicians for Social Responsibility (PSR). (2011). The Fukushima nuclear crisis. *Medicine and Global Survival*, *4–5*.

Reich, E. S. (2011). Fukushima disaster: some lessons already clear. *Nature News Blogs*. http://blogs.nature.com/news/2011/03/fukushima_disaster_some_lesson.html. Last accessed 10.04.11.

Shenckman, L. (2011). What effect will the radiation have on the Japanese? *Science*. http://news.sciencemag.org/scienceinsider/japan_quakes. Last accessed on 10.04.11.

Shrader-Frechette, K. (2002). *Environmental justice: Creating equality*. New York, Oxford: Reclaiming Democracy.

Shrader-Frechette, K. (2007). Trimming exposure data, putting radiation workers at risk. *American Journal of Public Health*, *97*(9), 1782–1786.

Shrader-Frechette, K. (2011). *What will work: Fighting climate change with renewable energy, not nuclear power*. New York: Oxford University Press.

Shrader-Frechette, K. (2012). Nuclear catastrophe, disaster-related environmental injustice, and Fukushima: prima facie evidence for a Japanese 'Katrina'. *Environmental Justice*, *5*(3), 133–139.

Spix, C. (2008). Do nuclear plants boost leukemia risk? *New Scientist*, *2642*, 6.

Suzuki, T. (2011). Deconstructing the zero-risk mindset: the lessons and future responsibilities for a post-Fukushima nuclear Japan. *Bulletin of the Atomic Scientists*, *67*(5), 9–18.

Tabuchi, H. (Apr. 2011). Braving heat and radiation for temp job. *New York Times*, *4*, 1–4A.

Tabuchi, H., & Pollack, A. (Apr. 2011). Strong aftershock jolts Japan; workers at nuclear plant take cover. *New York Times*, *8*, 14A.

Takahashi, H. (2011). Japan's nuclear disaster Severs Town's economic lifeline. *New York Times*. www.nytimes.com/2011/04/03/world/asia/03futuba.html?pagewanted=all. Last accessed 20.09.11.

The Economist. (2011). *Come Back in Ten Years' Time*. Vol. *398*(8726), p. 47–48. (26 Mar.– 1 Apr).

Tipton, E. (2008). Defining the poor in early twentieth-century Japan. *Japan Forum*, *20*(3), 361–382.

UNICEF. (2005). *State of the world's children*. New York: UNICEF.

US CFR. (2002). Standards, Title 10, Part 20, sections 1301 and 1402.

US EPA. (2005). 40 CFR Part 197.

Walker, S. (2000). *Permissible dose*. Berkeley: University of California Press. pp. 3–5, 37–38, 89, 51, 71, 75.

Chapter 13

Safety Culture and Safety Quality

Carl Rollenhagen
Department of Philosophy, Royal Institute of Technology, Teknikringen, Stockholm, Sweden
E-mail: carl.rollenhagen@vattenfall.com

Chapter Outline

13.1 Introduction 215
13.2 Personal and Organizational Values 219
13.2.1 Personal Values 220
13.2.2 Organizational Values and the Competing Values Framework 221
13.3 Safety Culture/Climate 223
13.3.1 Safety Values, Competing Values, and Bounded Rationality 224
13.4 Safety Quality and Human Reliability Analysis 226
13.5 Implications for Safety Management 227
13.5.1 Assessing Safety Culture and Safety Quality 227
13.5.2 Safety Culture Initiatives 229
13.6 General Discussion 230

13.1 INTRODUCTION

Nuclear and radiological risk prevention represents a specific knowledge domain in safety science and risk management. Nevertheless, this specific field is part of a more general knowledge domain associated with concepts and practices that have general applicability for most risk sensitive systems. The concept of safety culture has been one important innovation for enhancing safety in the nuclear field as well as in many other branches.

One of the strongest driving forces for research and practices departing from the concept of safety culture has been the observation that neither technical factors nor "human errors" are by themselves sufficient to explain major accidents (Pidgeon, 2010; Pidgeon & O'Leary, 2000). Consequently, it has been argued

Radioactivity in the Environment, Volume 19
ISSN 1569-4860, http://dx.doi.org/10.1016/B978-0-08-045015-5.00013-7

that to fully understand safety and risk, one should also explore the larger cultural and organizational context in which organizations operate. For example, the external environment including technological, institutional, and socioeconomic factors influences organizations policies through management decisions. This, in turn, has effects on systems for risk control, plans, policies, individual behavior, etc. (Rasmussen, 1997). The concept of safety culture has sometimes been used as a holistic term aiming for this system-oriented perspective. The nuclear domain has been one of the forerunners for applying the concept of safety culture but the concept is today used in many safety critical domains.

As a consequence of a broader concept of safety it is has become common to launch safety enhancement project subsumed under the labels of safety culture and safety climate[1]. Today, qualitative and quantitative indicators of safety culture/climate are found in many organizations. Differences in opinion do exist concerning what we ought to mean by safety culture. However, there seem to be at least some common understanding amongst researchers and practitioners arguing that safety culture refers to collectively held values, beliefs, knowledge and behavior of relevance for safety. For example, an influential definition of safety culture was provided by the Advisory Committee on the Safety of Nuclear Installations (ACSNI/HSE, 1993).

"The safety culture of an organization is the product of individual and group values, attitudes, perceptions, competencies and patterns of behavior that determine the commitment to, and the style and proficiency of, an organization's health and safety management. Organizations with a positive safety culture are characterized by communications founded on mutual trust, by shared perceptions of the importance of safety and by confidence in the efficacy of preventive measures".

How and to what extent attributes such as organizational structures, instructions, technology etc., also should be included in definitions of safety culture varies among scholars (Guldenmund, 2000, 2007): a risk of including almost everything in "safety culture" is that the concept becomes empty as an analytical construct (Rollenhagen, 2010). In any case, the value(s) people hold about safety seems to be a good and fairly noncontroversial position for many discussions about safety culture. The reason for this is commonsensical and departs from the assumption that if we value X, then we usually also strive for realizing the states of affairs associated with X. However, It is also easily recognized that people in organizations usually harbor values of many kinds; production, innovation, safety, organizational growth, etc. Sometimes, but not necessarily, such configurations of values oppose each other; production may conflict with safety and innovation values may conflict with safety values. In this paper, my intention is to take a closer look at what we could mean with safety as a value and how safety culture may be distinguished from other similar concepts such as quality. In particular I will argue that:

1. The concepts of safety culture and that of safety climate emanate from different research traditions. In this article the concepts will be used interchangeably.

H1: It is not the value of safety perceived as an absolute attribute in the matrix of organizational values that determines the strength of safety culture, but rather how the value of safety relates to the manifold of other values simultaneously strived for.

That the value of safety in some situations might oppose the value of good financial outcomes has been much discussed in safety science literature (Zohar & Luria, 2004) and is commonly found (more or less explicit) in the public discourse about risk and safety. However, the balance between safety and "production" is just one of many potential value conflicts facing organizations. Organizations are usually burdened by a huge amount of goals, projects, requirements etc., arising both from internal and external sources. In management rhetoric's, safety is often claimed to have priority over other ambitions, but such an espoused safety value may easily be compromized when meeting a complex reality. Nuclear power production is a good example of how many different values are at stake and the balance between these can be challenging.

I will argue that safety culture should be perceived as a relative construct defined as how people in an organization perceive and value safety in relation to other values and how the value of safety is implemented (the actions taken). Such a definition supports a research strategy that explores safety values in the context of other values (Colley, Lincolne, & Neal, 2013). I will argue that it is not only a matter of "balance" between different values: it is also important to understand the *extent* of other values i.e. the number of goal and values found on the agenda and how priorities are set among these: attention, after all, is a limited resource.

The above suggestion (H1) entails that an organization, in principle, could launch many seemingly good safety culture initiatives but the overall safety culture many nevertheless be weak. A main reason for this is that safety values, due to limited attention resources, risk to become "smeared out" in relation to many other values also embraced by an organization. To put a specific focus on values demands that our understanding of safety culture should be more narrowly defined than is usually suggested. In the following, I will elaborate on this topic by suggesting the following:

H2: It makes both theoretical and pragmatically sense to differentiate between safety culture and safety quality. Safety culture is a human centerd concept whereas safety quality basically focuses on those artifacts that support safe conduct.

H3: Safety competence can be viewed as an interface between values and safety quality. Thus safety competence can be perceived as a reflection of values but also as being a "support function" for safety quality.

Based on the above suggestions, a structural model of safety culture, safety competence and (safety) support functions (safety quality) are depicted in Figure 13.1.

Safety culture in Figure 13.1 is thus a concept assumed to consist of two parts: (1) shared values and (2) safety competence (declarative and procedural knowledge, including behavior).

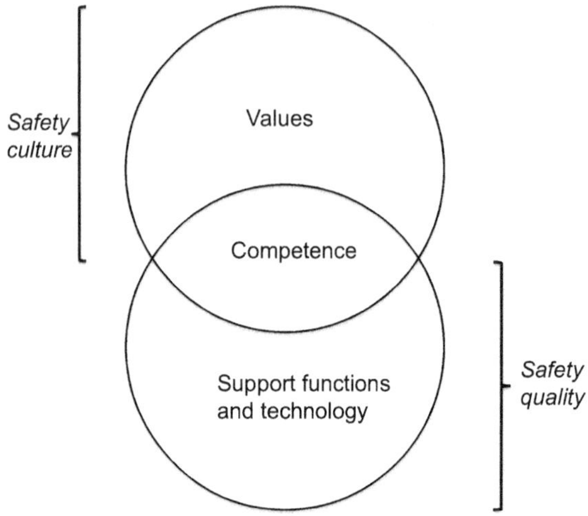

FIGURE 13.1 Safety culture, safety competence, and safety quality.

The concept of *safety quality* is here meant the perceived quality of those concrete arrangements used in an organization to support safety (including technology). For example, procedures, human machine interfaces, time resources, and training arrangements are objects that people perceive as more or less functional for supporting safe and reliable operations. In safety, quality is also included resources (manning, financial, and time) and core-production technology.

As indicated above, the concept of *competence* as it appears in Figure 13.1 should be interpreted as an interface between safety values and specific safety arrangements (safety quality). For example, if the human factors domain is valuated as being of high importance for safety, and that this value is salient in the competition with other values, then such a value focus should then be transferred (the "competence interface") into high-quality processes and products. The competence interface is recursive in the sense that it also applies to itself when transforming general competence values into working arrangements for developing competence in different domains. The concept of competence as it is applied here includes behavior, for example adherence to safety rules, communication, etc.

The nuclear industry, in which the concept of safety culture was born, has depending on its potential for environmental impact from radiological release, a very strong motive to develop methods and concepts for enhancing safety. Concepts such as safety culture and safety climate are currently used for this purpose. However, when such concepts are used it is likewise important to avoid confusion. The present text can hopefully contribute to some conceptual distinctions that may serve to stimulate further discussion about what we can mean by safety culture and how it relates to other similar concepts.

The remainder of this text elaborates on the above suggestions and their consequences for safety management and safety culture research. Already at the beginning, it should be remarked that the present discussion does not include an empirical test of the suggestions made. However, some anecdotal observations from the author's personal experience in the nuclear sector will be used as illustrations. Section 13.2 first introduces some background concepts associated with the concept of value in general and personal and organizational values in particular. In the third section, focus is on the concept of safety culture/climate and the need to prioritize and make tradeoffs in organizations. In the fourth section, the concept of safety quality is discussed in the context of human reliability analysis (HRA). In the fifth section, some consequences for safety management are suggested followed by a general discussion in Section 13.6.

13.2 PERSONAL AND ORGANIZATIONAL VALUES

The concept of value has been much discussed and its meaning varies among disciplines. Values are associated with many philosophical controversies, many of which are not particularly attended to (if at all) in the pragmatic (safety) use of the value concept. For example, a (controversial) distinction is often made between intrinsic values and extrinsic/instrumental values (in philosophy) but this is not often discussed in the safety literature. The former means that something is valued for itself and not as a mean to reach other values. If safety is perceived as an extrinsic value, then it is seen as something important for reaching something else (such as production). However, one could also argue that safety should be perceived as an intrinsic value i.e. something which should be endorsed as an end state rather as a means for reaching something else.

Both in psychology and in sociology, values are explored, for example, in research about the structure of personal- and organizational values and their interaction. In research about safety culture, the value of safety is surprisingly often discussed without much reference to what can be meant with the concept of *safety* in the first place. By comparison, discussions about safety culture have scrutinized the general concept of culture and organizational culture. This imbalance in safety culture research may at first sight seem surprising but is presumably a consequence of different research traditions.

The literature about values is extensive and the intention here is not to pursue a review of the subject. Instead, a more narrow departure in some selected research departing from Schwartz is adopted (Schwartz, 1992, 1994, 2006; Schwartz & Boehnke, 2004; Schwartz, Melech, Lehmann, Burgess, & Harris, 2001).

To generally approach (organizational) culture/climate from a value perspective has been suggested by several researchers (Hofstede, 1980; Howard, 1998; Parker, Brown, Child, & Smith., 1977; Rokeach, 1973; Rousseau, 1990; Sheridan, 1992). In this literature, values are perceived to serve as standards or criteria for evaluation of various states of affairs (people, events, objects etc.). Values represent beliefs and motivational constructs and are usually perceived as being more abstract in comparison with norms and attitudes (e.g. Allport, 1961; Feather, 1995; Inglehart, 1997; Kluckhohn, 1951; Kohn, 1969; Morris, 1956; Rokeach, 1973). Furthermore, values are generally assumed to influence people's attention and interpretation of situations (Schwartz, Sagiv, & Boehnke, 2000) as well as providing cues for goal setting, priorities, and behavior (Gollwitzer, 1996). For example, if safety is collectively perceived as having a high and longstanding priority in an organization, then we may assume that this value influences the interpretation of situations, goal and plans, safe behavior, and so forth. In conclusion, the concept of value has been influential as a descriptive and exploratory construct in many disciplines, including philosophy, psychology, sociology, and political science.

13.2.1 Personal Values

Schwartz has identified the following 10 personal values as being salient in cultures around the world: Power, Achievement, Hedonism, Stimulation, Self-direction, Universalism, Benevolence, Tradition, Conformity, and Security. Two orthogonal dimensions have been found to summarize these values (Schwartz, 1992, 1994). In the first dimension, power/achievement stand in opposition to universalism/benevolence (i.e. self-enhancement vs self-transcendence). In the other dimension, values of self-direction and stimulation are contrasted with security, conformity, and tradition (i.e. openness to change vs conservation).

How can we perceive the concept of safety culture in the framework suggested by Schwartz? First, safety culture is often associated with some conservatism with respect to change and innovation—changes should not be made without an analysis of possible risks associated with change. As a tentative hypothesis, we would thus expect a correlation between at least some aspects of positive safety culture and those personal values that focus on tradition and conformity. However, the concepts of "tradition" and "conformity" are obviously open to several interpretations. Under the assumption that we by "tradition" understand personal values associated with low risk taking, conservative decision making, and resistance to risky changes, then a positive safety culture would be associated with conservation values. By the same token, personal values of conformity could in the context of safety culture/climate be interpreted as adhering to safety rules and regulations as a positive ideal. However, such an interpretation can be challenged since a personal value of conformity could also be interpreted as something negatively for safety, for example in terms of group think being negative for safety. By and large, however, it seems reasonable that many of those personal values that often have been associated with a strong/good safety culture are "conservatively flavored" in comparison with values associated with "openness to change". If this is true, then we are faced with a difficulty because a strong/good safety culture must also be learning and adaptive culture open for innovation and change (Reason, 1997). This would imply, then, that a set of competing values may be present—to what extent they compete or cooperate depends on the situation.

The other main dimension in Schwartz model can, I suggest, be reinterpreted as an ethical dimension that on the one hand describes personal values associated with a strong self-focus, and, on the other hand, a set of more altruistically flavored ethical orientations—some of them surely important for safety. We would thus expect that at least some individuals holding strong personal values focusing on self-enhancement, power, and hedonism would contribute to a less developed safety culture in comparison with a moral stance that transcends self- and power interests. It should be evident to the reader that the discussion above represents a set of hypothesis in need of further elaboration and empirical research. For now it is suffice to say that the model by Schwartz represent one interesting candidate for further comparisons of how various personal values may interact with the more global construct of shared safety cultures in organizations.

The nuclear domain presents a particularly interesting case for safety value research. Radiological protection involves both concerns for those who directly handle radioactive material (personal safety) but of course also for many other stakeholders that can be affected by radiological releases.

13.2.2 Organizational Values and the Competing Values Framework

The competing values framework (CVF) emerged from studies of organizational effectiveness (Campbell, 1977). In an attempt to reduce the many criteria that existed in the literature about organizational effectiveness, Quinn and Rohrbaugh (1983) suggested a "competing values approach" to organizational analysis by application of expert judgments and multidimensional scaling. The results suggested that: "organizational researchers share an implicit theoretical framework, and, consequently, that the criteria of organizational effectiveness can be sorted according to three axes or value dimensions" (p.369). The first dimension was related to organizational focus "...from an internal, microemphasis on the wellbeing and development of people in the organization to an external, macroemphasis on the wellbeing and development of the organization itself" (p. 369). The second dimension was related to organizational structure, with, at the one end, emphasis on *stability* and, at the other end, emphasis on *flexibility*. The third dimension was related to means-end relationships—e.g. planning and goal setting vs outcomes (productivity).

The dimensions found were perceived as representing "competing values". As stated by Quinn and Rohrbaugh: "...flexibility versus stability reflects a basic dilemma of organizational life. The different viewpoints in considering order and control versus innovation and change are at the heart of the most heated debates in sociology, political science, and psychology. While many social theorists have emphasized authority, structure, and coordination, others have stressed diversity, individual initiative, and organizational adaptability" (1983:370). The next pair of competing values (external–internal) were commented as; "From the external view, the organization is a logically designed tool with the ultimate goal to of accomplishing its task and acquiring resources.... emphasis is on overall competitiveness of the organization in sometimes changing environments. From the internal perspective, the organization is a sociotechnical system. Participants have unique feelings, likes and dislikes, and require consideration, appropriate information, and stability in the workplace" (1983:370). Finally, the means-end dimension is discussed as an organizational dilemma, for example in the context of time. The different mind-sets that emerged from the CFV framework have been described as climates and cultures and related to different normative models of organizing (Quinn, 1988).

Comparing the CVF with that of Schwartz, it is evident that there appears to be some similarity between the dimensions of flexibility vs control in CFV and of openness to change vs conservation in Schwartz framework. Both models contrast the features of flexibility and openness with a more stable controlled state with less uncertainty/risk. As a critical remark speaking against the above interpretation, one may argue that the concept of "control" has several interpretations. For example, adaptive control is associated with flexibility. It would therefore be a mistake to suggest a too oversimplified dichotomy between control and flexibility. This is certainly a relevant objection. However, one should remember that the competing value framework was based on research reflecting a set of implicit assumptions among managers and organizational researchers rather than a theoretical analysis of various control perspectives.

With respect to the internal vs external dimension in the CVF, and the self-enhancement vs self-transcendence characteristics of Schwartz framework, there is less overlap. Of course, one could argue that the "internal" dimension in CVF has an abstract similarity to an internally (ego) focused ethics and that the "external dimension" has a correspondence in an ethics that stretches outward (universalism), but this is perhaps a too far-reaching conclusion and shall not be pursued further. This said, it seems however important to consider how ethics is related to safety culture—a neglected topic in safety culture research.

In summary; reviewing two previously suggested frameworks, both with empirical support, a set of questions in need of further elaboration and analysis emerge. Organizations with high safety demands must find a balance between stability and change. Since organizations face various internal and external pressures (Rasmussen, 1997) and strive to realize many goals and values, difficult priority problems may occur. We would also suspect that individual personal values may give rise to priority problems partly reflected by different ethical concerns.

Applied to the nuclear domain and radiological protection several issues are evoked by the above discussion. The potential value conflict between innovation/flexibility vs stability/control can clearly be perceived in the nuclear industry. For example, upgrading old nuclear power plant with modern technology is a difficult task (verification and validation of software can be very demanding). In general, values associated with flexibility/innovation and values associated with stability/control are not easily merged. People at nuclear power plants are expected to follow rules at the same time they are expected to be critical toward the rules etc.

13.3 SAFETY CULTURE/CLIMATE

As mentioned in the introduction, the concepts of culture and climate are both increasingly applied as descriptive, explanatory, and normative constructs for various states of affairs in organizations. For example, the concept of organizational culture is used in connection with organizational performance (Ogbonna & Harris, 2000); safety culture (INSAG, 1991) is applied in association with various risk and safety aspects, and ethical culture (Chen, Sawyer, & Williams, 1997) highlights moral aspects. Still another use of culture (and climate) focuses on innovative capacity and innovations (Tellis, Prabhu, & Rajesh, 2009). However, what is often lacking in research focusing on various specific organizational cultures/climates is an analysis of their interrelations. Specific cultures/climates associate with different values, goals, and priorities that may be in conflict—either more or less chronically, or in situations when organizations are facing problems and pressures (Zohar, 2008, 2010).

After the concept of safety culture was introduced in the aftermath of the Chernobyl accident, a wealth of research has contributed to new research topics and procedures including measurements of safety climate. Of earlier origin than the word safety culture is the concept of safety climate. This concept, with its cradle in organizational climate research, has often been associated with people's perception of safety priorities (Zohar, 1980) and with psychometric research strategies (scaling approaches) used to reveal the assumed generic dimensional structure of the construct (Collins & Gadd, 2002; Guldenmund, 2007;).

Whether safety climate should best be modeled as a uni-dimensional or a multidimensional construct has been debated (Neal, Griffin, & Hart, 2000; Zohar & Luria, 2005). Under the assumption that safety climate can be modeled with several dimensions (which is the most common notion), the quest for revealing a proper factor structure has resulted in a variety of dimensions (Brown & Holmes, 1986; Dedobbeleer & Beland, 1991; Zohar, 1980, 2002a,b). However, "management commitment to safety" appears as the most common factor found by researchers (e.g. Cheyne, Cox, Oliver, & Tomás, 1998; Cohen, 1977; Dedobbeleer & Beland, 1991; DeJoy, Schaffer, Wilson, Vandenberg, & Butts, 2004; Donald & Canter, 1994; Eiff, 1999; Hofmann & Stetzer, 1996; Ostrom, Wilhelmsen, & Daplan, 1993; O'Toole, 2002; Rundmo, 1996; Rundmo & Hale, 2003; Seo, Torabi, Blair, & Ellis, 2004; Simonds & Shafari-Sahrai, 1977; Smith, Cohen, Cohen, & Cleveland, 1978; Zohar, 1980; Zohar & Luria, 2005).

Some studies have suggested a core of safety climate dimensions that seem relatively stable across studies. For example, in a study by Johnson (2007), a three-factor model (e.g. Caring, Compliance, and Coaching) was able to predict safety outcomes. The proposed model was closely associated with a dimensional structure suggested by Zohar and Luria (2005); e.g. Active Practices (Monitoring–Controlling), Proactive Practices (Instructing–Guiding), and Declarative Practices (Declaring–Informing). However, a crucial question is to what extent general dimensions such as "management commitment" is unique for modeling safety climate or if such dimensions are common facilitators for most cultures/climates. In that case, nothing particularly interesting emerges from attempts to characterize various climates in isolation (safety, production, innovation). Even if the concentration is on attitudes, values and behavior rather than on more specific supportive functions (instructions etc.), as objects for evaluation, there is still little in current safety climate models that reveals the full scope of values that one may expect to find in organizations. I will thus conclude that much of this classical research has not addressed the right problem since a too limited scope of values has been addressed when exploring safety climate.

13.3.1 Safety Values, Competing Values, and Bounded Rationality

The discussion so far indicates that it is reasonable to find many different values and corresponding cultures/climates operate in organizations. Furthermore, it is also reasonable to assume that depending on different subcultures, some sets of values are more embraced than others. For instance, persons belonging to an upper management subculture may hold other assumptions, values and attitudes than those found in a maintenance subculture.

In a recent study by Colley et al. (2013), the role of different organizational values as an antecedent of perceived safety climate was explored. The authors refers to previous research by Zohar and Luria (2004) who argued that safety climate depends on the relative importance people place on safety relative production. Colley et al. (ibid.) recognize that several values are in focus in most organizations, and not only safety and production. To explore how a broad set of values might interact with safety climate, Colley at al. departed from the CVF.

A reasonable hypothesis is that depending on the profile of values (e.g. their relative strength to each other), different climate characteristics would emerge in an organization. For example, Quinn and Spreitzer (1991) argue that, in general, companies with a *balanced value profile* perform better than organizations with an unbalanced profile. Colley et al. (2013) takes these observations as a basis for a hypothesis that "individuals who perceive that their organization has a balanced value profile also will perceive that their organization has a more positive safety climate, and report fewer incidents in comparison with individuals that perceive that their organization has an imbalanced value profile" (p. 71). The rational for this hypothesis is expressed as; "A balanced profile suggests that people are valued, trained and supported (human relation focus), there are adequate and useful rules and procedures (internal focus), production goals and targets are appropriate and achievable (rational goal focus) and the system is adaptable, innovative and has up-to date technologies and equipment (open system focus)" p. 71.

Colley et al. suggest that individuals who perceive their organization as focusing either on human relations or open system models will perceive a better safety climate than those focusing on either internal processes or rational goal models. The rational behind this hypothesis is that previous research has found that if there is an overemphasis on production (rational goal model) people may feel that achieving production targets are more important than care for people (Wright, 1986) and if people believe that managers are overly focused on formal rules and procedures, managers might be perceived as more interested in compliance in itself rather that actually caring about people (Morgan, 1986). These assumptions and associated research can of course be questioned but appear to have face validity at least judging from my personal experience as a researcher and consultant in the nuclear industry. I have often heard critical voices regarding the influence of production pressures and the burdens of bureaucracy.

Colley et al. researched a sample of individuals working in high-risk industries in Australia. The best safety performance was found in cultures that were characterized by a Human Relation-Rational Goal Model and the worst performance was found in profiles biased toward the Internal Process and Rational Goal Model. However, the data did not support the hypothesis that a balanced profile is best for safety since none of the data profiles had that balanced characteristics. Interestingly, however, the data showed that profiles showing a weak safety performance were characterized by emphasis on the Internal Process Model (e.g. bureaucracy). On the other hand, profiles with better safety performance shared an emphasis on Human relations. Colley et al. interpret their results in the context of a previous research that found that overly focusing on control by rules and restrictions tend to limit motivation and learning creating a passive safety orientation (Parker, Turner, & Griffin, 2003; Turner, Chmiel, & Walls, 2005). An interesting finding from this research is that focusing on production (e.g. Rational Goal Model) was not in itself negative for safety given that production focus was balanced with a supportive and flexible human orientation. On the other hand, when the Rational Goal Model was strongly combined with focus on rules and procedures (e.g. Internal Process Model), this profile was found to be negative for safety.

Even though research such as the one illustrated above may be a step toward viewing a broader repertoire of values in relation to safety, many methodological problems remain. First, value focus may change depending on context so there is a difference between situational value focus and a more stable general value profiles. Secondly, what people say about organizational values may not reflect organizational behavior. It is furthermore questionable if subjective data in terms of questionnaires actually are able to handle these problems of value diagnostics. To conclude, understanding organizational values and how they associate with safety is, despite many years of general organizational and safety culture research, still a challenge. However, here is another research tradition that more directly has addressed human and organizational issues in relation to safety that we shall now consider namely HRA. The traditions exploring safety culture and safety climate has unfortunately been much separated from the human reliability tradition (HRA).

13.4 SAFETY QUALITY AND HUMAN RELIABILITY ANALYSIS

Whereas writings about safety culture and safety climate often take a departure in theories about organization and general culture research, the HRA tradition has been closely associated with risk analysis and system engineering (Aven, 2003; Melnick & Everitt, 2008). Reliability analysis found in the tradition of system engineering and risk analysis set out to assess failures in technical systems and their consequences. HRA focuses particularly on the human component as a risk in such assessments. French, Bedford, Pollard., and Soane (2011) distinguish between two related ways to apply HRA. The first use (referred to as summative analysis) gives a quantitative estimate of a human failure probability that is then used for supporting decisions about choice of technology, retrofits, maintenance strategies, etc. A second use of HRA (formative use) is to support decisions about how to improve a system and its supporting functions (procedures, man–machine interfaces, organizational structures, etc.). The most common practice related to HRA has been to focus on well-defined event sequences at the sharp end of operation and maintenance. By means of the general strategy to decompose event sequences into well-defined tasks and investigate supporting human performance shaping factors (PSF), many versions of HRA method have been suggested and applied (for overviews see Adhikari et al., 2008; Gertman & Blackman, 1994; Hollnagel, 1993, 1998; Kirwan, 1998, 2008; Swain & Guttman, 1983). Various criticisms have been directed toward HRA, for example that these methods not have been able to incorporate culture and other contextual factors to a sufficient extent in the models (French et al., 2011; Hollnagel, 1998). But to be fair, some previous research on risk analysis has in fact tried to incorporate organizational factors in some models (Boring, 2007). The concepts of organizational culture, safety culture, and safety climate sometimes (but not always explicitly) take a position in these frameworks. Some earlier attempts to incorporate organizational factors into risk analysis (such as probabilistic risk analysis) are represented by, for example; Davoudian et al. (1994a,b), Embrey (1992), and Pate-Cornell and Murphy (1996). More recent models are represented by, for example: Mohaghegh and Mosleh (2009). These scholars provide a set of theoretical foundations and principles for incorporating organizational factors in probabilistic risk assessment. Critique has however been directed toward the often-used strategy to model organizational factors by means of conventional event sequences and fault trees. Several authors have expressed a need for more dynamic approaches (Leveson, 2004). In a model suggested by Mohaghegh and Mosleh (2009) this criticism is to some extent handled since they use several different modeling strategies. An observed characteristic associated with some recent complex models is that these become very complex and therefore less usable for more practically oriented purposes (Roelen et al., 2011). Building on these observations, as an analytical strategy it would be pragmatically sound to make a conceptual distinction between safety culture and safety quality.

It should also be mentioned in this context that some research that is close to the HRA tradition has sleeked to develop methods for analysis of safety management. One example is the work by Kennedy and Kirwan (1998) who developed what they call Safety Culture and Operability approach (SCHAZOP), a tool for diagnostics of safety management practices based on a similar philosophy as the well-known method of Hazard and Operability study approach (Kletz, 2006).

To conclude, the HRA tradition provides tools and concepts that motivate a separation between safety culture and safety quality. We shall now discuss what this implies for safety management.

13.5 IMPLICATIONS FOR SAFETY MANAGEMENT

13.5.1 Assessing Safety Culture and Safety Quality

It has become rather common to attempt to asses safety culture and safety climate and try to use the result as an indicator for safety. Such assessments are usually made by means of interviews and/or questionnaires in order to reach a diagnostic summary in qualitative or quantitative terms. The predictive value of such diagnostics has been researched (Arboleda, Morrow, Crum, & Shelley, 2003; Diaz & Cabrera, 1997; Garavan & O'Brien, 2001; Glendon & Stanton, 2000; Lee & Harrison, 2000; Mearns, Whitaker, & Flin, 2001; Niskanen, 1994; O'Toole, 2002; Silva, Lima, & Baptista, 2004; Varonen & Mattila, 2000; Vredenburgh, 2002; Zohar, 2000). Moderate to poor predictive validity is usually obtained (but also see Zohar, 2002a,b).

In assessments of safety culture/climate, there is sometimes a mix between value-oriented objects (e.g. safety vs production values) on the one hand, and measures of more specific arrangements that support safe behavior on the other hand (i.e. what is here referred to as safety quality). I believe that this mixture is unfortunate. It would be better, I believe, to more clearly attempt to separate safety culture interpreted as values from concrete safety arrangements interpreted as safety quality. In the safety climate tradition, the strong focus on management commitment toward safety do indeed reflect value focus (what I here have called safety culture) but such summative measures fail, in my view, largely to scrutinize also other general value dimensions. As mentioned, research based on the CVF has started to address this issue (Colley et al., 2013; Reiman & Rollenhagen, 2012; Silva et al. 2004) but more has to be accomplished in this direction. For example, more efforts should be made about diagnosing how organizations need to compromise and tradeoff between values, goals, and ambitions facing organizations. One route of research would be to investigate organizational pressures by studying *the number of goals* that simultaneously exist in organization and relate this to safety attention and practices. Such studies may reveal true safety priorities. Also using free text data in the context of safety climate questionnaires can reveal safety priorities. I have used such data in the context of safety climate studies and in my experience the free text data are often of more interest that the quantitative estimates of different safety climate dimensions.

With respect to what I call safety quality, the general philosophy found behind HRA has significance for general safety management to a much higher extent that is often being realized in everyday-safety management practices (Kennedy & Kirwan, 1998). The HRA tradition is different from most attempts to measure safety culture in the sense that HRA is considerably more situational specific. HRA addresses particular processes and the quality of supporting operator functions. However, HRA is often perceived as a specialist domain in the context of risk analysis (Probabilistic Safety Assessment (PSA)) rather than a general methodology for quality assessment. Although attempts have been made to generalize the HRA approach and to include elements from safety culture/climate research, the models produced are often so complex that it is doubtful in they will be used as practical safety management tools. However, by making use of the general philosophy found in the risk analytical and safety engineering traditions, one could eventually develop practical and relatively easy-to-use analytical instruments. Such a strategy constitutes a "middle way" between holistic safety culture/climate assessments and HRA assessments. Some projects along this line can be found by Kennedy and Kirwan (1998) and Kongsvik, Almklov, and Fenstad, (2010) as well as in other more qualitative analytical approaches.

The framework suggested here may be used to suggest some general rules of the thumb to enhance safety culture and safety quality in organizations.

First, based on the present framework it is important to explore the general organizational values found in an organization and to position safety in the context of these values. As said previously, this entails both finding the strength of various values and also the nominal extension of values (i.e. how many goals and values are the organization attempting to reach within a given time frame). Various configurations of values are distributed unevenly in organizations depending on different functions. This context sensitivity of values must be taken into account in trying to obtain a summative result of value distribution. Also, to ask about the value of "safety" as a summative value seems problematic without also considering safety contribution from other values. For example, rather than try to position the value of, say, innovation or human resources in relation to safety, one has to explore in greater detail how innovation and human resource management may also contribute to safety and how they might contribute to other nonsafety-related states of affairs.

Secondly, by assessing specific processes and use various performance-shaping factors identified in the context of HRA research, it is possible to obtain a diagnostics of safety quality. But instead of the highly context specific sequences usually assessed in HRA assessments, safety quality assessments could also be performed at a higher level of abstraction e.g. a group of activities (maintenance, operation, etc.) could be modeled as a generic process.

13.5.2 Safety Culture Initiatives

The framework suggested here can be used to initiate safety culture initiatives in terms of value discussions performed in different functions (operation, maintenance, etc.). Such discussion could depart from a generic set of value dimensions—for instance those found in the competing framework as well as those found in the research by Schwartz. By highlighting different potential tradeoffs and dilemmas, people participating in such discussions can hopefully develop an increased awareness about safety aspects inherent also in values that are not obviously safety related but nevertheless may influence safety.

The general problem of how to promote safety interventions in organization is of course of crucial importance and numerous different suggestions can be found in the literature. I shall not here attempt to review this work but just pick one example of research which I find particularly relevant in the present context.

Hale, Guldenmund, van Loenhout, and Oh (2010) reviewed safety culture and safety management interventions to improve safety. They found that the overall success rate of 17 reviewed projects was about 50%. Although methodological problems make it difficult to generalize, some general findings are worth noticing. First, successful projects had twice as many independent interventions as the unsuccessful projects, Hale et al., interpret this finding as the importance of creating a critical mass of "input energy" to withstand general resistance associated with organizational change. In terms of the framework discussed here this finding could also be interpreted to mean that safety values must be recognized across the whole organization to have a real influence in the competition from other values. Secondly, successful projects (several performance measures were used including accident data, unsafe behavior etc.) had very strong support (managers, dedicated coordinators, safety professionals, etc.). Interestingly, the research of Hale et al. also found indications that the amount of pressure facing the organizations (investment stops, reorganizations, economic climate, etc.) did not discriminate between successful and nonsuccessful safety projects. Such a tendency, if valid in larger samples, would perhaps speak against the hypothesis introduced in the beginning of this chapter i.e. that more pressures, goal, values, etc. would make it more difficult to focus on safety values in organizations. On the other hand, the research by Hale et al. also indicated that the successful projects were characterized by a planned and systematic approach: "So, although it seems important to have many interventions, it is also important not to let them tumble over each other and create confusion. Order and breaking the safety problem into manageable chunks to drive action ensures that actions can be coordinated and followed up (p. 1034)". Also worth noting from this research is that the training efforts that were most successful had focus on upper management training.

In conclusion, the research performed by Hale et al. points to the importance of management support that opened up for dialogues about safety with the workforce. In this context, a free communication about experienced risks was found to be important. In view of the framework suggested here, the results indicate that safety values should be endorsed by management and spread to the workforce by means of active communication and empowerment—something that is perhaps self evident but nevertheless worth repeating. What I call "competence" (for safety) must comprise both management and workforce and upper management training seems as a necessary (but not sufficient) factor for spreading safety values in organization. A free flow of communication and information regarding safety-related matters creates what Reason (1997) refers to as an informed and a reporting culture. Information about what I call safety quality is also one important piece of information that should attract management attention. This means, that concrete reports about how the safety support is experienced in different processes/functions must reach management attention which, in turn, opens for dialogues about what can be done to improve weak spots in the safety support. But instead of burying such factors in general holistic safety climate measures it seems more pragmatic to develop specific safety quality tools for different professional groups.

13.6 GENERAL DISCUSSION

Radiological safety is deeply interwoven with ethics. Ethics has to do with values. Safety culture and values are related. The meaning of the term "safety culture", I believe, is usually intuitively understood as reflecting some shared property among people concerning assumptions, values, norms, attitudes, and behavior associated with safety. But because the concept of safety itself is seldom defined more precisely, but instead taken for granted in much discussion about safety culture, together with a "systemic ambition" in much safety culture research, then one might question many safety culture assessment strategies as being rather vague. Moreover, an important issue concerns if safety culture really should be conceived of as a "systemic" concept? Some would surely answer affirmative, other are not so sure. Those leaning toward the systemic side usually back up their position with platitudes such as: "culture tells what an organization IS and not what it HAS". But what does that mean? If the concept of organizational culture includes everything imaginable associated with organizations, then what is not culture? Some organizational culture frameworks have tried to escape this trap by suggesting "layers" of which some are perceived as representing more fundamental culture than others. For example, in Schein's model (Schein, 1985) organizational culture is discussed in terms of what he calls basic assumptions at the bottom and deepest level, over what he calls espoused values and finally artifacts at the surface and most visible level.

Indeed, the framework suggested here share a similarity with the framework suggested by Schein. Both frameworks seek to abstract different facets of the more general concept of "organizational culture". However, Schein's model, it seems to me, has sometimes been misunderstood in the sense of attributing what he calls artifacts as necessarily reflecting a specific organizational culture. This is not what Schein means—various artifacts may or may not reflect a specific organizational culture, it depends. For instance, technology itself is an artifact but its design may rather reflect state of affairs (beliefs. values, norms etc.) of the designer's organization rather of those that operate the technology. What Schein calls basic assumptions is not applied in the present framework but such assumptions may nevertheless associate with values. Here I think it might be appropriate to differentiate between the *ground* and the *basis* for values (Kaplan, 1964). A value has a ground in the sense that a given value may be justified by various reasons but this is not the same as the basis for a given value. The basis is about what causes it. In Schein's model this basis is described as "basic assumptions" e.g. those taken-for-granted (and often unconscious beliefs), which causes people to hold various values. In fact, Shein makes a point in that people generally do not consciously know what causes values to exist in an organization.

It seems to me, though, that there might be many causal explanations for embracing certain values, some of these causes might well be in the realm of basic more or less conscious assumptions but others might reflect moral considerations of a much more deliberate nature. Some values might for example be better explained as a result of deliberate risk-taking and/or a certain stance with respect to moral issues rather than be understood by some deep taken-for-granted collective cognitive belief structures. In any case, what is important in the here suggested framework is that there seems to be reasons for separating a value dimension from a more concrete quality dimension when using the concept of safety culture. One may argue that this is what has motivated the distinction between safety culture and safety climate where the later concerns people's perceptions and expressed attitudes of various objects in their work environment (including management attention to safety, safe behavior, training, meetings, policy, resources etc.). Culture on the other hand is often assumed to reflect some deeper level of beliefs and basic values. However, in the literature this separation between safety culture and safety climate is seldom clear. By lifting out safety quality as a separate concept this can lead to a more focused discussion of safety in terms of values and in terms of supporting (safety) functions.

Competence for safety can be perceived as an interface between values and quality—it is by knowing the risks and acting upon them that safety may be strengthened. Also by being explicit about (safety) competence, this important dimension becomes more salient in safety culture talk. For example, observing that people do not follow rules could mean that safety values have low priority, that a low safety quality of some reason forces people to take shortcuts, that training is less than adequate, etc. A score on a safety climate assessment investigating "perceptions" or "attitudes" to rule following may be a starting point for an analysis of such deeper causes, but there is also a risk that various behavior modification techniques are used to promote change but without fostering a deeper understanding of the reasons for particular behaviors. Also, the conceptual separations made between safety culture, safety competence, and safety quality may make it easier to avoid collective moralistic attributions (Rollenhagen, 2010).

In conclusion: rather than endless debates over what safety culture and safety climate really means one can reduce some of the problem by using the distinction between safety culture and safety quality and let the competence factor serve as an interface between the two. Moreover, more research needs to be done that explore ideas around values (at different levels; national, branch, organizational, personal) and how these values interact to influence competence and safety quality.

Finally, let us now return to the hypothesis (H1) presented at the beginning of this paper, that is:

H1: It is not the value of safety perceived as an absolute attribute in the matrix of organizational values that determines the strength of safety culture, but rather how the value of safety relates to the manifold of other values simultaneously strived for.

What has the discussion above led us regarding H1? First, there is evidence confirmed with common observations, that people in organizational cultures harbor many different values and some of these may be in conflict. To understand safety culture as a phenomenon (in terms of values and competence) in solitude from other values is therefore a misconception and misuse of the concept of safety culture. Secondly, due to the existence of many values and goals an organization faces, we need some strategy to *protect safety values* from a too strong influence from some other values that may cannibalize on the safety values. How can this be achieved? Here are some suggestions:

- Development of a professional safety ethics for personnel in safety critical organizations may be a one step to protect safety values.
- Strong independent safety departments have the benefit of presenting an organizational platform for protecting safety values.
- Indicators can be developed that measure *the amount* of nonsafety-directed goals, projects, etc. that exist at a certain time in an organization. Under the assumptions that an organization has a very large amount of such goals, projects, etc. in operation, and that these may draw attention away from safety values, we then might have a useful (but paradoxical it may seem) *safety indicator.*
- When accident investigations are made, care should be taken to explore how many simultaneous goals, values, etc. were in operation when the event happened.

REFERENCES

Adhikari, S., Bayley, C., Bedford, T., Busby, J. S., Cliffe, A., Devgun, G., et al. (2008). *Human reliability analysis: a review and a critique*. Booth Street West, Manchester M15 6PB: Manchester Business School.

Advisory Committee on the Safety of Nuclear Installations (ACSNI). (1993). *Study group on human factors, third report: Organising for safety*. London: HMSO.

Allport, G. W. (1961). *Pattern and growth in personality*. New York: Holt, Rinehart & Winston.

Arboleda, A., Morrow, P. C., Crum, M. R., & Shelley, M. C., II (2003). Management practices as antecedents of safety culture within the trucking industry: similarities and differences by hierarchical level. *Journal of Safety Research*, *34*(2), 189–197.

Aven, T. (2003). *Foundation of risk analysis and decision oriented perspective*. Chichester: John Wiley and Sons.

Boring, R.L., (2007). Dynamic human reliability analysis: benefits of simulating human performance. European Safety and Reliability Conference (ESREL 2007). INL/CON-07-12773. Idaho National Laboratory.

Brown, R. L., & Holmes, H. (1986). The use of a factor analytic procedure for assessing the validity of an employee safety climate model. *Accident Analysis and Prevention*, *18*(6), 455–470.

Campbell, J. P. (1977). On the nature of organizational effectiveness. In P. S. Goodman & J. M. Pennings (Eds.), *New perspectives on organizational effectiveness* (pp. 13–55). San Francisco: Jossey-Bass.

Chen, A. Y. S., Sawyer, R. B., & Williams, P. F. (1997). Reinforcing ethical decisions through corporate culture. *Journal of Business Ethics*, *16*, 855–865.

Cheyne, A., Cox, S., Oliver, A., & Tomás, J. M. (1998). Modelling employee attitudes to safety. *Work and Stress*, *12*, 255–271.

Cohen, A. (1977). Factors in successful occupational safety programs. *Journal of Safety Research*, *9*, 168–178.

Colley, S.K., Lincolne, J., Neal, A., (2013). An examination of the relationship amongst profiles of perceived oraganizational values, safety climate and safety outcomes.

Collins, A. M., & Gadd, S. (2002). *Safety culture: A review of the literature*. Sheffield: Health and Safety Laboratory, Human Factors Group.

Davoudian, K., Wu, J. S., & Apostolakis, G. (1994a). Incorporating Organizational Factors into Risk Assessment Through the Analysis of Work Processes. *Reliability Engineering and System Safety*, *45*, 85–105.

Davoudian, K., Wu, J. S., & Apostolakis, G. (1994b). The work process analysis model (WPAM). *Reliability Engineering & System Safety*, *45*, 107–125.

Dedobbeleer, N., & Beland, F. (1991). A safety climate measure for construction sites. *Journal of Safety Research*, *22*, 97–103.

DeJoy, D. M., Schaffer, B. S., Wilson, M. G., Vandenberg, R. J., & Butts, M. M. (2004). Creating safer workplaces: assessing the determinants and role of safety climate. *Journal of Safety Research*, *35*, 81–90.

Diaz, R. I., & Cabrera, D. D. (1997). Safety climate and attitude as evaluation measures of organizational safety. *Accident Analysis and Prevention*, *29*(5), 643–650.

Donald, I., & Canter, D. (1994). Employee attitudes and safety in the Chemical Industry. *Journal of Loss Prevention in the Process Industries*, *7*, 203–208.

Eiff, G., 1999. Organizational Safety Culture. Proceedings of the Tenth International Symposium on Aviation Psychology, Columbus.

Embrey, D. E. (1992). Incorporating management and organizational factors into probabilistic safety assessment. *Reliability Engineering and Systems Safety*, *38*, 199–208.

Feather, N. T. (1995). Values, valences and choice: the influence of values on the perceived attractiveness and choice of alternatives. *Journal of Personality and Social Psychology*, *68*, 1135–1151.

French, S., Bedford, T., Pollard, S. J. T., & Soane, E. (2011). Human reliability analysis: a critique and review for managers. *Safety Science*, *6*, 753–763.

Garavan, T. N., & O'Brien, F. (2001). An investigation into the relationship between safety climate and safety behaviors in Irish organizations. *Irish Journal of Management*, *22*(1), 141–170.

Gertman, D. I., & Blackman, H. S. (1994). *Human reliability and safety analysis data handbook*. New York: John Wiley.

Glendon, A. I., & Stanton, N. (2000). Perspectives on safety culture. *Safety Science*, *34*(3), 193–214.

Gollwitzer, P. M. (1996). The volitional benefits of planning. In P. M. Gollwitzer & J. A. Bargh (Eds.), *The psychology of action*. New York: Guilford.

Guldenmund, F. W. (2000). The nature of safety culture: a review of theory and research. *Safety Science*, *34*, 215–257.

Guldenmund, F. W. (2007). The use of questionnaires in safety culture research – an evaluation. *Safety Science*, *45*, 723–743.

Hale, A. R., Guldenmund, F. W., van Loenhout, P. L. C.H., & Oh, J. I. (2010). Evaluating safety management and culture interventions to improve safety: effective intervention strategies. *Safety Science*, *48*, 1026–1035.

Hofmann, D. A., & Stetzer, A. (1996). A cross-level investigation of factors influencing unsafe behaviours and accidents personnel. *Psychology*, *49*, 307–339.

Hofstede, G. (1980). *Culture´s consequences: International differences in work related values*. Beverly Hills, CA: Sage.

Hollnagel, E. (1993). *Human reliability analysis: Context and control*. London: Academic Press.

Hollnagel, E. (1998). *Cognitive reliability and error analysis method – CREAM*. Oxford: Elsevier Science.

Howard, L. W. (1998). Validating the competing values model as a representation of organizational cultures. *International Journal of Organizational Analysis*, *6*(3), 231–250.

Inglehart, R. (1997). *Modernization and postmodernization*. Princeton, NJ: Princeton University Press.

INSAG. (1991). *Safety culture*. IAEA. Safety Series No 75-INSAG-4.

Johnson, S. E. (2007). The predictive validity of safety climate. *Journal of Safety Research*, *38*(5), 511–521.

Kaplan, A. (1964). *The conduct of inquiry: Methodology for behavioral science*. San Francisco, CA: Chandler. p.428.

Kennedy, R., & Kirwan, B. (1998). Development of hazard and operability-based method for identifying safety management vulnerabilities in high risk system. *Safety Science*, *30*, 249–274.

Kirwan, B. (1998). Safety management assessment and task analysis – a missing link? In M. Baram & A. Hale (Eds.), *Safety management: The challenge of change* (pp. 67–91). Pergamon.

Kirwan, B. (2008). *Human reliability assessment. Encyclopedia of quantitative risk analysis and assessment*. John Wiley and Sons, Ltd.

Kletz, T. (2006). *Hazop and hazan* (4th ed.). Taylor & Francis.

Kluckhohn, C. (1951). Values and value-orientations in the theory of action: an exploration in definition and classification. In T. Parsons & E. Shils (Eds.), *Toward a general theory of action* (pp. 388–433). Cambridge, MA: Harvard University Press.

Kohn, M. L. (1969). *Class and conformity*. Homewood, Il: Dorsey Press.

Kongsvik, T., Almklov, P., & Fenstad, J. (December 2010). Organisational safety indicators: some conceptual considerations and a supplementary qualitative approach. *Safety Science*, *48*(10), 1402–1411.

Lee, T., & Harrison, K. (2000). Assessing safety culture in nuclear power stations. *Safety Science, 34*(1–3), 61–97.

Leveson, N. (2004). A new accident model for engineering safer systems. *Safety Science, 42*(4), 237–270.

Mearns, K., Whitaker, S. M., & Flin, R. (2001). Benchmarking safety climate in hazardous environments: a longitudinal, interorganizational approach. *Risk Analysis, 21*.

Melnick, E. L., & Everitt, B. S. (2008). *Encyclopedia of quantitative risk analysis and assessment.* Chichester: John Wiley and Sons.

Mohaghegh, Z., & Mosleh, A. (2009). Incorporating organizational factors into probabilistic risk assessment of complex socio-technical systems: principles and theoretical foundations. *Safety Science*. http://dx.doi.org/10.1016/j.ssci.2008.12.008.

Morgan, G. (1986). *Images of organisations*. London: Sage Publications.

Morris, C. W. (1956). *Varieties of human value*. Chicago: University of Chicago Press.

Neal, A., Griffin, M. A., & Hart, P. M. (2000). The impact of organizational climate on safety climate and individual behaviour. *Safety Science, 34*, 99–109.

Niskanen, T. (1994). Safety climate in the road administration. *Safety Science, 17*(4), 237–255.

Ogbonna, E., & Harris, L. (2000). Leadership style, organizational culture and performance: empirical evidence from UK companies. *International Journal of Human Resources Management, 11*(4), 766–788.

Ostrom, L., Wilhelmsen, C., & Daplan, B. (1993). Assessing safety culture. *Nuclear Safety, 34*, 163–172.

O'Toole, M. (2002). The relationship between employees' perceptions of safety and organizational culture. *Journal of Safety Research, 33*(2), 231–243.

Parker, S. R., Brown, R. K., Child, J., & Smith, M. A. (1977). *The sociology of industry* (3rd ed.). London: Georg Allen and Unwin.

Parker, S. K., Turner, N., & Griffin, M. A. (2003). Designing healthy work. In D. A. Hoffman & L. E. Tetrick (Eds.), *Health and safety in organisations: A multilevel perspective* (pp. 91–130). California: Jossey-Bass.

Pate-Cornell, M. E., & Murphy, D. M. (1996). Human and management factors in probabilistic risk analysis: the SAM approach and observations from recent application. *Reliability Engineering and System Safety, 53*(2), 115–126.

Pidgeon, N. (2010). System thinking, culture of reliability and safety. *Civil Engineering and Environmental Systems, 27*(3), 211–217.

Pidgeon, N., & O'Leary, M. (2000). Man-made disasters: why technology and organizations (sometimes) fail. *Safety Science, 34*, 15–30.

Quinn, R. E. (1988). *Beyond rational management: Mastering the paradoxes and competing demands of high performance*. San Francisco, CA: Jossey-Bass.

Quinn, R. E., & Rohrbaugh, J. (1981). A competing values approach to organizational effectiveness. *Public Productivity Review, 5*, 122–140.

Quinn, R. E., & Rohrbaugh, J. (1983). A spatial model of effectiveness criteria: towards a competing values approach to organizational analysis. *Management Science, 29*, 363–377.

Quinn, R. E., & Spreitzer, G. M. (1991). The psychometrics of the competing values culture instrument and an analysis of the impacts of organisational culture on quality of life. *Research in Organizational Change and Development, 5*, 115–142.

Rasmussen, J. (1997). Risk management in a dynamic society: a modelling problem. *Safety Science, 27*(2–3), 183–213.

Reason, J. (1997). *Managing the risks of organizational accidents*. Aldershot: Ashgate Publishing Limited.

Reiman, T., & Rollenhagen, C. (2012). Competing values, tensions and trade-offs in management of nuclear power plants. *Work*, *41*, 722–729.

Roelen, A. L. C., Lin, P. H., & Hale, A. R. (January 2011). Accident models and organisational factors in air transport: the need for multi-method models. *Safety Science*, *49*(1), 5–10.

Rokeach, M. (1973). *The nature of human values*. New York: Free Press.

Rollenhagen, C. (2010). Can focus on safety culture become an excuse for not rethinking design of technology? *Safety Science*, *48*, 268–278.

Rousseau, D. M. (1990). Assessing organizational culture: the case of multiple methods. In B. Schneider (Ed.), *Organizational climate and culture* (pp. 17–43). San Francisco, CA: Jossey-Bass.

Rundmo, T. (1996). Associations between risk perception and safety. *Safety Science*, *24*, 197–209.

Rundmo, T., & Hale, A. (2003). Managers' attitudes towards safety and accident prevention. *Safety Science*, *41*, 557–574.

Schein, E. H. (1985). *Organizational culture and leadership. A dynamic view*. San Francisco: Jossey-Bass.

Schwartz, S. H. (1992). Universals in the content and structure of values: theory and empirical tests in 20 countries. In M. Zanna (Ed.), *Advances in experimental social psychology* (Vol. 25, pp. 1–65). New York: Academic Press.

Schwartz, S. H. (1994). Are there universal aspects in the content and structure of values? *Journal of Social Issues*, *50*, 19–45.

Schwartz, S. H. (2006). Basic human values: theory, measurement, and applications. *Revue Française de Sociologie*.

Schwartz, S. H., & Boehnke, K. (2004). Evaluating the structure of human values with confirmatory factor analysis. *Journal of Research in Personality*, *38*(3), 230–255.

Schwartz, S. H., Melech, G., Lehmann, A., Burgess, S., & Harris, M. (2001). Extending the cross-cultural validity of the theory of basic human values with a different method of measurement. *Journal of Cross-Cultural Psychology*, *32*, 519–542.

Schwartz, S. H., Sagiv, L., & Boehnke, K. (2000). Worries and values. *Journal of Personality*, *68*, 309–346.

Seo, D. C., Torabi, M. R., Blair, E. H., & Ellis, N. T. (2004). A cross validation of safety climate scale using confirmatory factor analytic approach. *Journal of Safety Research*, *35*, 427–445.

Sheridan, J. E. (1992). Organizational culture and employee retention. *Academy of Management Journal*, *35*(5), 1036–1056.

Silva, S., Lima, M. L., & Baptista, C. (2004). OSCI: an organizational and safety climate inventory. *Safety Science*, *42*(3), 205–220.

Simonds, R. H., & Shafari-Sahrai, Y. (1977). Factors apparently affecting injury frequency in eleven matched pairs of companies. *Journal of Safety Research*, *9*, 120–127.

Smith, M. J., Cohen, A., Cohen, H. H., & Cleveland, R. S. (1978). Characteristics of successful safety programs. *Journal of Safety Research*, *10*, 5–15.

Swain, A. D., & Guttman, H. E. (1983). *Handbook of human reliability analysis with emphasis on nuclear power plant application*. USNRC. NUREG/CR-1278.

Tellis, G. J., Prabhu, J. C., & Rajesh, K. C. (2009). Radical innovation across nations: the preeminence of corporate culture. *Journal of Marketing*, *73*(1), 3–23.

Turner, N., Chmiel, N., & Walls, M. (2005). Railing for safety: job demands, job control and safety citizenship. *Journal of Occupational Health Psychology*, *10*(4), 504–512.

Varonen, U., & Mattila, M. (2000). The safety climate and its relationship to safety practices, safety of the work environment and occupational accidents in eight wood-processing companies. *Accident Analysis and Prevention*, *32*, 761–769.

Vredenburgh, A. G. (2002). Organizational safety: which management practices are most effective in reducing employee injury rates? *Journal of Safety Research*, *33*(2), 259–276.

Wright, C. (1986). Routine death: fatal accidents in the oil industry. *Sociological Review*, *34*(1), 265–289.

Zohar, D. (1980). Safety climate in industrial organizations: theoretical and applied implications. *Journal of Applied Psychology*, *65*(1), 96–102.

Zohar, D. (2000). A group-level model of safety climate: testing the effect of group climate on microaccidents in manufacturing jobs. *Journal of Applied Psychology*, *85*(4), 587–596.

Zohar, D. (2002a). Modifying supervisory practices to improve subunit safety: a leadership-base intervention model. *Journal of Applied Psychology*, *87*(1), 587–596.

Zohar, D. (2002b). The effects of leadership dimensions, safety climate, and assigned priorities on minor injuries in work groups. *Journal of Organizational Behaviour*, *23*, 75–92.

Zohar, D. (2008). Safety climate and beyond: a multi-level multi-climate framework. *Safety Science*, *46*, 376–387.

Zohar, D. (2010). Thirty years of safety climate research: reflections and future directions. *Accident Analysis and Prevention*, *42*, 1517–1522.

Zohar, D., & Luria, G. (2004). Climate as social-cognitive construction of supervisory practices: scripts as proxy of behaviour patterns. *Journal of Applied Psychology*, *89*, 322–333.

Zohar, D., & Luria, G. (2005). A multilevel model of safety climate: cross level relationships between organization and group-level climates. *Journal of Applied Psychology*, *90*, 4,616–4,628.

Part IV

Proliferation and the Nuclear Fuel Cycle

Chapter 14

The Legacies of Soviet Nuclear Testing in Kazakhstan

Fallout, Public Health and Societal Issues

Susanne Bauer*[a], Boris Gusev[b], Tatyana Belikhina[b], Timur Moldagaliev[b,c], and Kazbek Apsalikov[b]

[a]*Goethe University Frankfurt, Department of Social Sciences, Frankfurt/Main, Germany*
[b]*Research Institute for Radiation Medicine and Ecology, Semei, Kazakhstan*
[c]*Semei Medical University, Kazakhstan*
**Corresponding author: E-mail: bauer@soz.uni-frankfurt.de*

Chapter Outline

14.1 The Semipalatinsk Nuclear Test Site 244
14.2 Research into Radiation Effects of Nuclear Testing Near Semipalatinsk during Soviet Time 245
14.3 Radiation Risk Research in Kazakhstan during the Early Post-Soviet Years 248
14.4 Addressing Nuclear Legacies in the New Economy of Kazakhstan 253
14.5 Conclusions 255

This chapter reports on the legacies of nuclear testing at the Semipalatinsk nuclear test site, Kazakhstan. Drawing on selected on-going local studies, we discuss current frameworks and conditions for the documentation of fallout effects in Kazakhstan. Research into the local health impact of nuclear fallout in the populations living in adjacent areas has remained scarce for most nuclear test sites worldwide. After 1991 the new independent states of the former Soviet Union implemented programs to address and document the impact of radiation exposures resulting from Soviet nuclear activities. In the Chelyabinsk area, Southern Urals, there was considerable contamination as a result of the Kyshtym accident in 1957 and due to the Mayak plutonium production (Burkart & Kellerer, 1994; Polikarpov & Aarkrog, 1993). The nuclear industries in Southern Urals have led not only to exposures among workers, but also to unprecedented internal exposures due to radioactive

Radioactivity in the Environment, Volume 19
ISSN 1569-4860, http://dx.doi.org/10.1016/B978-0-08-045015-5.00014-9

discharges of the plant into the Techa river, which served as the water supply to downstream villages.

The development of the Soviet nuclear program after World War II involved uranium mining through the Wismut AG in East Germany. From there, the uranium ores were transported to the Soviet Union and, in part, processed at the Mayak plutonium facilities in Southern Urals. Nuclear warheads were developed in Arzamas-16 and prepared for nuclear testing at the two major test sites Semipalatinsk and Novaia Zemlia. One end of this Cold War nuclear production chain was the Semipalatinsk polygon (test site) located in the steppe areas of the Soviet Republic of Kazakhstan. Even though the Central Asian steppe was sparsely populated, numerous settlements and a few smaller towns with 10,000s of inhabitants were within a few hundreds of kilometers distance to the epicenters. The population of Semipalatinsk city, located 170 km east of the polygon, was about 116,000 in 1949 (Gusev, Sekerbaev, Rozenson, Tchaijunusova, & Apsalikov, 1998). Moreover, numerous summer and winter pastures of collective farms that had replaced the formerly nomadic livestock economies were located in the areas that were heavily exposed due to atmospheric nuclear testing between 1949 and 1962.

In the early years after the dissolution of the Soviet Union, formerly classified information on nuclear issues became available for the first time. In addition to a veritable information boom in the public sphere, governments of the new independent states began to retrospectively estimate cumulative fallout doses of the populations in the exposed areas. With the end of the Cold War division that had shaped nuclear science west and east of the iron curtain, much of this research was practiced in novel constellations of international cooperation and shifted to issues of management and mitigation of nuclear legacies. Since there were different practices and assumptions of what counted as proof of exposure in post-Cold War radiation science, epidemiological risk assessments in particular were done in different ways in Soviet and western traditions. Epidemiology was not the only discipline involved in risk assessment, as a host of physical and biological methods of dose reconstruction and biomarker measurement had evolved.

The study of health effects of radiation takes place through large-scale epidemiological studies, including studies of medical and occupational exposures. Largely however, this line of research has built on studies of the atomic bomb survivors in Hiroshima and Nagasaki. The follow-up of the atomic bomb survivors was the largest epidemiological study at the time. Its core aim was to derive quantitative radiation risk estimates and study the dose–response relationship of effects in this exposed population. Since the late 1940s until the mid-1970s, these studies were conducted first by the Atomic Bomb Casualty Commission (ABCC) and then by the successor organization, the Radiation Effects Research Foundation (RERF) in Hiroshima. In the 1950s and 1960s medical studies on the bomb survivors in Japan focused on genetic effects of radiation (Lindee, 1994) and later on increases in cancer risk among the bomb survivors, as for instance in Pierce, Shimizu, Preston,

Vaeth, and Mabuchi (1996). These studies have become standard references for any investigation of radiation risk including those conducted on Soviet nuclear legacies after 1991. Moreover, current regulations and dose limits for occupational exposures are based largely on the results from the studies of atomic bomb survivors.

When information on the nuclear facilities became accessible to international research, western researchers hoped to complement and validate current risk estimates through epidemiological studies of the exposures in the Soviet Union. Especially with regard to nonuniform exposures in the intermediate and low dose range, scientists emphasized the uncertainties of extrapolations from the high acute external exposures of the atomic bomb survivors to the lower doses. Thus, the importance of "learning" from exposures in the Soviet Union was stressed, framing these projects as "unique opportunities" for radiation biology (Burkart, 1996). Of particular interest to radiation research were the exposures involving chronic internal exposures and thus different from Hiroshima and Nagasaki. Largely with the rationale of improving current risk estimates and regulation, funding for collaborative research was made available by the European Union as well as by the US National Institutes of Health, Japanese agencies, and as part of many bilateral agreements during the 1990s. At first, international studies of nuclear exposures in the Soviet Union concentrated on the health impact of Chernobyl, before scientists also trained their attention to exposures due to the Southern Urals nuclear facilities and the fallout from nuclear testing near Semipalatinsk.

This chapter seeks to describe the current documentation efforts of the public health impact in contemporary Kazakhstan as well as post-Soviet mitigation projects following the closing of the Semipalatinsk test site in 1991. While little information on radiation and health was openly available on the radiation exposures in the Soviet Union until the late 1980s, research into health effects took place since the 1950s in closed, military institutions, the Institute of Biophysics and related institutions in Southern Urals and in Semipalatinsk. Today's Institute of Radiation Medicine and Ecology in Semei (formerly Semipalatinsk) is based on one of these institutes, the Dispanser No. 4, as the institute was called when founded in 1957, in charge of ecological and health monitoring in the areas surrounding the nuclear test site. Focusing on research of the Institute of Radiation Medicine and Ecology, we review studies on local health effects and describe the implementation of epidemiological research in Soviet and the early post-Soviet era as well as in today's Kazakhstan. Rather than conducting a review of risk assessments, we attend to the changing conditions of knowledge production in the Soviet and post-Soviet context. Thus, this chapter provides an insight into current assessments of the long-term impacts, including societal issues and psychosocial effects and the local public health consequences of radiation exposure in the past.

14.1 THE SEMIPALATINSK NUCLEAR TEST SITE

Construction works for the first nuclear test site began after a Decree of the Council of Ministers of 19 June 1947. On the northeastern end of the future test site in the central Asian steppe areas west of Semipalatinsk, a closed nuclear weapons research city was built. It was named Kurchatov, after the Soviet atomic physicist and engineer; in command of Stalins's atomic bomb project was the chief of the secret police Lavrentii Beria (Gordin, 2009). The town of Kurchatov did not appear on maps except for a train station named "konechnaia" (terminal). The test site (polygon) comprised an area of approximately 19,000 km^2 west of the city of Semipalatinsk (now: Semei), situated in the northeastern part of Kazakhstan, bordering the Altai region, the Mongolian Republic, and China. More than 110 atmospheric and 340 underground tests were conducted near Semipalatinsk between 1949 and 1989 (Iakubovskaia, Nagibin, & Suslin, 1998; Mikhailov et al., 1996; Grosche 2002). A total of 715 Soviet nuclear tests were carried out between 1949 and 1990, with 456 and 130 near Semipalatinsk and on Novaia Zemlia, respectively[1]. In addition to the nuclear tests at the two test sites, 129 nuclear detonations took place outside these test sites—in the Russian Federation (91), in Kazakhstan (33), in Ukraine (2), in Uzbekistan (2), and in Turkmenistan (1) (Mikhailov et al., 1996). Some of these nuclear explosions were conducted for nonmilitary purposes, for instance for construction and the planned large-scale irrigation projects.

The first nuclear device, an above ground fission bomb that was exploded on August 28, 1949, resulted in fallout northeast of the Semipalatinsk test site. Due to weather conditions prevailing at the time of the explosions, among the administrative regions adjacent to test site, the Semipalatinsk and Ust-Kamenogorsk regions of the Kazakh Soviet Republic were affected most. Fallout also extended to the southwestern parts of the Altai region, an administrative unit, now part of the Russian Federation, located north-east at approximately 200 km from the borders of the test site.[2] Atmospheric nuclear tests conducted from 1949 until 1963 caused considerable radiation exposures in the proximity of the test site. In particular, the explosions in 1949, 1951, 1953, 1956, and 1962 contributed major fractions of the cumulative radiation dose in villages adjacent to the test site (Bauer, Gusev, Pivina, Apsalikov, & Grosche, 2005; Gusev,

1. At the second nuclear test site on Novaia Zemlia atmospheric nuclear tests were conducted between 1955 and 1962, with the most contaminating nuclear detonation on September 7, 1957 resulting in exposure of reindeer herds and consumers of reindeer meat (Burkart et al., 1999; UNSCEAR, 2000).

2. In the early 1990s the Russian Ministry of Emergency Situations set up a Federal Program "Semipalatinsk Test Site/Altai". The program included the compilation of health data and dose reconstruction as well as a registry of the rural population of Altai region settlements (Shoikhet, Kiselev, Algazin, et al., 2002; Shoikhet, Kiselev, Loborev, et al., 2002; Shapiro, Kiselev, & Zaitsev, 1998; Shoikhet et al., 1999). These exposures northeast of the Kazakhstan territory were largely due to fallout from the first Soviet nuclear test in 1949.

Rozenson, & Abylkassimova, 1998; Gusev, Sekerbaev, et al., 1998). The first Soviet thermonuclear bomb (yield: 400 kt TNT equivalent) of August 12, 1953 was the only time when a part of the downwind settlements were evacuated to other nearby areas, a measure that did not prevent exposure. After a preliminary moratorium on atmospheric nuclear tests of 1963, underground testing continued through 1989, with the last explosion carried out on October 17th, 1989 (Mikhailov et al., 1996). Several underground explosions resulted in craters, e.g. those on January 15 and October 14, 1965, and also produced a radioactive cloud and near-surface contamination (Izrael et al., 2000).

The nuclear test site was officially closed by Decree of president Nursultan Nazarbaev on August 29, 1991, 42 years after the first Soviet atomic bomb test. The scientific institutes in Kurchatov on the former test site, with several research reactors, were integrated into a reorganized National Nuclear Center that has continued operations now focusing on nuclear energy, nuclear technologies, radiological monitoring, and risk assessment.

14.2 RESEARCH INTO RADIATION EFFECTS OF NUCLEAR TESTING NEAR SEMIPALATINSK DURING SOVIET TIME

While there was secrecy around the nuclear program and radiation risks, epidemiological research in the Soviet Republic of Kazakhstan was conducted mostly by the "Dispanser No. 4", a medical research unit in Semipalatinsk that reported to the Institute of Biophysics. Yet, the first biomedical investigation of health effects was carried out by the Institute for Regional Pathology through what was later referred to as "Balmukhanov-Atchabarov expeditions" (Balmukhanov, Gusev, & Balmukhanov, 2002). The study was carried out between 1956 and 1960 after an "unknown intoxication" had been reported in cattle of the "Thelmann Kolkhoz" located close to the epicenter of nuclear testing (Academy of Sciences of the Soviet Republic of Kazakhstan, 1958). The comprehensive study included measurements of alpha- and beta-activity in soil, plants, and food products, veterinary examination of livestock animals as well as a comparative medical survey of the health status of the populations in rural areas in the Almaty, Karaganda, and Semipalatinsk regions. While in the Alma-Ata region, levels of radiation were found within the range of background values, the situation was different in the Abaiski district of the Semipalatinsk region. Here, radiation levels in soil samples were reported to substantially exceed the Soviet dose limits of 10^{-7} Ci/kg. Further, this early expedition found pathological organ alterations in sheep and goats in areas close to the test site (Academy of Sciences of the Soviet Republic of Kazakhstan, 1958).

The medical part of the study was aimed at a description of the characteristics of regional pathology in terms of disease prevalence. The physicians examined 1635 people aged 4–90 years in five districts of the Semipalatinsk and Almaty regions. The report for the population of the Semipalatinsk region showed high prevalence of hematological alterations and a range of vegetative disorders, disorders of the gastrointestinal tract and respiratory system (Academy of Sciences of the Soviet Republic of Kazakhstan, 1958). These conditions were described as mass phenomena, found in healthy individuals, and also among young people. The report's authors came to the conclusion, after ruling out brucellosis, helminthosis, or vitamin C deficiency as causes for the observed differences between the areas, that the observed pathology could point to "chronic radiation disease" (Academy of Sciences of the Soviet Republic of Kazakhstan, 1958). The report remained classified and Soviet authorities did not support the report's conclusions (Balmukhanov et al., 2002).

In 1957 the Dispanser No. 4—a specialized center for the study of radiation effects—was opened in Semipalatinsk (Nugent, Zhumadilov, Gusev, & Hoshi, 2000). This Dispanser No. 4—often code-named "brucellosis hospital"—was specialized in oncology and radiation medicine. Its main tasks included radiation monitoring, assessment of the health impact due to fallout, and medical follow-up of the population living in areas affected by fallout. From 1962, staff at the Dispanser No. 4 comprised 120 employees, including 42 medical doctors and nurses. In addition to the clinical sections of a ward with 20 beds, physicians carried out regular examinations in eight districts of the Semipalatinsk region. At the time, this work was classified "top secret" and any contact with the military institutions and research on the test site were strictly prohibited. Thus, these activities of health monitoring were completely separated from the military activities of measuring radiation doses throughout the 1960s and 1970s. Focusing on the study of medical effects, the institute broadened its research activities in radiation biology, cytogenetics, and medical follow-up during the 1970s. A first internal report compiled in 1981 compared the highly exposed population (determined by age at exposure) of six heavily exposed settlements (Sarzhal, Kainar, Karaul, Dolon, Kanonerka, Mostik) (5600 persons) with the remaining residents of these settlements (12,000), reporting increased solid cancer rates in the highly exposed group.

It was only with the late 1980s that more information became available. An interdepartmental commission was set up by the Soviet Ministry of Health in 1989 to address public concerns regarding the impact of radiation exposures. This study provided epidemiological documentation for exposed settlements and measured radioactive strontium and cesium in soil samples collected in areas downwind of the test site. It also reported cytogenetic studies of chromosome aberrations among three groups—people living in settlements close to the test site, students at the Medical Academy who had recently moved to the area, and students and faculty born in Semipalatinsk. These cytogenetic examinations carried out at the Institute of Medical Radiology, Obninsk confirmed increased frequencies of aberrations in villages near the test site (Sevan'kaev et al., 1995). These results, together with other early reports of increased rates of cancer and congenital malformations, played an important role in negotiations about further research and compensation programs.[3] Radiation measurements in soil confirmed the presence of alpha-emitters. The commission's report was finalized during Soviet time and recognized that in particular the 1949 and 1953 nuclear tests resulted in considerable contamination due to fallout in areas of Kazakhstan and in the Altai region and that external doses to the adjacent population exceeded Soviet radiation safety limits (Balmukhanov et al., 2002). All these findings would lend support to calls by the government of Kazakhstan for international assistance in the management of nuclear legacies.

Researchers from the Soviet Centre for Oncology of the Medical Academy and the Soviet Health Ministry published a descriptive study of the incidence of malignant tumors in the Semipalatinsk region in 1991 (Bul'bulian & Tokareva, 1991). They compared age-standardized incidence rates by administrative units, according to official records to general rates of the Soviet Union and the Soviet Republic of Kazakhstan between 1959 and 1988. Age-specific incidence rates of malignant tumors were found higher in the Semipalatinsk region than in the Soviet Union or Kazakhstan as a whole, mainly due to oesophageal cancer. Incidence rates of malignant tumors in the lymphatic and hematopoietic system were lower in the Semipalatinsk region in the 1960s and 70s, but reached the level of the Soviet Union and exceeded Kazakhstan levels in the mid-1980s. The observed general increase of malignant neoplasms in the general population was considered to be partly due to improvement of cancer diagnostics and registration standards (Bul'bulian/Tokareva, 1991).

A study of childhood cancer incidence in the administrative regions bordering the test site was conducted by Zaridze, Li, Men, and Duffy (1994). Scientists collected documentation of cancer cases of children aged up to 14 years between 1981 and 1990 from all hospitals in the three regions (Semipalatinsk, Pavlodar, and Karaganda), using census data of 1979 and 1989 as a reference. Statistical tests were conducted for association between incidence and distance to the epicenter of nuclear tests (air and underground test sites, atomic lake). A significant inverse trend with distance to test epicenter was shown for acute leukemia and for all cancers combined. Using distance to the test site as exposure proxy, rates of childhood cancer were found increased closer to the test sites in the Semipalatinsk and (former) East Kazakhstan regions. Yet, some regional variations in the Pavlodar and Karaganda regions, especially for leukemia incidence could not be explained by distance to the test site. Here, Zaridze et al. (1994) discussed an influence of urban/rural status, exposure to chemical carcinogens from Karaganda industrial plants and other factors.

Public debates over radiation risk due to fallout in the late 1980s were followed by a veritable information boom after the closure of the nuclear test site in 1991 and numerous calls to assess the public health impact. Despite the amount of health research conducted over the decades and first attempts to link exposure and disease data, securing epidemiological proof for radiation effects

3. Increased cancer mortality rates (by 40%) were reported in Sarzhal and Kainar when compared to Kokpekty, an area with minimal exposure (Balmukhanov et al., 2002).

remained a challenging task for researchers in Kazakhstan, in particular given the lack of infrastructure and resources for research during the immediate post-Soviet years as well as the inconsistency of information and the difficulties in obtaining to primary documents for dosimetry.

14.3 RADIATION RISK RESEARCH IN KAZAKHSTAN DURING THE EARLY POST-SOVIET YEARS

According to the 1949 census, 1.5 millions of inhabitants were living in the administrative regions adjacent to the territory of the test site at the time of atmospheric nuclear testing. Delineating the areas of exposure after decades of nuclear exposure and reconstructing doses posed challenges: The exposure depended on weather conditions, wind directions that formed the trajectories of the radioactive cloud as well as on radionuclide deposition, shielding, season, and accumulation in the food chain. During Soviet time, estimates of effective doses to the population had been reported in classified internal reports of the Institutes of Biophysics (Stepanenko et al., 2011). In the open literature, the first publication that gave dose estimates (Tsyb et al., 1990) estimated maximal effective doses at 1.6 Sv in Dolon, the settlement with highest exposure in 1949. Most of the early dose reconstructions were based on measurements of gamma–radiation that the military conducted during and after each nuclear test, or on measurements of soil samples conducted that now could be used as additional data for dose estimation.

For decades, these data had also been kept separate from the health studies done by the physicians and their reports to the authorities in Moscow. Apart from descriptive comparisons by exposure status or area, epidemiological linkage of dose estimates and disease rates had not been possible until the 1990s. When the test site was closed, however, much of the primary information including military data were transferred to the Central Physical Technical Institute in Sergiev-Posad near Moscow and access to primary data for dose reconstruction was limited for scientists in Kazakhstan (Balmukhanov et al., 2002).

The staff of the Dispanser No. 4 (now Institute of Radiation Medicine and Ecology) conducted a first exposure mapping in 1991, using soil samples collected in the 1960s and measured for residual radionuclides. As identified by the interdepartmental commission's study, administrative districts were assigned to radiation risk zones based on information on two major dose-contributing detonations of 1949 and 1953 (Gusev, Rozenson, et al., 1998; Gusev, Sekerbaev, et al., 1998). Aimed at a first estimation of the exposed population numbers—relevant for social and public health programs in the Semipalatinsk region—three district-based exposure zones[4] were defined (Gusev, Sekerbaev, et al., 1998).

4. After administrative reforms in 1997, the Semipalatinsk region became part of the East Kazakhstan *oblast'* with the center Ust-Kamenogorsk. Here, we refer to exposed territories and demographic data of the former Semipalatinskaia *oblast'* divided into 15 administrative districts (*raiony*).

This assessment ranked the population of the Abaiskii, Abralinskii, Zhana-Semeiskii, and Beskaragaiskii districts (situated closest (40–100km) to the test site) at highest risk, followed by the districts situated at 150–250km to the test site as "zones of intermediate exposure" and districts situated at distances of 250–800km from the testing area as exposed to low or negligible doses. The first definitions of three zones of radiation risk were carried out in order to plan further research and develop a compensation policy.

The radiation risk zoning was also used to estimate the number of exposed people: According to the first mapping of "radiation risk zones" of the Semipalatinsk region, among a total population of 294,000 in 1949, 46,000 people lived in high exposure districts. 159,000 people lived in areas with intermediate exposure levels and 89,000 people in districts with low radiation exposure (Gusev, Rozenson, et al., 1998; Gusev, Sekerbaev, et al., 1998; Rakhypbekov et al., 1999; Nugent et al., 2000). In the late 1950s and early 1960s, the population of the Semipalatinsk region increased due to immigration from bordering territories of China and relocations from other regions of the former USSR to Kazakhstan, in particular with Khruchchev's agricultural reform that set out to cultivate large steppe areas of central Asia through irrigation. By the end of 1962, the total population of the Semipalatinsk regions had increased to 471,500, with radiation exposures depending on age at exposure and, for migrants to these areas, on the time of their relocation. While most high exposure areas were located in the Semipalatinsk region, areas in the Ust-Kamenogorsk region (now merged into the East Kazakhstan region) needed to be considered as well.

It was this first risk zoning that also entered the "Law on the Social Protection of the Citizens and Victims of the Semipalatinsk Nuclear Test Site" issued in 1992 (Iakubovskaia et al., 1998; Nugent et al., 2000). The law introduced four exposure groups (with estimated dose ranges of >1 Sv, 350mSv–1 Sv, 70–350mSv, 1–70mSv), each of which were assigned different levels of compensation and benefits. The latter allotted small direct payments (depending on exposure category and calculated by a coefficient related to the minimum social welfare), additional paid holidays and retirement at younger age as well as free medical treatment and rehabilitation if indicated. The implementation of the program was postponed throughout the 1990s and amended by additional regulations. As a later result however, people living in exposed areas could demand information on their exposure status as well as claim respective medical care and rehabilitation in case of a radiation-induced disease (such as cancer, thyroid disease, diseases of the hematological system, cardiovascular disease, and congenital malformations).[5]

A retrospective cancer-incidence study compared an exposed area consisting of eight exposed villages and five nonexposed villages of a comparison area in

5. Decree No. 34 "On the approval of the list of diseases associated with exposure to ionizing radiation and rules to establish causal relations between disease and exposure to ionizing radiation" Government of the Republic of Kazakhstan (13 Jan 2004).

the Kokpektinskii district using five year cumulative incidence information of the period from 1949 to 1990 (Gusev, Rozenson, et al., 1998; Gusev, Sekerbaev, et al., 1998). These cross-sectional comparisons describe the temporal development of solid cancer incidence and found a marked increase in cancer rates in the exposed areas at 20 and 40 years after exposure. The study described the shifts in proportions of the different cancer localizations, using a multiple cross-sectional design with inclusion of young age groups at each 5-year interval. The differences between groups were significant for esophagus cancers and for stomach and liver cancers combined. Until 1970, esophagus was the most frequent cancer site with declining rates from the 1970s onward, whereas the incidence of stomach and liver cancers increased over time. The second peak of excess cancer rates was mainly related to high rates of lung, breast and thyroid carcinoma (Gusev, Rozenson, et al., 1998; Gusev, Sekerbaev, et al., 1998). While such local studies both in Kazakhstan as well as in the Altai region focused on cancer patterns in the population at a descriptive level, standard approaches of western cancer epidemiology aim at analytical studies, using individual-level instead of area data to avoid aggregation bias that epidemiologists termed "ecological fallacy". One of the challenges for western epidemiologists working with these data was that they did not fit their standard formats of cohort designs with a person-years approach or of case–control studies.

After the independence of the Kazakhstan Republic and the official closure of the test site, the Kazakhstan government asked for assistance in the assessment and management of its nuclear legacies. The UN General Assembly adopted a resolution on Semipalatinsk at its 52nd session in December 1997, asking for the support the international community in the assessment of the economic, ecological, and humanitarian consequences of nuclear testing in this region. In 1998, a UN needs-assessment mission developed an integrated program of priority projects on health, humanitarian aid, ecology, economic development, and information dissemination. The report to the 53rd session of the General Assembly led to a second resolution on Semipalatinsk, stating the need for international attention, cooperation and coordination of responses to the legacies of the Semipalatinsk test site (United Nations General Assembly, 1998). Since then, international efforts have been intensified that assess the radiological situation on the test site, also involving IAEA and WHO. Many of these international efforts defined as their first step to narrow down and confirm actual doses, which shifted priorities away from health effects to validation of the dosimetric data. Also a range of analytical epidemiological studies were begun in the mid-1990s, striving to enroll and incorporate into the western epidemiological formats the data collected or available from the Soviet records of the public health system and the Dispanser No. 4 (Grosche et al., 2002; Bauer et al., 2005; Bauer, Gusev, Pivina, Apsalikov, & Grosche, 2006; Land et al., 2008; Mudie et al., 2007, 2010; Grosche et al., 2011).

These research projects conducted in international collaboration analyzed the data that had been collected since the 1960s—two historical cohorts set up already during the 1960s under supervision of the Institute of Biophysics. These studies worked with a person-years at-risk approach and individualized dose estimates to examine the association between radiation dose and cancer mortality, taking into account age at exposure, attained age, and time since exposure. Follow-up data (1960–1999) for this historical cohort included 19,545 people in 10 exposed and six nonexposed settlements (Bauer et al., 2005; Grosche et al., 2011). The rate ratios showed substantially increased cancer rates in the exposed group and a significant dose–response effect for all solid cancers. To assess potential bias due to selection or other differences between the two groups and due to differential emigration levels, analyses were also restricted to the exposed cohort and to different times of follow-up to account for losses to follow up, resulting in still significant but lower excess risks (Bauer et al., 2005).

Subsequently, the WHO funded a reproductive health study that was aimed to assess reproductive health outcomes in different exposure categories. For this study, a part of the historical cohort (for high exposure villages and selected birth cohorts) was extended and a complete list of people living in the exposed settlements was compiled, using *kolkhoz* books (*pokhoziaistvennye knigi*) and birth and death certificates (Mudie et al., 2007, 2010). One of the challenges of retrospective analytical studies is to trace the exposure groups, for instance complete birth cohorts in the exposed villages born before and during the time of nuclear testing. The ideal epidemiological constellation of complete follow-up though is difficult to achieve, as the retrospective tracing of complete migration histories and mortality follow-up is hardly feasible given the migration within and out of the former Soviet Union. With a large part of the population, especially those of Russian and German descent, leaving after the dissolution of the USSR, the follow-up of the "exposed cohort" under study would not only extend to Kazakhstan or the Altai region, but also to western Europe, the Middle East, and elsewhere. Efforts to perform vital status follow-up for a retrospectively established exposure registry in the Altai region proved feasible only for approximately 40% of the initial population (Shoikhet et al., 1999). To some extent, mechanisms for epidemiological studies among emigrants from the Soviet Union in Germany were developed (Ronellenfitsch, Kyobutungi, Becher, & Razum, 2006).

Collaboration for the thyroid study reported by Land et al. (2008) was distinct in that by conducting a medical screening of the exposed population, this collaboration generated its own outcome data through examinations conducted jointly by teams from the US National Cancer Institute and the Institute of Radiation Medicine and Ecology. For the first time, this study worked with individualized dose estimates based on a joint Russian/US methodology for the estimation of external doses and internal doses to the thyroid based on the trajectory of fallout deposition and ingestion of radioactive iodine with food products, based on questionnaires on dietary habits. The study showed clear associations of the thyroid nodule prevalence with radiation dose; the size of the effects corresponded to what would be expected based on previous studies on the atomic bomb survivors and medical exposures (Land et al., 2008). This dosimetry system was also used for a later assessment of cardiovascular disease in the historical cohort (Grosche et al., 2011). While validated to a greater extent, the drawback of using this Russian-US-dosimetry-system for other than thyroid studies, however, was its limitation to external doses as an approximation of the total dose due to the complexity of the reconstruction of the internal exposure.

Different from the local research conducted in Kazakhstan and its close connection to a social program, much of the international health studies had little or no connection to mitigation. To funding agencies, international relevance emerged on an entirely different register. Beyond the assessment and documentation of the local health impact, radioepidemiologic studies in the Soviet Union were expected to improve current radiation risk estimates derived from studies in Japanese atomic bomb survivors, occupational studies, e.g. in uranium miners exposed to chronic radiation, and studies on therapeutic radiation at high-dose fractions (UNSCEAR, 2000, 2008). This was relevant due to considerable uncertainties as to the risks due to nonuniform exposure in the intermediate and lower dose ranges. Furthermore, there is heterogeneity between the results that come from different research scales, such as molecular biology or epidemiology.

At the societal level, epidemiological studies translate into public health by indicating possibilities of primary, secondary, and tertiary preventions. In the case of radiation contamination, this implicates to adapt health services to specific needs of people living in areas affected by nuclear fallout. Rural areas adjacent to the polygon had long been underserved in terms of medical facilities, with only very basic healthcare available to its residents. In this situation, epidemiology should be linked to a broader public health perspective aimed at improving healthcare services, counseling and information, particularly in rural and deprived areas. Current concerns range from radioecological safety of farming, implications for the livestock economies and long-term health effects. Environmental health risk assessments need to be generated in a way that takes up local knowledge and is accessible to practitioners, physicians, teachers, educators, social workers, and the general public. Following Coughlin (1996) and Coughlin and Beauchamp (1996), research benefits and burdens need to be distributed equally and, therefore, epidemiological studies should be designed to obtain knowledge that benefits the group of which the epidemiological study subjects are representative. In other words, this requires that those bearing the burden receive an appropriate benefit. Transforming "from a Cold War production mission to that of a peace-time, environmental restoration and cleanup mission" (Reed, Lemak, & Hesser, 1997: p. 633) applied to nuclear weapons cleanup and waste management, as well as to research into health effects of nuclear testing. In that sense, health research should work toward the goal of environmental and transitional justice.

14.4 ADDRESSING NUCLEAR LEGACIES IN THE NEW ECONOMY OF KAZAKHSTAN

Since 2002 additional government funding by the Ministry of Health for a socio-economic program, rehabilitation and health promotion in the areas that were exposed due to nuclear testing enabled the Institute for Radiation Medicine and Ecology to open a clinic and gradually expand clinical work, diagnostics, and research activities and develop monitoring and screening programs. The continued study of radiation and other anthropogenic risk factors and a prognosis of the radiological situation in the regions of Kazakhstan, the reconstruction of effective radiation doses based on empirical models and residual radionuclides in the environment have remained core research directions. Moreover, current activities include development and implementation into practice of diagnostics, therapeutic, and preventive measures against radiation associated disease; screening, treatment, and rehabilitation of the population exposed to radiation and other environmental factors, expert consulting on causal association of diseases with radiation exposure; conducting disease prevention and health promotion among the exposed population. During the past 10 years, the clinical work of the institute has broadened, including methods of individual dosimetry for the calculation of individual effective doses. In the most heavily affected areas of the East-Kazakhstan region, screening examinations have been conducted in order to identify healthcare needs and provide rehabilitation, if indicated.

One of the core projects has been the implementation of an automated registry of the exposed population that has served as the basis for dose assessment as to the legal status with regard to compensation policies. From 2003 to 2005 with support by the Ministry of Health of the Republic Kazakhstan, the institute has implemented and maintains a State Scientific Automated Medical Registry of the Kazakhstani Population Exposed to Radiation. This central database serves as resource for epidemiological research projects. Recently, new lines of research into health effects in descendants born to irradiated parents have been added to the previously conducted epidemiological and dosimetric studies. These include studies among descendents of those exposed to fallout between 1949 and 1962. In East Kazakhstan only, there are about 250,000 descendants—the second and third generation after the directly exposed population. Moreover, studies have been added that extend somatic effects of radiation to include psychological consequences, which so far have remained virtually unstudied. The psychological aspects that are part of living with nuclear legacies—such as the sensual imperceptibility of the danger, expectation of delayed effects on health, uncertainty of individual prognosis—have been conceptualized as challenging the adaptative capacities of the individual.

Attempts to objectify and rank psychological consequences in situations of radiation exposure in Southern Urals and the Chernobyl nuclear power station can serve as comparisons for the Semipalatinsk results, when taking into account the differences in dose formation and demographics of radiation risk groups. Using the exposure registry mentioned above, psychological effects were examined in the second generation of the exposed population. Inclusion criteria were confirmed presence in territories contaminated by radioactive residuals with a dose estimate >200 mSv in the first generation. The control group comprised migrants to exposed areas after termination of the nuclear weapons testing. For this study, information on psychological effects was retrieved from medical documents between 2007 and 2011 showing individual diagnoses for an exposed and a comparison group of 1350 and 1010 individuals, respectively. Among an additional group of people with confirmed functional disorders, a survey was conducted using standard questionnaires on social well-being. Results indicated that key issues of living with nuclear legacies, unlike wars and events of nature, were the potential long-term consequences of exposure in terms of uncertain health and genetic effects. This is a chronic psychological burden that brings about experiences of "not guaranteed" or "hopeless" futures. While preliminary, we consider that those findings can contribute to develop ways to extend rehabilitation and prevention also to psychosocial issues of living with nuclear legacies.

At present, much of the scientific research activities and the entire social and health sector of the region deal with nuclear test consequences. A whole region and its health infrastructure thus has been coshaped by its nuclear legacy, involving a rehabilitation and compensation program, infrastructures for medical research, radiation monitoring and health monitoring. Retrospective assessment and reconstruction of doses among population living in adjacent areas will be continued in order to develop a comprehensive strategy for mitigation of the health impact for the population living in areas with radioecological risks. Key research concerns include the study of somatic and genetic effects among the second generation as well as the implementation of follow-up mechanisms of deterministic effects among descendents of exposed parents and the implementation of their medical-social rehabilitation.

Future research areas will comprise the study of mechanisms of low dose effects on human organism, and quantification of the health impact on the population due to nuclear testing, as well as the potential risks of further nuclear developments in Kazakhstan. A major issue in addition to the epidemiology of fallout exposures due to nuclear weapons testing near Semipalatinsk are the uranium miners in Kazakhstan, where 50% of the former Soviet uranium ores and about 25% of the world's reserves are reportedly located (Kazymbet & Seisebaev, 2002). While much of the radiation research focuses on dealing with the legacies of Soviet exposures due to the Cold War nuclear race, one can observe a continued rise of new and different nuclear economies, especially in uranium mining for export.

14.5 CONCLUSIONS

Epidemiological investigations and risk assessments are heterogeneous and contingent undertakings with different agendas, ranging form the generation and improvement of universalized risk estimates to the local documentation and mitigation of radiation exposure. Throughout the past decades, the conditions of researching the public health impact of the legacies of nuclear testing in Semipalatinsk have tremendously changed. Historically, a specific mode of risk research evolved as part of the nuclear programs during the Cold War and continues to shape the infrastructures of the nuclear present. The early years and the collecting of health and exposure data by Soviet scientists provide the infrastructures that have enabled and to some extent prestructured radioepidemiological research. Since 1991, despite a shared Soviet history of exposure due to nuclear testing, retrospective investigations of the health impact of nuclear testing at the Semipalatinsk test site have taken place in two countries—in the Republic of Kazakhstan and the Russian Federation—with different institutional and economic conditions for research and compensation. The epidemiological studies already conducted bear witness of the heterogeneity of research practices as well as the differences in what is scientifically recognized as evidence, and the different traditions of reporting and articulating health effects.

The cleanup of the legacy of nuclear weapons productions involved major changes within the nuclear complex. Both in the US and the USSR (Reed et al., 1997) this meant to shift from a top-down military organization toward a new structure involving different stakeholders and public debate.[6] While some documentation on nuclear programs has been made available for other test sites, for instance those in Nevada and the Marshall Islands, in most cases by far more studies were conducted among the veterans than in populations downwind to the test sites (Bauer, 2006). For the Semipalatinsk test site, exposures of the local population may exceed those in other nuclear weapons test sites, in terms of numbers of exposed people and collective doses. In the post-Soviet states, comprehensive research programs into health effects or screening of the entire population potentially exposed to local fallout from nuclear testing in most nuclear test sites of the western world have not been conducted. In that sense, the Semipalatinsk test site stands out not only as to the severe population exposures but also in the mobilization of funds, medical programs, and the public into the study of fallout effects. Thus, the nuclear program and the nuclear legacies are features that have implicated on and reconfigured much of contemporary society and everyday life in these areas.

6. For issues of public involvement and radiation research on atmospheric nuclear testing in the US, see Hoffmann, Apostonaei, & Thomas, 2002.

REFERENCES

Academy of Sciences of the Soviet Republic of Kazakhstan. (1958). *Radioactivity of the environment, health status of the population and cattle in central Kazakhstan, Almaty*. Published in 1997, Nagasaki Association for Hibakushas' Medical Care (NASHIM). (in Russian and Japanese).

Balmukhanov, S. B., Gusev, B. I., & Balmukhanov, T. S. (2002). *Radioactivity and population health status around the Semipalatinsk Nuclear Test Site*. Almaty.

Bauer, S. (2006). *The health impact of atmospheric nuclear testing. Cancer epidemiology in areas adjacent to the Semipalatinsk nuclear test site, Kazakhstan*. Frankfurt a.M. New York: Peter Lang.

Bauer, S., Gusev, B. I., Pivina, L. M., Apsalikov, K. N., & Grosche, B. (2005). Radiation exposure due to local fallout from Soviet atmospheric nuclear weapons testing in Kazakhstan: solid cancer mortality in the Semipalatinsk historical cohort, 1960–1999. *Radiation Research*, *164*(4), 409–419.

Bauer, S., Gusev, B. I., Pivina, L. M., Apsalikov, K. N., & Grosche, B. (2006). Esophagus cancer and radiation exposure due to nuclear test fallout: an analysis based on the data of the Semipalatinsk historical cohort, 1960–1999. *Radiatsionnaia Biologiia Radioekologiia*, *46*(5), 611–618.

Bul'bulian, M. A., & Tokareva, G. D. (1991). Deskriptivnoe epidemiologicheskoe issledovanie zlokachestvennykh novoobrazovanii v Semipalatinskoi oblasti Kazakhskoi SSR (Descriptive epidemiological study of malignant neoplasms in the Semipalatinsk district of the Kazakh S.S.R.). *Vestnik Akadademii Meditsinskikh Nauk SSSR*, *7*, 59–63.

Burkart, W. (1996). Radioepidemiology in the aftermath of the nuclear program of the former Soviet Union: unique lessons to be learnt. *Radiation and Environmental Biophysics*, *35*, 65–73.

Burkart, W., & Kellerer, A. (Eds.), (1994). *Radiation exposure in the Southern Urals. A collection of invited papers providing a first assessment of radiation doses and health effects in workers and communities exposed since 1948 in Southern UralsThe Science of the Total Environment*, *142*(Special issue).

Burkart, W., Kellerer, A., Bauer, S., Harrison, J., Land, C., Shoikhet, YaN., et al. (1999). Health effects. In F. Warner & R. J. C. Kirchmann (Eds.), *Nuclear test explosions: Environmental contaminations and health effects* (pp. 179–228). Chichester.

Coughlin, S. S. (1996). Environmental justice: the role of epidemiology in protecting unempowered communities from environmental hazards. *The Science of the Total Environment*, *184*, 67–76.

Coughlin, S. S., & Beauchamp, T. L. (Eds.). (1996). *Ethics and epidemiology*. Oxford: New York.

Gordin, M. D. (2009). *Red cloud at dawn. Truman, Stalin, and the end of the atomic monopoly*. New York: Farrar Strauss and Giroux.

Grosche, B. (2002). Semipalatinsk nuclear test site: Introduction. *Radiation and Environmental Biophysics*, *41*, 53–56.

Grosche, B., Lackland, D. T., Land, C. E., Simon, S. L., Apsalikov, K. N., Pivina, L. M., et al. (2011). Mortality from cardiovascular diseases in the Semipalatinsk historical cohort, 1960–1999, and its relationship to radiation exposure. *Radiation Research*, *176*(5), 660–669.

Grosche, B., Land, C., Bauer, S., Pivina, L. M., Abylkassimova, Z. N., & Gusev, B. I. (2002). Fallout from nuclear tests: health effects in Kazakhstan. *Radiation and Environmental Biophysics*, *41*, 75–80.

Gusev, B. I., Rozenson, R. I., & Abylkassimova, ZhN. (1998). The Semipalatinsk nuclear test site: a first analysis of solid cancer incidence (selected sites) due to test-related radiation. *Radiation and Environmental Biophysics*, *37*, 209–214.

Gusev, B. I., Sekerbaev, A. K., Rozenson, R. I., Tchaijunusova, N. J., & Apsalikov, K. N. (1998). *Late effects of exposure to ionizing radiation. Studies in the resident population in the Semipalatinsk area. Edited by The Public Health Committee, Ministry of Education, Culture and Public Health of the Republic of Kazakhstan.* Almaty: Kazakh Scientific Research Institute for Radiation Medicine and Ecology.

Hoffmann, F. O., Apostonaei, A. I., & Thomas, B. A. (2002). A perspective on public concerns about exposure to fallout from the production and testing of nuclear weapons. *Health Physics*, *82*, 736–748.

Iakubovskaia, E. L., Nagibin, V. I., Suslin, V. P. (1998). *Semipalatinskii iadernyi poligon—50 let (The Semipalatinsk nuclear test site—50 years).* Novosibirsk.

Izrael, YuA., Stukin, E. D., Petrov, V. N., Anspaugh, L., Doury, A., Kirchmann, R. J. C., et al. (2000). Nuclear explosions and their environmental contamination. In F. Warner & R. J. C. Kirchmann (Eds.), *Nuclear test explosions* (pp. 33–98). Chichester: Environmental and Human Impacts.

Kazymbet, R. K., & Seisebaev, A. T. (2002). Problems of the complex assessment of radiobioecological situation and public health in uranium-extraction regions in Kazakhstan. *Radiatsionnaia Biologiia Radioekologiia*, *42*, 750–753.

Land, C. E., Zhumadilov, Z., Gusev, B. I., Hartshorne, M. H., Wiest, P. W., Woodward, P. W., et al. (2008). Ultrasound-detected thyroid nodule prevalence and radiation dose from fallout. *Radiation Research*, *169*, 373–383.

Lindee, M. S. (1994). *Suffering made real. American science and the survivors at Hiroshima.* Chicago: Chicago University Press.

Mikhailov, V. N., Andryshin, I. A., Bogdan, V. V., Vashchinkin, S. A., Zelentsov, S. A., Zolutkin, G. E., et al. (1996). *USSR nuclear weapons tests and peaceful nuclear explosions 1949 through 1990.* Moscow: Ministry of the Russian Federation for Atomic Energy. The Ministry of Defense of the Russian Federation.

Mudie, N. Y., Gusev, B. I., Pivina, L. M., Schoemaker, M. J., Rijinkova, O. N., Apsalikov, K. N., et al. (Nov, 2007). Sex ratio in the offspring of parents with chronic radiation exposure from nuclear testing in Kazakhstan. *Radiation Research*, *168*(5), 600–607.

Mudie, N. Y., Swerdlow, A. J., Gusev, B. I., Schoemaker, M. J., Pivina, L. M., Chsherbakova, S., et al. (2010). Twinning in the offspring of parents with chronic radiation exposure from nuclear testing in Kazakhstan. *Radiation Research*, *173*(6), 829–836.

Nugent, R. W., Zhumadilov, Z. S., Gusev, B. I., Hoshi, M. (2000). *Health effects of radiation associated with nuclear weapons testing at the Semipalatinsk nuclear weapons test site*. New York: Semipalatinsk/Hiroshima.

Pierce, D. A., Shimizu, Y., Preston, D. L., Vaeth, M., & Mabuchi, K. (1996). Studies of the mortality of atomic bomb survivors. Report 12, Part I. Cancer: 1950–1990. *Radiation Research*, *146*, 1–27.

Polikarpov, G. G., & Aarkrog, A. (1993). The radioecological problems of Eurasia and the sources of radioactive environmental contamination in the former USSR. *Radiobiologiia*, *33*(1), 15–24.

Rakhypbekov, T. K., Gusev, B. I., & Sekerbaev, A. Kh. (1999). *Nuclear weapons testing at the Semipalatinsk and Chinese polygons and their consequences for the population of the East-Kazakhstan region of Kazakhstan.* Report to the Ministry of Health, Semipalatinsk: Education and Sport of the Kazakhstan Republic and the Scientific Research Institute for Radiation Medicine and Ecology.

Reed, R., Lemak, D. J., & Hesser, W. A. (1997). Cleaning up after the Cold war: management and social issues. *The Academy of Management Review*, *22*(3), 614–642.

Ronellenfitsch, U., Kyobutungi, C., Becher, H., & Razum, O. (2006). All-cause and cardiovascular mortality among ethnic German immigrants from the Former Soviet Union: a cohort study. *BMC Public Health*, *26*(6), 16.

Sevan'kaev, A. V., Ankina, M. A., Golub, E. V., Zhloba, A. A., Zavitaeva, T. A., Kozlov, V. M., et al. (1995). The results of cytogenetic studies of persons from the settlements adjacent to the Semipalatinsk testing ground. *Radiatsionnaia Biologiia Radioekologiia*, *35*, 596–607.

Shapiro, C. S., Kiselev, V. I., & Zaitsev, E. V. (Eds.), (1998). *Nuclear tests: Long term consequences in the Semipalatinsk/Altai region. NATO ASI Series*. (Vol. 36). Berlin/Heidelberg.

Shoikhet, YaN., Kiselev, V. I., Algazin, A. I., Kolyado, I. B., Bauer, S., & Grosche, B. (2002). Fallout from nuclear tests: health effects in the Altai Region. *Radiation and Environmental Biophysics*, *41*, 69–74.

Shoikhet, YaN., Kiselev, V. I., Loborev, V. M., Sudakov, V. V., Zelenov, V., & Azarov, L. (2002). Fallout from nuclear tests: dosimetry in the Altai region. *Radiation and Environmental Biophysics*, *41*, 57–60.

Shoikhet, YaN., Kiselev, V. I., Zaitsev, E. V., Kolyado, I. B., Konovalov, BYu., Bauer, S., et al. (1999). A registry for exposure and population health in the Altai region affected by the Semipalatinsk nuclear tests. *Radiation and Environmental Biophysics*, *38*, 207–210.

Stepanenko, V. F., Skvortsov, V. G., Ivannikov, A. I., Dubov, D. V., & Tsyb, A. F. (2011). Methods of individual retrospective physical dosimetry in the problem of estimation of consequences of accidental irradiation. *Radiatsionnaia Biologiia Radioekologiia*, *51*(1), 168–177.

Tsyb, A. F., Stepanenko, V. A., Pitkevich, V. A., Ispenkov, E. A., Sevankaev, A. V., Orlov, MYu., et al. (1990). Around the Semipalatinsk testing ground: radioecological situation and exposure of population in the Semipalatinsk region (based on the materials of the Interagency Commission report). *Meditsinskaia Radiologiia*, *35*, 1–12.

United Nations General Assembly. (1998). *International cooperation and coordination for the human and ecological rehabilitations and economic development of the Semipalatinsk region of Kazakhstan. Report of the Secretary General, 23 September 1998 A/53/424*. New York.

United Nations Scientific Committee on the Effects of Atomic Radiation (UNSCEAR). (2000). *Sources and effects of ionising radiation. Report to the General Assembly. With scientific annexes*. New York.

United Nations Scientific Committee on the Effects of Atomic Radiation (UNSCEAR). (2008). *Sources and effects of ionising radiation. Report to the General Assembly. With scientific annexes*. New York.

Zaridze, D. G., Li, N., Men, T., & Duffy, S. W. (1994). Childhood cancer incidence in relation to distance from the former nuclear testing site in Semipalatinsk, Kazakhstan. *International Journal of Cancer*, *59*, 471–475.

Chapter 15

Moral Dilemmas of Uranium and Thorium Fuel Cycles

Behnam Taebi
Department of Values and Technology, Delft University of Technology, Delft, The Netherlands
E-mail: b.taebi@tudelft.nl

Chapter Outline

15.1 Introduction 260
15.2 Existing Nuclear Fuel Cycle: Uranium 261
15.2.1 Safety 262
15.2.2 Security 264
15.2.3 Sustainability 265
15.2.4 Economic Viability 266
15.3 The Closed Fuel Cycle and Intergenerational Justice Dilemmas 267
15.3.1 Short-term Safety Compromised, while Long-term Safety is Enhanced 268
15.3.2 Additional Security Concerns in Conjunction with Plutonium 268
15.3.3 Resource Durability Enhanced, while Short-term Environmental Friendliness is Compromised 269
15.3.4 Less Economic due to Expensive Reprocessing 270
15.3.5 Important Assumptions Concerning Reprocessing 270
15.4 Is Thorium a Viable Substitute or Supplement for Nuclear Fuel? 273
15.4.1 Proliferation Resistance: Using Thorium to Produce Less Plutonium 274
15.4.2 Waste-Management Benefits of Using Thorium in Molten Salt Reactors 275
15.4.3 Challenges and Shortcomings of Thorium Cycles 275
15.5 Conclusions 277

Radioactivity in the Environment, Volume 19
ISSN 1569-4860, http://dx.doi.org/10.1016/B978-0-08-045015-5.00015-0

15.1 INTRODUCTION

Recent developments in Germany's energy policy aptly symbolize the controversies surrounding the nuclear power debate. In August 2010, after months of debate and consideration, Chancellor Merkel's administration decided to extend the lifetime of Germany's 17 nuclear power plants. Surprisingly, Germany was also the first country in the world to revise its nuclear power policy, following the disastrous Fukushima-Daiichi events of March 2011. Simultaneously, the worldwide debate on the extent to which nuclear power should have a role in supplying our energy demands continues.

While increasingly more states are being swayed by the fact that nuclear power can enhance domestic energy security, boost energy levels, and reduce greenhouse gas byproducts, critics point to the continued risk of reactor accidents—Fukushima made this issue painfully clear—the dangers surrounding nuclear fuel transportation, fears of proliferation and the vexing problem of how to deal with long-lived nuclear waste as reasons why it should be curtailed.

But as politicians, energy experts and the general public weigh up the pros and cons, one key element linked to harnessing energy from the atom is being neglected: the relationship between the different nuclear power producing methods (i.e. fuel cycles) and the different safety, security, and economic considerations that each method brings. The technical choices made today will not only determine the extent of the risk posed today, but they will also seriously affect the burden faced by humanity in the form of contaminated byproducts that can remain radiotoxic for hundreds of thousands of years. Rather than reducing nuclear power to a simple yes/no, good/bad dichotomy, we need to first focus on the advantages and disadvantages of each nuclear energy production method, including the burdens and benefits posed now and in generations to come.

This will not, of course, answer the thorny question of whether we should go nuclear on a larger scale or retain our current nuclear reactors. We can only answer this question if we consider nuclear energy in relation to other energy sources. We first have to distinguish between the different nuclear power production methods in terms of the different moral considerations they bring. Such analysis could help us to establish a *desirable energy mix*. Not only does this lend more accuracy to the debate but it also enables an ethically informed discussion to take place on the ideal future energy mix and the possible role of nuclear energy. We need to include the new technology prospects and reflect on the desirability of future fuel cycles, the aim being to support research and development paths that could culminate in the industrialization of a certain desired technology. This chapter will take up this challenge.

In Section 15.2, I will first specify what is morally at stake in nuclear power production. I will start by discussing the open fuel cycle, the most straightforward and common type of nuclear power production. I will identify several of the moral values at stake. Values are things worth striving for if we are to achieve what we deem to be "good". Precisely what constitutes *good* in nuclear power production has implications for different groups of people that are spatially or temporally distinct. I will discuss the four main values that play a key part in nuclear power production and waste management: safety (the public health impacts), security (sabotage and proliferation), sustainability (the environmental impacts and energy resource availability), and economic viability (embarking on new technology and its continuation).[1] I will operationalize these values for different fuel cycles, by elaborating on how they relate to each step of the fuel cycle. Section 15.3 extends these moral considerations by including the closed fuel cycle that extends the open fuel cycle by recycling spent fuel after irradiation in a reactor. Section 15.4 focuses on thorium as a possible nuclear fuel for the future. Even though the relevance of thorium as a nuclear fuel has been acknowledged from the early days of nuclear power production, no thorium cycle yet exists. Nevertheless, there is renewed interest in thorium because of its resource durability, but also because of the security enhancing and waste-management benefit prospects. It is therefore important to include the future prospects of thorium as a substitute or as a complementary fuel and to contemplate its ethical considerations. Section 15.5 summarizes the findings made in this chapter.

15.2 EXISTING NUCLEAR FUEL CYCLE: URANIUM

In this section I will first discuss the open fuel cycle[2] type common in the U.S., Sweden, Canada, and many other countries. I will then identify the four important values at stake and elaborate on how each step affects these values.

The open fuel cycle consists of five main steps. In Step 1, natural uranium is mined and milled, this process is similar to the mining of other metals, with the difference that uranium and its decay products emit ionizing radiation. Step 2 involves the chemical purification and enrichment of uranium. Natural uranium consists of the two main isotopes ^{235}U and ^{238}U. Only the first isotope (^{235}U) is *fissile* and deployable as a fuel in currently operational Light Water Reactors (LWR).[3] However, this fissile uranium only constitutes 0.7% of all natural uranium. In order to produce a type of fuel that can be efficiently used in LWRs, we need to increase the content of this isotope to 3–5%; this process is known as *enrichment*. Enriched uranium[4] is converted into uranium dioxide and used to fabricate fuel (Step 3), which can be used in an LWR (Step 4). A typical fuel assembly will remain in the reactor for about four years; the remainder that is

---

1. For a detailed discussion of these values and how they have played a role in the history of nuclear reactor design, please see (Taebi & Kloosterman, in press).
2. The detailed information and figures about the open fuel cycle is based on two MIT reports (MIT, 2003, 2011).
3. Light Water Reactors are Generation II reactors of the type comprising the vast majority of the currently operable nuclear power plants.
4. It is actually the gaseous uranium hexafluoride that will be used to fabricate fuel.

discharged from the reactor is called *spent* fuel. Spent fuel is not necessarily a waste, but in the open fuel cycle it is disposed of as waste. Before final disposal underground and in deep geological repositories (Step 5), spent fuel must be temporarily stored and cooled in storage facilities for several decades (Step 4). In the remainder of this section, I will present a moral analysis of the open fuel cycle by first introducing the four main values at stake in nuclear energy production; I will then *operationalize* these values by elaborating on the effect of each step in the open fuel cycle.

15.2.1 Safety

The IAEA et al. (2006, p. 5) defines public safety as "the safety of nuclear installations, radiation safety, the safety of radioactive waste management and safety in the transport of radioactive material." Safety as a value refers here to those concerns that pertain to the exposure of the human body to radiation and to the subsequent health effects.[5] In radiation health, we distinguish between different types of radiation (α, β, γ and neutron radiation) and the various health effects. It is both the nature of the radiation, the type of exposure (i.e. inhalation, ingestion etc.) and the period of exposure that determines the radiotoxic effects of any radiation (Smeesters, 2008). It is therefore important to include all types of ionizing radiation in our moral analysis. The general philosophy of radiation protection is "to reduce exposure to all types of ionizing radiations to the lowest possible level" (ICRP, 1959, p. 10).

In all phases of the open fuel cycle, there is an ionizing radiation that has to be coped with. Though natural uranium emits fairly limited amounts of ionizing radiation, it is important to consider the steps 1 to 3 safety risks, because workers will be continuously exposed to such low-dose radiation.[6] Furthermore, the disposal of uranium tailings in uranium mines, and depleted uranium around any uranium enrichment facility forms a major source of nuclear waste. It is particularly the long-lived isotopes of radium (^{226}Ra) and the gaseous decay product radon (Rn) that constitute health concerns for radiation workers.[7] The high-dose radiation in the reactor is of a different type and also cause serious risks, due to the strong radioactive decay in the fuel; this radiation is shielded in the reactor. The radiation emitted from spent fuel in the interim storage period

5. The IAEA et al. (2006, p. 5) defines safety as "the protection of people and the environment against radiation risks." The radiation consequences for the environment and for nonhuman life will be returned to when the matters of sustainability and environmental friendliness are examined.
6. When referring to low-dose radiation, we really mean the probabilistic radiation effects: i.e. the probability of severe consequences (e.g. cancer) as a result of long-term exposure to such radiation (de Saint-Georges, 2008). For an overview of the occupational health standards, see (Hansson, 1998).
7. I owe this suggestion to Gilbert Eggermont and Jean Hugé who criticized my earlier publications on this issue, in which I overlooked the health concerns of uranium mines and depleted uranium; see (Eggermont & Hugé, 2011, p. 45).

(Step 4) also has to be carefully isolated, especially since serious decay and heat production occur during the first decades after fuel has been discharged from the reactor. This is why repositories will not be permanently closed until 80–100 years after the emplacement of last loads of spent fuel. The health effects of the alpha, beta, and gamma radiations in spent fuel are fairly known but what is less known is the effects of neutron radiation.[8]

In conjunction with the longevity of nuclear waste, safety is a value that specifically relates to future generations as well. The safety of future generations has been one of the concerns from the early days of nuclear power production (NRC, 1966). How we should protect future generations from the harmful effects of radiation remains a subject of an ongoing discussion, both in technical literature (how and where to build repositories that best guarantee long-term protection), and in policy-related documents. The prevailing notion is that our responsibility to future generations will diminish over time which is why we do not have an obligation to offer the same level of protection to all future generations. In the US, this culminated in a policy to introduce a two-tiered standard in order to distinguish between short-term and long-term radiological protections (EPA, 2005). However, this distinction lacks solid moral justification (Shrader-Frechette, 1993, 2005; Taebi, 2012).

As stated above, only ^{235}U is fissile and deployable in light water reactors. The major constituent of the fuel (^{238}U) is *fertile*, meaning that upon absorbing neutrons it converts to fissile ^{239}Pu. In addition to the unused ^{235}U, ^{238}U and ^{239}Pu, spent fuel comprises other fissile and nonfissile plutonium isotopes, *minor actinides*, namely americium (Am), curium (Cm) and neptunium (Np), and fission products. Essentially, spent fuel poses a radiation risk throughout the period of dangerous radioactive decay, something referred to as the *waste lifetime*, dominated by the presence of plutonium and americium. In general, spent fuel is believed to be radiotoxic for a period of about 200,000 to one million years. Precisely how we determine the waste lifetime remains a matter of definition depending on the point of reference we choose. Generally, when determining the waste lifetime, spent fuel is compared with the same amount of natural uranium, or uranium ore; it is the period after which the radiotoxicity of emitted radiation from spent fuel will reach the same radiotoxicity of that emitted by the same amount of natural uranium; see the dotted line in Figure 15.1.

There have been at least two criticisms leveled at the way in which the radiotoxicity of nuclear waste has been compared to that of natural uranium ore. Firstly, spent fuel consists of different chemical components to natural uranium, thus meaning that the effects on health of emitted radiation are not necessarily similar (Eggermont & Hugé, 2011, p. 46). Secondly, natural radiation can

8. The energetic alpha particles released during the decay of the element americium (Am) could knock out neutrons from lighter elements such as oxygen (O). Furthermore, the spontaneous fission of curium (Cm) could produce neutrons at an even higher rate (Wallenius, 2011). Americium and curium are two of the minor actinides produced during the irradiation of uranium fuel.

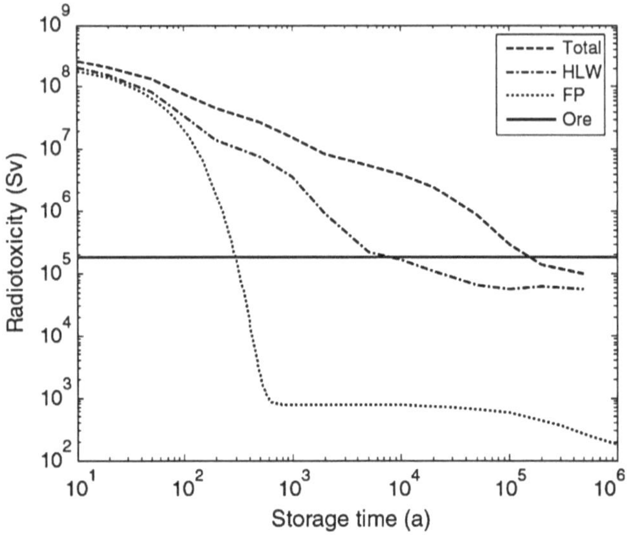

FIGURE 15.1 The radiotoxicity of spent fuel, high-level waste (HLW), and fission products, compared to the radiotoxicity derived from the same amount of uranium ore.

also cause serious health problems. Therefore, referring to naturally caused radiation does not offer sufficient justification for tolerating comparable levels of man-made radiation (Shrader-Frechette, 2005).

The waste lifetime denotes the period that nuclear waste needs to be isolated from the environment. So in addition to the uranium ore line, one can also use the peak-dose criterion. In the US, a period of one million years was proposed by the National Academy of Science, which suggested that in terms of the nuclear waste produced in the US, the peak-dose will occur after 750,000 years. The American regulator has endorsed one million years as the period of time necessary for the isolation of American waste (EPA, 2008).[9]

15.2.2 Security

In the IAEA's Safety Glossary, nuclear security is defined as "any deliberate act directed against a nuclear facility or nuclear material in use, storage or transport, which could endanger the health and safety of the public or the environment" (IAEA, 2007, p.133). Even though safety and security apparently do overlap to an extent, I shall keep the value of "security" separate so as to be able to distinguish between unintentional and intentional harm. In an open fuel cycle, proliferation threats arise from the enrichment of uranium. Uranium needs to be enriched to 3–5% (and in some reactor types 20%) for power production purposes in reactors. Highly Enriched Uranium (HEU) is produced by allowing

9. For an elaborate discussion of this issue see Chapter 8 in Vandenbosch and Vandenbosch (2007).

the enrichment process to exceed 70%, a level only required for the manufacture of nuclear weapons. The Hiroshima bomb dropped in 1945 was created from HEU. When the enrichment exceeds 20%, the application can only be for nuclear arms; the IAEA has well-developed inspection methods to detect such activity in any facility under their control.

Serious contention concerning the expansion of civilian nuclear technology among new members is the matter of whether each country should have its own enrichment facility. A good example is Iran, which insists on having its own enrichment facility. While the Non-Proliferation Treaty gives all member states (including Iran) the right to follow through all steps of the nuclear fuel cycle, the existence of an enrichment facility clearly increases proliferation risks. There are currently a few countries that operate enrichment facilities. The countries that enrich impose limitations on the importing countries in order to avoid proliferation issues; the plutonium currently present in spent fuel constitutes a considerable proliferation risk too. More will be said about this in Section 15.3.

15.2.3 Sustainability

Sustainability is one of the most frequently discussed and perhaps contested notions in all the literature on nuclear power. It is not my intention to enter into those discussions here and I certainly do not intend to assess the sustainability of nuclear power. One common and influential definition concerning sustainable development is the Brundtland definition in which the ability of present generations to meet their own needs without compromising the ability of future generations to meet their needs is emphasized (WCED, 1987). In nuclear power production and nuclear waste management, this definition relates to at least two specific issues, namely the state of the environment bequeathed by us to posterity—referred to as *environmental friendliness*—and the availability of natural (nonrenewable) energy resources on which future well-being of generations relies, referred to as *resource durability*.

15.2.3.1 Environmental Friendliness

The value of *environmental friendliness* relates to the accompanying radiological risks to the environment. Radiological risks, as perceived in this chapter, express the possibility or rather the probability that radioactive nuclides might leak into the biosphere and harm both humans and the environment. Issues that relate to the harming of human beings have already been subsumed under the heading safety. The effect of the same radiation on the environment and nonhuman animals is included here under the heading of environmental friendliness. Whether we should protect the environment for its own sake or for what it means to human beings is a longstanding discussion that is still continuing in environmental philosophy. I do not intend to take a stance on this matter here. I prefer to preserve the value of "environmental friendliness" as a separate value in order to allow a broader number of views to be reflected through this set of values. Those who adhere to the anthropocentric approach will then simply merge this value with the value of "safety", while those who adhere to the nonanthropocentric approach will explicitly include in their analysis those risks and burdens other specifies will be exposed to as a result of humanity's nuclear power production and consumption. The latter could drastically change the ethical analysis.

15.2.3.2 Resource Durability

If we now consider the period from the time of the industrial revolution up until the present, it would be fairly straightforward to conclude that the availability of energy resources has played a key role in augmenting and sustaining people's well-being. The appropriate consumption of nonrenewable natural resources over time is one of the central issues of sustainability; "later generations should be left no worse off [...] than they would have been without depletion" (Barry, 1989, p. 519). The value of *resource durability* is defined as the availability of natural resources for the future or as the providing of an equivalent alternative (i.e. compensation) for the same function. In an open fuel cycle, we intend to use uranium and nuclear fuel only once. The remaining spent fuel then officially has to be disposed of underground for a very long period of time. Spent fuel contains various isotopes including uranium and plutonium that could also be used as fuel; this aspect will be discussed in Section 15.3.

15.2.4 Economic Viability

The next value that we shall discuss in relation to sustainability is that of *economic viability*. One might question whether economic issues have inherent moral relevance and whether it is justified to present economic durability as a moral value. We can safely assume that the safeguarding of the general well-being of society (also, for instance, including health care issues) has undeniable moral relevance. However, in my interpretation of economic viability in this chapter I do not refer to the general well-being but only to those aspects of well-being that have to do with nuclear energy production and consumption. With this approach, economic aspects are not of inherent moral relevance; it is rather what stands to be achieved from such economic potential that makes it morally worthy.

This is why the value of economic durability is presented in conjunction with other values. First and foremost, economic viability should be considered in connection with resource durability. In that way, it relates to the economic potential for the initiation and continuation of an activity that produces nuclear energy. As we shall see in the following sections, certain future nuclear energy production methods may well require serious R&D investments for further development; particularly new methods that are based on new types of reactors will require serious investment prior to industrialization. Economic viability could also become a relevant notion when efforts are made to safeguard the safety and security of posterity by introducing new technology designed to reduce the lifetime of nuclear waste. In general, economic viability is defined here as the economic potential to embark on a new technology and to safeguard its continuation in order to uphold all the other values.

Since the open fuel cycle is the *shortest* cycle, in other words the one necessitating least nuclear activity compared to the closed fuel cycle, we can argue that its economic burdens are low compared to other fuel cycles where spent fuel is further recycled. It should, however, be noted that since we do not yet have any geological repositories for the disposal of spent fuel, we cannot yet accurately estimate the costs of such a repository. The legal requirements attached to building such repositories could, for instance, impose additional technical criteria therefore making it more expensive than anticipated.

15.3 THE CLOSED FUEL CYCLE AND INTERGENERATIONAL JUSTICE DILEMMAS

In Section 15.2, I briefly assessed the open fuel cycle on the basis of the four values of safety, security, sustainability, and economic viability. In this section, I will assess the main alternative for the open cycle, namely the closed fuel cycle, in terms of the same moral values. With this cycle, spent fuel is no longer viewed as waste and the idea is to recycle it. As stated above, less than 1% of the uranium ore consists of the fissile isotope ^{235}U. The major isotope of uranium (^{238}U) is not fissile and it must be converted into fissile plutonium (^{239}Pu) that is deployable for energy production. In the closed fuel cycle, spent fuel will undergo a chemical process to extract the useable elements, including plutonium. Such recycling treatment is referred to as *reprocessing*. During reprocessing the uranium and plutonium isotopes in the spent fuel are isolated and recovered; the remaining materials are put into a glass matrix to be immobilized; this is known as High-Level Waste.

There are two rationales to the closed fuel cycle. Firstly, taking radiotoxicity as a criterion, it could reduce the waste lifetime to c. 10,000 years; simultaneously the volume of the remaining High-Level Waste (HLW) could be reduced by two-thirds. Secondly, it will enable the more efficient use of nuclear fuel since recycled uranium could be added to the beginning of the fuel cycle. The extracted plutonium must be used for manufacturing *Mixed Oxide Fuel* (*MOX*), a nuclear fuel based on uranium and plutonium oxide. MOX fuel is deployable in the existing LWRs.

Reprocessing is a technology as old as nuclear weapons themselves. The first reprocessing plant was built as part of the Manhattan project in the US during the Second World War. Its primary purpose was to extract plutonium from irradiated uranium fuel for use in nuclear weapons; that was to culminate in the Nagasaki bomb. Worldwide, there are only five commercial reprocessing plants operable: namely in France, the U.K., Russia, India, and Japan. Japan is the only nonweapon state that was building a reprocessing plant.

15.3.1 Short-term Safety Compromised, while Long-term Safety is Enhanced

Reprocessing seems to have important long-term benefits, since it reduces the period of necessary care by a factor 20 (from 200,000 years to 10,000 years). It does, however, introduce various short-term safety risks when compared to the open fuel cycle. Firstly, reprocessing is a chemical process with radiological risks. Secondly, reprocessing produces more short-lived (but less radiotoxic) waste that needs to be disposed of as well. Thirdly, reprocessing plants are only available in a handful of countries and the transporting of radiotoxic material introduces additional risks. Just to elucidate: the closed fuel cycle is the favored cycle in West Europe. European countries that opted for the closed fuel cycle method need to transport their spent fuel to the reprocessing plants in La Hague (France) or Sellafield (UK); HLW and the separated uranium and plutonium must then be sent back to the country of origin because each country is responsible for disposing of its own waste. Alternatively, uranium and plutonium could also be sent to a third country for further use as fuel.

It should be noted that the risk of large quantities of radioactive waste being released during the transport of spent fuel and HLW is small; the countries concerned have extensive experience both with sea transport and rail transport in Europe. In view of the latter point, a 2006 U.S. National Research Council report emphasized however that the vulnerability of this transport to terrorist attack need to be examined (NRC, 2006). As stated above, proliferation relates both to the dissemination of knowledge and technology on the manufacturing of nuclear weapons and to sabotage with radiotoxic materials.

15.3.2 Additional Security Concerns in Conjunction with Plutonium

In closed fuel cycles, the remaining uranium in spent fuel, along with different isotopes of plutonium will be removed so that it can be reused as fresh fuel, but of course the extracted plutonium also carries proliferation threats. Proliferation is by far the most important concern when reprocessing. Security is one of the main reasons why the US, which has about one fourth of all the world's nuclear reactors, does not reprocess. The major stockpiles of plutonium in the US derive from the nuclear warheads that were dismantled after the Cold War era. The idea of producing more plutonium is generally considered to be highly undesirable in the US.

To illustrate the seriousness of these proliferation risks, 8 kg of *weapon-grade* plutonium (^{239}Pu) is sufficient to produce a bomb with the devastation potential of the Nagasaki bomb. The kind of plutonium emerging from a power reactor under normal circumstances consists of different isotopes including ^{238}Pu, ^{240}Pu and ^{239}Pu; Figure 15.2 shows the buildup of different plutonium isotopes during fuel irradiation. For destructive purposes, plutonium must contain as much as

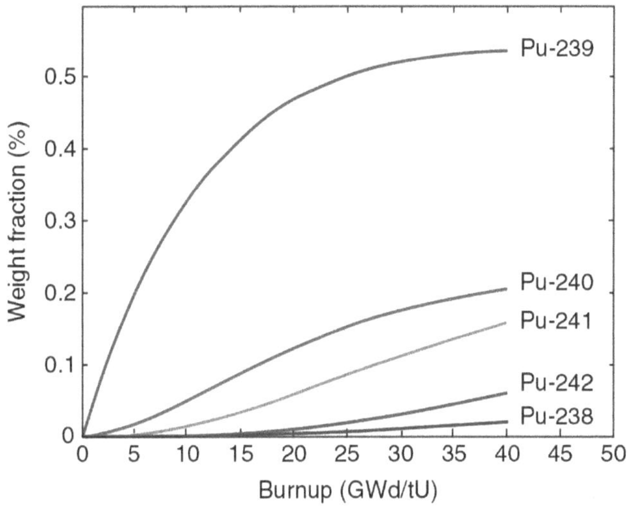

FIGURE 15.2 The burning up of different plutonium isotopes in an LWR. *Source: (Taebi, 2012.).* (For color version of this figure, the reader is referred to the online version of this book.)

possible ^{239}Pu; this corresponds to the relatively short burn up time, as illustrated in Figure 15.2. Removing nuclear fuel after a couple of weeks of burning up could thus be taken as evidence of ill intent. While *civilian* plutonium does not have the yield of weapon-grade plutonium it does carry serious security risks.

To conclude, the closed fuel cycle seems to increase proliferation concerns in the short term. On the other hand, it reduces proliferation concerns in the long run because the material potentially deployable for proliferation (plutonium) will not be retained in the spent fuel. Current spent fuel inventories are safeguarded as well as military facilities, since this spent fuel contains plutonium.

15.3.3 Resource Durability Enhanced, while Short-term Environmental Friendliness is Compromised

In Section 15.2, I distinguished between two aspects of sustainability, namely resource durability and environmental friendliness. The environmental aspects are closely related to the safety risks. The closed fuel cycle seems to have serious long-term safety (and thus environmental) benefits, but it brings with it various short-term safety and environmental concerns. Regarding resource durability, reprocessing seems to create important benefits as well. In the early days of nuclear energy production and after Eisenhower's "Atoms for Peace" speech of 1953, reprocessing was promoted as the technology that could lead to sufficient supplies of nuclear fuel. Instead of using the uranium fuel once, it would be used more efficiently. Both the remaining uranium in spent fuel and the plutonium produced could be reused. In the first years of commercial nuclear power developments in the 1960s all countries considered reprocessing.

With the knowledge we currently have on the abundance of uranium resources, the resource durability argument seems to have become less persuasive. Yet, resource durability still plays an important part in countries that do not have access to other types of resources. Japan, for instance, that operates one eighth of all the power plants in the world, favors reprocessing from the resource durability perspective.[10] Japan does not have any oil or gas resources, nor does it have uranium; efficient use of nuclear fuel is therefore very important. Japan is building its own reprocessing plant in Rokkasho but it is not yet operational.

15.3.4 Less Economic due to Expensive Reprocessing

Reprocessing plants are extremely expensive facilities. Along with the proliferation concerns, the cost aspect is the main reason why the US is not in favor of the closed cycle option. This is also the reason why small producers of nuclear power that do adhere to the closed cycle system prefer to transport their spent fuel to other commercial reprocessing plants in other countries than to build their own plants (WNA, 2013).

A Harvard University study concludes that even if there is a substantial growth in nuclear power (which would make reprocessing plants more economically viable), it is more likely that the open cycle will remain significantly cheaper, at least for the coming 50 years.[11] Another study carried out by the Massachusetts Institute of Technology upholds the same view on the economic aspects of reprocessing, concluding that according to certain assumptions and under the present US conditions, the closed cycle method will be four times as expensive as the open cycle system (MIT, 2003). Reprocessing will only be competitive if the uranium prices substantially increase, but considering the fact that the identified uranium sources have grown over the last couple of years (IAEA-NEA, 2011) that is unlikely. It seems reasonable to conclude that reprocessing creates additional short-term economic burdens (compared to the open fuel cycle option).

15.3.5 Important Assumptions Concerning Reprocessing

It may thus be concluded that the open fuel cycle seems to be preferable from the perspective of the present generation, since it creates fewer safety and security risks and is less costly. The closed fuel cycle is, on the other hand, more beneficial from the point of view of future generations, because it reduces the

10. The number of operational power reactors in Japan before the Fukushima Earthquake and Tsunami was 55 (WNA, 2011), while on January 1st 2013 there were 50 reactors operable (WNA, 2013).

11. The key findings of this study were later published (Bunn, Fetter, Holdren, & Van der Zwaan, 2005).

long-term safety concerns attached to waste disposal while helping to extend nonrenewable resources further into the future. At the same time, the closed cycle creates greater short-term safety and security concerns as well as economic burdens. This results in an intergenerational justice dilemma. Do we, the present generation, have an obligation to reduce future burdens as much as possible? If so, to what extent and under what conditions are the additional burdens[12] for present generations justifiable? The decision to choose a given fuel cycle should thus be viewed from the point of view of what justice to future generations requires from us.[13]

The long-term safety benefits of the closed fuel cycle rely however on three important assumptions that need to be further scrutinized:

> Assumption 1: *The extracted uranium and plutonium will be reused for energy purposes;*
> Assumption 2: *The waste lifetime will be substantially reduced (by a factor of 20); and*
> Assumption 3: *The waste volume will be substantially reduced (by a factor of 3).*

Assumption 1 states that all the separated uranium and plutonium will be reused in energy reactors. The separated uranium could be easily reintroduced to the beginning of the fuel cycle. In the case of plutonium, however, a strong case must be made because without reusing (or destroying) all the separated plutonium neither the waste-management benefit assumptions 2 & 3 nor the resource durability assumption 1 make any sense.

In other words, if we fail to make it plausible that extracted plutonium will eventually be fissioned, we create more de facto risks—both in the short-term and the long-term—and that will substantially alter the analysis. Critics of reprocessing emphasize that "about half of the plutonium being separated [worldwide] is simply being stockpiled at reprocessing plants along with the associated high-level waste from reprocessing" (Von Hippel, 2007, p. 3). This only creates additional proliferation concerns.

It seems evident that countries pursuing the closed fuel cycle option need to take these considerations into account. A 2011 report issued by the Royal Society in the UK presented several recommendations designed to respond to these proliferation concerns of reprocessing, for instance "[s]pent fuel should be reprocessed only when there is a clear plan for its reuse" (The-Royal-Society, 2011). These recommendations furthermore assert that separated plutonium should be converted to MOX fuel as soon as possible, since it is the extracted (pure) plutonium that is proliferation sensitive; mixed with uranium it will lose its military usefulness. By the same token, new

12. Burden in this context entails technological risks as a result of reprocessing, the transport of spent fuel and HLW etc. as well as economic costs for the closed fuel cycle.
13. Together with Jan Leen Kloosterman, this argument has been extensively discussed elsewhere (Taebi & Kloosterman, 2008).

technological solutions have been presented to keep uranium and plutonium mixed during reprocessing. Even though plutonium could still be separated from such a mixture (Von Hippel, 2007, p. 4), it could make reprocessing more proliferation sensitive. Furthermore, global governance proposals have been drawn up, particularly in the US, the most notable of which is the "Global Nuclear Energy Partnership" (GNEP) proposal, in which it is stated that weapon states and Japan are expected to provide reprocessing services for other states.[14]

When substantiating the waste-management assumptions (2 and 3), there is an important technical problem. The existing method for fissioning plutonium is to mix it with depleted uranium and insert it into LWRs as MOX fuel. However, the presence of the major isotope of uranium (^{238}U) in MOX means that during operation we both fission and breed (i.e. create) plutonium. This means that some neutrons will cause the fissioning of existing plutonium in MOX while others will be absorbed by the fertile ^{238}U, which will cause the production of more plutonium. On balance, spent MOX fuel (not to be confused with the spent fuel of uranium dioxide in an open cycle) contains 30% less plutonium that the fresh MOX fuel (Wallenius, 2011, p. 21). If one examines the radiotoxic inventory of MOX spent fuel, one can argue that disposing of spent MOX fuel poses at least the same long-term risks.[15]

This means that the current closed fuel cycle with single MOX fuel in Light Water Reactors does not really live up to expectations. Assuming that all plutonium can be extracted from spent fuel,[16] we can argue that the radiotoxicity of the remaining HLW will be substantially reduced. However, the remaining MOX spent fuel possesses the same long-term radiotoxicity. This seriously calls assumption 2 into question. By the same token, assumption 3 seems to be problematic as well. Likewise, the question here is whether the volume of the waste emanating from MOX spent fuel has been taken into account in the claim that reprocessing reduces the volume of waste to one third of the original volume of the spent fuel.[17]

This means that it will not be possible to remove all plutonium in the existing closed fuel cycles. Alternative fuel cycles should be considered for maximum actinide recycling, if a long-term waste-management benefit is to be effected. Full plutonium recycling is, for instance, possible with the "fully closed fuel

14. Information retrieved from the website of the World Nuclear Association (WNA) http://www.world-nuclear.org/info/inf117_international_framework_nuclear_energy_cooperation.html (retrieved on February 12 2013). The former GNEP has further been developed to The International Framework for Nuclear Energy Cooperation (IFNEC).

15. This is because there will be a higher concentration of long-lived transuranic elements (e.g. plutonium americium etc.) (Wallenius, 2011, p. 21).

16. A recovery rate of 99.88% for plutonium has been achieved in La Hague (Wallenius, 2011, p. 18).

17. Unfortunately I cannot find credible figures regarding volumes of spent MOX fuel.

cycle": a once-through cycle in an LWR after which spent fuel is recycled in a fast reactor (MIT, 2003, Figure 4.3). Fast reactors use fast neutrons in contrast to the thermal neutrons characteristic of LWRs; fast neutrons possess high kinetic energy and are therefore capable of fissioning more actinides. In such a fully closed cycle, only fast reactor fuel will be reprocessed; the extracted plutonium involved in the reprocessing of spent fast reactor fuel will be used as fresh fuel; the remaining waste is supposed to be free of plutonium (MIT, 2003, p. 33).[18] Alternatively, we can use thorium to get rid of plutonium. This will be further discussed in the next section.

15.4 IS THORIUM A VIABLE SUBSTITUTE OR SUPPLEMENT FOR NUCLEAR FUEL?

Since the beginning of nuclear energy era, thorium has been considered a viable nuclear fuel. Especially because in the early years of the development (i.e. 1950s to 1970s) the nuclear industry was rapidly growing and uranium resources were considered to be insufficient to accommodate that growth (Lung & Gremm, 1998). Especially Germany and the US were interested in utilizing thorium as an alternative fuel; with the new discovery of uranium deposits this enthusiasm faded away in those same countries in 1970s (IAEA, 2005). Nevertheless, for two reasons there is still a serious interest in thorium. Firstly, thorium is 3–4 times more abundant than uranium. Secondly, the global distribution of thorium is different from uranium. Countries with large thorium and limited uranium deposits, such as India, therefore have a vested interest in thorium fuel cycles. In addition to the wish to increase the availability of nuclear energy fuel, there have been other historical reasons for such an interest in thorium, reasons linked to the perspective of enhancing security (i.e. reducing proliferation concerns), and the benefits for the nuclear waste management (MIT, 2011, p. 181).

Using thorium instead of uranium does, however, present a technical problem. Natural thorium is not fissile and therefore not directly deployable in a reactor. Much like the major uranium isotope (^{238}U) that also first needs to be converted into a fissile material (i.e. plutonium), thorium must first be converted into fissile, thus deployable, ^{233}U. This is why we can only deploy thorium in conjunction with another fissile material. Depending on the perspective from which thorium is deployed, different fissile materials could be mixed with thorium, including the fissile uranium isotopes (^{233}U) produced during the earlier irradiation of thorium, enriched uranium (^{235}U) of the type used in LWRs, extracted plutonium from LWR spent fuel, Highly Enriched Uranium (HEU) or the weapon-grade plutonium extracted from dismantled warheads after the Cold War.

18. Together with Andrew Kadak, the intergenerational consequences of several future fuel cycles have been thought through. This cycle is analyzed in that paper and it is called there the transmuter fuel cycle; (Taebi & Kadak, 2010, pp. 1351–1352).

Since thorium fuel is very much in an experimental stage, there is no operational thorium cycle that can be discussed. Here below I will highlight the basic characteristics of two possible thorium cycles that might be suitable for reducing proliferation and for promoting waste-management benefits.

15.4.1 Proliferation Resistance: Using Thorium to Produce Less Plutonium

The main reason why thorium made such a comeback in the 1990s was because it could contribute to making fuel cycles more proliferation resistant. One of the leading approaches was to construct a fuel assembly consisting of fuel rods with reactor grade uranium (with no more than 20% fissile ^{235}U) and natural thorium (^{232}Th); this fuel would be deployable in LWR. The special design makes it possible to refuel uranium and thorium at a different pace; uranium needs to be refueled more frequently. It has been estimated that this design would reduce the amount of plutonium produced to somewhat more than one third of the level of a typical uranium open fuel cycle in an LWR.[19]

Producing less plutonium is of course a security improvement in its own right, but there are two more factors that further support the proliferation-resistant aspect of this fuel. Firstly, the compilation of different plutonium isotopes of this Th-U fuel is such that a bomb made from this plutonium would probably not have much explosive yield (Hargraves & Moir, 2010).[20] Secondly, it would be more difficult to separate plutonium from spent Th-U fuel than to remove plutonium from an LWR spent fuel. This requires further explanation. During irradiation of Th in the reactor, not only fissile ^{233}U will be produced but also another uranium isotope, namely ^{232}U. This isotope has a very short half-life (73 years) and it is a strong gamma emitter (IAEA, 2005, pp. 82–83). This means that heavily shielded facilities and automated equipment that could be remotely operated would be required to separate plutonium from the remaining fuel (IAEA, 2005; Kazimi, 2003). It has been argued this would necessitate well-funded resources at national program level and that enriching uranium to a high degree or separating plutonium from spent uranium fuel (which is the more *conventional* way of manufacturing nuclear explosives) would probably be less problematic (Hargraves & Moir, 2010, p. 312).

Thorium could also be deployed in combination with plutonium. This has two advantages: (1) less plutonium (and other long-lived actinides) will be produced and (2) existing stockpiles of plutonium (derived from reprocessed spent

19. This figure is based on IAEA estimations. There are two ways to mix uranium and thorium, namely homogenously and heterogeneously. From the proliferation resistance perspective it is more beneficial to mix them heterogeneously. This estimation shows a reduction in plutonium from 250 kg (per GWh per year) in a uranium cycle to 70–90 in a heterogeneous uranium thorium cycle (IAEA, 2005, p. 82).

20. See Section 15.3.2 for more information on the proliferation potentials of plutonium isotopes.

fuel or dismantled nuclear weapons) could then be burned. The Th–Pu fuel cycle is believed to consume twice as much plutonium as the current uranium plutonium MOX fuel (IAEA, 2005, p. 16).

15.4.2 Waste-Management Benefits of Using Thorium in Molten Salt Reactors

Thorium fuel could also be used in Molten Salt Reactors (MSR), which use liquid rather than solid fuel. All the power reactors currently operational in the world make use of solid fuel. It is a combination of molten thorium fluoride and uranium fluoride[21] that will be used in this type of reactor; the molten salt stream will serve both as a fuel and a coolant. The salt circulates and fissioning takes place during the process. There are at least two important safety features that characterize this approach. Firstly, the coolant (molten salt) is not pressurized and cannot therefore explode. This does away with the need to contain the reactor in thick-walled reactor pressure vessels designed to prevent radioactive release following pressure-induced explosion. Secondly, MSR has at the bottom a plug of salt (a different kind of salt than the fuel/coolant) that is kept at temperatures below its freezing point. If the molten salt rises to a critical temperature, the plug will melt and the fuel/coolant will be poured into a catch basin; this is a passive safety feature (Hargraves & Moir, 2010, p. 310).

More importantly, MSR claims to have attractive waste-management benefits. In thorium and uranium fluoride cycles, the waste stream of Th-U fuel could result in the production of much lower quantities of plutonium and long-lived minor actinides (i.e. americium, neptunium, and curium) and more or less the same amount of fission products (Gruppelaar & Schapira, 2000). Fission products are short-lived isotopes in the waste that will generally decay to nonradiotoxic levels within 300 years. After that period, thorium uranium spent fuel is estimated to be 10,000 times less radiotoxic than spent fuel derived from a uranium open fuel cycle (Hargraves & Moir, 2010, p. 309). On the other hand, other long-lived radionuclides will be produced, including other isotopes of thorium (^{229}Th) and protactinium (^{231}Pa).

15.4.3 Challenges and Shortcomings of Thorium Cycles

It seems that there are no technological barriers to the development of different thorium fuel cycles. However, compared to uranium cycles, the databases and experiences with thorium fuels are very limited (IAEA, 2005). In relation to this technological immaturity, the UK National Nuclear Laboratory (NNL) holds the position that "the thorium fuel cycle does not currently have a role to play in the UK context, other than its potential application for plutonium management

21. This combination is actually added to a third carrier salt.

in the medium to long term" (NNL, 2010). An additional reason for the UK being averse to thorium cycles is the uncertain waste-management benefits since the serious reduction in the waste stream of thorium cycles is based on a "self-sustaining thorium cycle" while "[m]ore realistic studies that take into account the effect of the U-235 or Pu-239 seed fuels required to breed the U-233 suggest the benefits are more modest" (NNL, 2010). Yet the NLL concedes that the substantial reduction in radiotoxicity promised by full thorium recycling does provide significant justification for continuing R&D investment.

A recent study conducted by the Norwegian government reached more or less the same conclusion; also Norway sees every reason for continuing with R&D but no reason for immediately developing a thorium cycle (Kara, 2008). Norway is an interesting case because of its vast thorium resources (i.e. 170,000 tons, equivalent to 100 times all the oil extracted from the Norwegian fields (Kara, 2008, p. 1)). Yet Norway is not operating any uranium cycles either, nor does it have plan to do so.

As discussed above, one important incentive for wanting to develop thorium cycles is to increase resistance to proliferation. However, like uranium, thorium has its origins in military applications. In 1955, in Operation Teapot, the US detonated a device with ^{233}U; the yield of this bomb was less than anticipated. In 1998 India also detonated a very small device based on ^{233}U.[22] Ultimately, it seems reasonable to presume that the strong gamma emitting ^{232}U makes thorium cycles more proliferation-resistant than uranium cycles. To sum up, thorium cycle seems to be more proliferation-resistant since it enables us to use up the currently existing reserves of plutonium and it produces less plutonium during operation. We should not, however, forget that some thorium cycles use enriched uranium (up to 20%) as the fissile material to keep the chain reaction going; see Section 15.4.1. This means that the proliferation-sensitive enrichment facilities will still remain necessary in these scenarios.

Finally, the nonproliferation feature of the thorium cycle does pose additional safety risks and economic burdens. As stated above, the presence of ^{232}U is a guarantee of nonproliferation; the separation of ^{233}U would be problematic because of the strong gamma emissions. By the same token, thorium spent fuel has to be processed with more caution and in remotely controlled facilities that would raise the costs of processing. Furthermore, possible radiation leakage during the operation of the reactor or waste reprocessing would pose serious radiation risks.

Despite the fact that there seems to be no serious interest in thorium among the current leading nuclear power countries in Europe or in North America, the new emerging economies of India and China have already shown a definite interest in thorium. China sees thorium as a possibility in its future shift

22. This information has been retrieved from the web site of the WNA: http://www.world-nuclear.org/info/inf62.html retrieved on 18 January 2013.

toward nuclear power as its primary energy source. India, on the other hand, has already been working on a thorium cycle since the 1950s. India has major thorium resources and the country's main aim is to develop an energy system based on domestic resources and domestic technology. These developments in China and India could be decisive for the future of thorium because together they will possess more than one-third of all future power plants (if one includes those under construction, planned or proposed).[23]

15.5 CONCLUSIONS

In this chapter I have first presented four moral values at stake in nuclear power production, namely safety, security, economic viability, and sustainability; the latter can be divided into resource durability and environmental friendliness. Using these values, I clarified some of the moral dilemmas presented by the nuclear fuel cycle. I first discussed two existing fuel cycles, the open fuel cycle that uses uranium once (the spent fuel has to be disposed of underground for 200,000 years) and the closed fuel cycle that extends the open cycle by reprocessing spent fuel (there the waste lifetime is reduced and the still deployable materials in spent fuel can be reused). I argued that the open uranium fuel cycle seems to be preferable from the perspective of the present generation, since it creates fewer safety and security risks and it is less costly. The closed fuel cycle, on the other hand, seems to be more beneficial from the point of view of future generations, because it reduces the long-term safety concerns of waste disposal while helping to make nonrenewable resources last longer. At the same time, the closed cycle creates more short-term safety and security concerns and economic burdens. This leads to the intergenerational justice dilemma. Does the present generation have an obligation to reduce future burdens as much as possible? If so, to what extent and under what conditions are the (additional) burdens for present generations acceptable?[24] The choice of a given fuel cycle should thus be made on the basis of what justice to future generations requires from us.[25]

The long-term safety benefits of the closed fuel cycle are, however, based on the assumption that all plutonium extracted during reprocessing will be used for energy purposes. I have scrutinized this assumption and have gone on to argue that current closed fuel cycles based on MOX fuel in LWR do not live up to that expectation. A substantial reduction in the waste lifetime could, however, be achieved with new (not yet developed) fuel cycles based on fast reactors

23. Worldwide there are 65 reactors under construction (29 in China and 7 in India), 167 reactors on order or planned (51 in China and 18 in India) and 317 proposed reactors (120 in China and 39 in India) (WNA, 2013).

24. One important issue is how the additional burdens are being distributed among people living today. See for a brief discussion of this issue (Taebi, 2011, pp. 187–188).

25. Together with Jan Leen Kloosterman, this argument has been extensively discussed elsewhere (Taebi & Kloosterman, 2008).

and full plutonium recycling. Thorium cycles are also being developed with a view to the waste-management considerations and proliferation concerns. I have highlighted these features in two propositions for possible future thorium fuel cycles.

Firstly, a combination of thorium oxide and uranium oxide was proposed to reduce security concerns; the amount of plutonium produced in such a cycle will be one third of that produced in a conventional closed fuel cycle using MOX. Secondly, MSRs were proposed because of their waste-management benefits. This would represent a revolutionary change in reactor design and it would use a combination of molten thorium fluoride and uranium fluoride as fuel; such a combination would, it is believed, substantially reduce the production of plutonium and other long-lived actinides. In short, thorium seems to offer promising prospects for the future of nuclear power, but it is a technology that is still several decades away from industrialization. When developing new thorium fuel cycles (and the reactors they require) the reasons for their development must be clear. In addition to that we need to further address the moral dilemmas attached to each cycle. I will illustrate this by giving the following example.

The nonproliferation benefits of thorium cycles are partly based on the fact that irradiated thorium contains two isotopes of uranium: namely ^{233}U and ^{232}U. The former isotope can be used for energy purposes but it has also destructive potential. What could impede the separation of this proliferation-sensitive uranium is the strong gamma emissions of the latter isotope, ^{232}U. A potential proliferator needs substantial funds and fully remotely operated facilities. However, the same nonproliferation feature would create additional processing costs for energy production. More importantly, there are also serious radiation risks resulting from the possible leakage of MSRs during operation. There seems thus to be a conflict between security on the one side and safety and economic viability on the other side.

To sum up, when deciding about the desirability of any new nuclear fuel cycle we need to appropriately address the moral dilemmas. What constitutes desirable depends on how we make trade-offs between different values. Such analysis has at least three advantages. Firstly, it adds a more nuanced perspective to currently dominating analyses solely based on economic perspectives. Secondly, this analysis could support research and development paths that could well culminate in the industrialization of a certain technology. Thirdly, it is a first step toward a broader discussion on the desirable energy mix and the possible role that should be played by nuclear energy (either uranium or thorium).

ACKNOWLEDGMENTS

I wish to thank the editors and Jan Leen Kloosterman for their useful comments. Of course, the usual disclaimer with regard to authorial responsibility applies.

REFERENCES

Barry, B. (1989). *The ethics of resource depletion democracy, power and justice, essays in political theory.* (pp. 511–525), Oxford: Clarendon Press.

Bunn, M., Fetter, S., Holdren, J. P., & Van der Zwaan, B. (June, 2005). The economics of reprocessing versus direct disposal of spent nuclear fuel. *Nuclear Technology, 150*, 209–230.

Eggermont, G., & Hugé, J. (2011). *Nuclear energy governance, deliverable 4.1, SEPIA project.* Brussels: Belgian Science Policy 2011 (Research Programme Science for a Sustainable Development).

EPA. (2005). *Public health and environmental radiation protection standards for Yucca Mountain.* 40 CFR Part 197, Part II, Washington D.C.: Office of Radiation and Indoor Air U.S. Environmental Protection Agency.

EPA. (2008). *Public health and environmental radiation protection standards for Yucca Mountain; final rule.* 40 CFR Part 197, Part III, Washington D.C.: Office of Radiation and Indoor Air U.S. Environmental Protection Agency.

Gruppelaar, H., & Schapira, J. P. (2000). *Thorium as a waste management option.* Final report EUR 19142EN, Brussels: European Commission.

Hansson, S. O. (1998). *Setting the limit: occupational health standards and the limits of science.* New York: Oxford University Press.

Hargraves, R., & Moir, R. (2010). Liquid fluoride thorium reactors. An old idea in nuclear power gets reexamined. *American Scientist, 98*(4), 304.

IAEA. (2005). *Thorium fuel cycle—Potential benefits and challenges.* Austria: IAEA.

IAEA. (2007). *IAEA safety glossary, terminology used in nuclear safety and radiation protection.* Vienna: IAEA.

IAEA-NEA. (2011). *Uranium 2011: Resources, production and demand.* Paris: IAEA and NEA-OECD.

IAEA, Euratom, FAO, IAEA, ILO, IMO., et al. (2006). *Fundamental safety principles.* Vienna: A joint publication of Euratom, FAO, IAEA, ILO, IMO, OECD-NEA, PAHO, UNEP, WHO.

ICRP. (1959). ICRP Publication 1, *Recommendations of the International Commission on Radiological Protection: Revised December 1954*(Vol. 1). Oxford: Pergamon Press.

Kara, M. (2008). *Thorium as an energy source: Opportunities for Norway.* Oslo: The Thorium Report Committee.

Kazimi, M. (2003). Thorium fuel for nuclear energy. *American Scientist, 91*(5), 408–415.

Lung, M., & Gremm, O. (1998). Perspectives of the thorium fuel cycle. *Nuclear Engineering and Design, 180*(2), 133–146.

MIT. (2003). *The future of nuclear power: An interdisciplinary MIT study.* Cambridge, MA: Massachusetts Institute of Technology (MIT).

MIT. (2011). *The future of nuclear power: An interdisciplinary MIT study.* Cambridge, MA: Massachusetts Institute of Technology (MIT).

NNL. (2010). *The thorium fuel cycle. An independent assessment by the UK National Nuclear Laboratory.* UK National Nuclear Laboratory (NNL).

NRC. (1966). *Understanding risk: informing decisions in a democratic society.* Washington DC: National Research Council (NRC), National Academy Press.

NRC. (2006). *Going the distance? The safe transport of spent nuclear fuel and high-level radioactive waste in the United States.* Washington D.C.: National Research Council.

de Saint-Georges, L. (2008). The radiological risk: uncertainty related to low dose exposure. In G. Eggermont & B. Feltz (Eds.), *Ethics and radiological protection* (pp. 47–58). Louvain-la-Neuve: Academia-Bruylant.

Shrader-Frechette, K. (1993). *Burying uncertainty: Risk and the case against geological disposal of nuclear waste*. Berkeley: University of California Press.

Shrader-Frechette, K. (2005). Mortgaging the future: dumping ethics with nuclear waste. *Science and Engineering Ethics*, *11*(4), 518–520.

Smeesters, P. (2008). Health effects of exposure to ionizing radiation: overview, justification principle, precaution and uncertainties. In G. Eggermont & B. Feltz (Eds.), *Ethics and radiological protection* (pp. 13–28). Louvain-la-Neuve: Academia-Bruylant.

Taebi, B. (2011). The morally desirable option for nuclear power production. *Philosophy & Technology*, *24*(2), 169–192.

Taebi, B. (2012). Intergenerational risks of nuclear energy. In S. Roeser, R. Hillerbrand, P. Sandin & M. Peterson (Eds.), *Handbook of risk theory. Epistemology, decision theory, ethics and social implications of risk* (pp. 295–318). Dordrecht: Springer.

Taebi, B., & Kadak, A. C. (2010). Intergenerational considerations affecting the future of nuclear power: equity as a framework for assessing fuel cycles. *Risk Analysis*, *30*(9), 1341–1362.

Taebi, B., & Kloosterman, J. L. (2008). To recycle or not to recycle? An intergenerational approach to nuclear fuel cycles. *Science and Engineering Ethics*, *14*(2), 177–200.

Taebi, B., & Kloosterman, J. L. Designing for nuclear safety, security & sustainability: a philosophical discourse of reactor design. In J. van den Hoven, I. Van de Poel & P. Vermaas (Eds.), *Handbook of ethics and values in technological design*. Dordrecht: Springer, in press.

The-Royal-Society. (2011). *Fuel cycle stewardship in a nuclear renaissance*. London: The Royal Society Science Policy Centre report 10/11.

Vandenbosch, R., & Vandenbosch, S. E. (2007). *Nuclear waste stalemate: Political and scientific controversies* (Vol. 61). Salt Lake City: The University of Utah Press.

Von Hippel, F. N. (2007). *Managing spent fuel in the United States: The illogic of reprocessing: A Research Report of the International Panel on Fissile Materials (IPFM)*.

Wallenius, J. (2011). *Transmutation of nuclear waste*. Märsta: Leadcold Books.

WCED. (1987). *Our common future*. Oxford: World Commission on Environment and Development (WCED).

WNA. (2011). *World Nuclear Power Reactors & Uranium Requirements 2011, Information Paper (2 March 2011)*. Retrieved from: http://www.world-nuclear.org/info/reactors.html.

WNA. (2013). *World Nuclear Power Reactors & Uranium Requirements (1 January 2013)*. Retrieved from: http://www.world-nuclear.org/info/reactors.html.

Part V

Public Participation

Chapter 16

Social Identities and Public Uptake of Science

Chernobyl, Sellafield, and Environmental Radioactivity Sciences

Brian Wynne
CSEC, Department of Sociology, Lancaster University, Lancaster, UK
E-mail: b.wynne@lancaster.ac.uk

Chapter Outline

16.1 Introduction 283
16.2 Sheep Farmers, Scientists, and Radiation Hazards: The Background 285
16.3 Scientific Knowledge and Social Identities 288
16.4 Public Belief and Private Dissent 291
16.5 Credibility: The Social Dimension 299
16.6 Conclusions: Lay Reflexivity and Social Identities 303

16.1 INTRODUCTION

The Chernobyl experience represents a major example of the general point that public responses to risks and risk information are rationally based upon their experience and judgment of the credibility and trustworthiness of the institutions, which claim to be in charge (Wynne, 1980). The 1986 Chernobyl nuclear accident and its far-flung radioactive fallout provided a richly endowed laboratory for identifying the factors affecting the public credibility of science, and for examining the relationship between that and the "public understanding of science" issue.

As the 1985 London Royal Society report on public understanding of science made abundantly clear, much of the impetus for the current interest in this subject stems from a broad anxiety among scientists and policy makers about what they see as the public's inability or unwillingness to understand "correct" messages about risks as given to them by the experts (The Royal Society of London, 1985). This self-defeating scientistic conception of the public understanding of science problem has been criticized before (Collins, 1987), and attributed to the

Radioactivity in the Environment, Volume 19
ISSN 1569-4860, http://dx.doi.org/10.1016/B978-0-08-045015-5.00016-2

neurosis of scientific institutions about their public credibility. I have noted the irony that this formulation of the problem only encourages more public alienation, hence justifying and consolidating the neurosis (Wynne, 1992).

A more interpretive framing of the public understanding of science problem, and of the closely associated risk perception issue, starts from the observation that public experiences of risks, risk communications or any other scientific information is never, and can never be, a purely intellectual process, about reception of knowledge per se (Douglas, 1986). People experience these in the form of material social relationships, interactions, and interests, and thus they logically define and judge the risk, the risk information, or the scientific knowledge as part and parcel of that "social package". A corollary of this is that people do not simply not understand science when they are seen to disregard it; they do not recognize it, or identify with it, morally speaking.

In other words, it is increasingly accepted that the issues of public understanding of science, and of public risk-perceptions, are not so much about public capabilities in understanding technical information, but about the trust and credibility they are prepared to invest in scientific spokespersons or institutions. The unduly cognitive emphasis of much public understanding of science and risk perception work has been undermined and shifted through this social transformation, which is still far from fully developed. (Several main sociological perspectives are represented in Krimsky & Golding, 1992.)

In this paper, I draw upon fieldwork from the Chernobyl issue to attempt to take this sociological transformation one step further. This fieldwork involved in-depth interviews with hill sheep farmers in the Lake District of Northern England who received intensive expert information and advice about environmental hazards from radioactive cesium isotopes deposited as a fallout from the Chernobyl accident. This fallout caused restrictions upon the free movement and sale of sheep in an area dominated economically and culturally by this particular traditional way of life (Wynne, 1989).

The conventional framing of the public understanding of science issue misleadingly reifies scientific knowledge, as if it were objective and context-free. The more recent recognition that trust and credibility are the basic dimensions in public "understanding", now also risks reifying these concepts, which would be just as misleading (Renn, 1991). Trust, or trustworthiness, and credibility are relational terms, about the nature of the social relationships between the actors concerned. They are not intrinsic to either actor nor to the information said to be transmitted between them (Luhmann, 1990).

I disavow theoretical commitments to "essentialist" models of beliefs, values, and trust, and of the relationships between them (Unger, 1976). Most research in public understanding of science involves observing or measuring what people believe after they have been exposed to scientific information of some kind(s). This is true of both large-scale, often quantitative survey research, and of more small-scale, qualitative research. An assumption tends to be made in both approaches, though it is not essential to the latter, that the beliefs or "understandings" are internally coherent, with a primary existence in the sense that they are not derivatives of other factors. Whether the beliefs are measured quantitatively or qualitatively, and whether they are about electrons, genes, or trustworthiness of social actors such as scientists, they are taken to be unambiguously real.

In this analysis of the reactions of hill-farmers to the Chernobyl crisis I wish to go not one, but two steps beyond the cognitivist approach (Lave, 1986), to show that the best explanatory concepts for understanding public responses to scientific knowledge and advice are not trust and credibility per se, but the social relationships, networks and identities from which these are derived. If we view these social identities as incomplete, and open to continual (re)construction through the negotiation of responses to social interventions such as the scientists represented, we can see trust and credibility more as contingent variables, influencing the uptake of knowledge, but dependent upon the nature of these evolving relationships and identities.

In the case described below, our interviews revealed the complex and multidimensional social basis of trust and credibility as a central factor in the reception or "understanding" of scientific advice by the farmers. They monitored and constructed evidence on this trustworthiness factor from a far wider range of behavior and demeanor of the expert institutions, including long-past behavior on related issues, than would be recognized by cognitivist approaches.

However, the same interviews also indicate ambivalence in relation to trust, and to what - and who- they actually believed about the sources of the radioactive cesium which had damaged their livelihoods. This ambivalence of belief and trust reflects the multifaceted and plural social networks and identities which the sheep farmers inhabit, contradicting simple notions of an unreflexive traditional lay culture. The evidence suggests that the beliefs the farmers construct, including their beliefs about the credibility and trustworthiness of different scientific and other social actors, are functions of the social networks with which they identify. There is nothing intrinsically different in this to the basic structure of scientific belief and commitment (Knorr-Cetina, 1989). "Understanding" or knowledge, its precision and resilience, is a function of social solidarity, mediated by the relational elements of trust, dependency, and social identity—constructing that "intellectual" understanding should be seen as a process of social identity-construction.

16.2 SHEEP FARMERS, SCIENTISTS, AND RADIATION HAZARDS: THE BACKGROUND

The hill-sheep farmers near the Sellafield (formerly Windscale) nuclear fuels reprocessing complex in the Lake District of Cumbria, Northern England, have more than a personal health interest in radiation risk information. Their economic viability depends totally upon rearing a large crop of lambs each spring, and selling them in the late summer and autumn, before they run out of their meager valley grazing due to the temporary overpopulation of lambs. The UK lamb industry exports heavily to continental Europe. Any public perception of radioactive blight on its product would destroy the industry, especially the hill sheep-farming sector, which is a key early part of the breeding cycle, but which is economically more fragile and offers the farmer few or no alternatives compared to lowland sheep farming.

The upland hill farming region in the Lake District is one of the few locations of relative solidarity and distinctive traditional cultural identity left in industrial Britain. Although (as shown later) this should not be overstated, these communities share an unusually demanding livelihood as a way of life; they occupy a distinct and sought-after geographical location, and have common historical traditions, linguistic dialects, and recreational pursuits. They also share particular "external" socioeconomic threats such as subordination to tourism, landlords and authorities who appear to be more and more concerned with meeting environmental and urban recreational demands on the country than with sheep farming. All of this was an important context of the post-Chernobyl crisis.

In May 1986, following the Chernobyl accident, upland areas of Britain suffered heavy but highly variable deposits of radioactive cesium isotopes, which were rained out by localized thunderstorms. The effects of this radioactive fallout were immediately dismissed by scientists and political leaders as negligible, but after six weeks, on 20 June 1986, a ban was suddenly placed on the movement and slaughter of sheep from virtually the whole of Cumbria.

Although this shock was mitigated somewhat by the confident scientific reassurances that the elevated levels of cesium in sheep, and hence the ban, would only last about three weeks, at the end of this period the restrictions were instead extended indefinitely. The confident dismissal of any effects at all only two months earlier, had now changed to the possibility of wholesale slaughter and complete ruin of hill sheep farms at the hands of a faraway stricken nuclear plant. At that time, over four thousand British farms were restricted. The initially wide restricted area in Cumbria (which included about five hundred farms) was whittled down within three months to a central crescent covering one hundred and fifty farms (see Figure 16.1). These farms remained restricted for many years, contrary to all the scientific assertions of the time. A later scientific review indicated that they could remain so for years, overturning completely the scientific basis upon which the previous policy commitments were made (Howard & Beresford, 1989).

Very close to this recalcitrant central "crescent" of longer-term radioactive contamination, almost suggesting itself as its focal point, is the Sellafield-Windscale nuclear' complex. The stories of Sellafield-Windscale and Chernobyl are intertwined in ways which I now unravel.

Sellafield-Windscale is a huge complex of fuel storage ponds, chemical reprocessing plants, nuclear reactors, defunct military piles, plutonium processing and storage facilities, and waste processing and storage silos. It has developed from its original role in the early 1950s of producing purely weapons-grade plutonium into a combined military and commercial reprocessing facility which stores and reprocesses thousands of tonnes of UK and foreign spent fuel. It is by far the biggest employer in the area, with a regular workforce of some five thousand swollen by a construction workforce of nearly the same size. It dominates the whole area not only economically, but also socially and culturally.

Sellafield has been the center of successive controversies, accidents, and events relating to its environmental discharges and workforce radiation doses, with increasingly powerful criticisms not only of allegedly inadequate

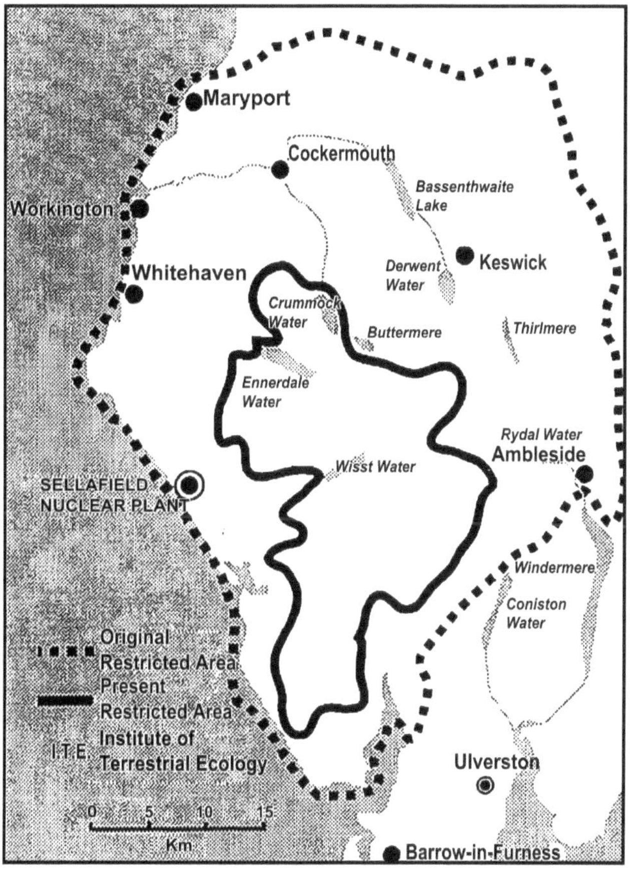

FIGURE 16.1 Map showing the restricted areas of Cumbria, from June 1986 (the original area), and from September 1986 (the long-term persistent one).

management and regulation, but also of poor scientific understanding of its environmental effects, and of the economic irrationality of the recycling option in nuclear fuel-cycle policy. In the early 1980s, the plant was alleged to be the center of excess childhood leukemia clusters; these excesses were confirmed by an official inquiry chaired by Sir Douglas Black, which nevertheless expressed agnosticism as to the cause. This controversy continues, with every scientific report exhaustively covered in the local and national media (Gardner et al., 1990). The plant operators were later shown to have misled the Black inquiry, inadvertently or not, over earlier levels of environmental discharge of radioactivity. In 1984, the operators were accused by the environmental group Greenpeace of contaminating local beaches above legal discharge levels, and were subsequently prosecuted; and in 1986 they were threatened with closure after another incident and an ensuing formal safety audit by the Health and Safety Executive. Despite huge investments in public relations, they have suffered a generally very poor public image for openness and honesty over the years.

Before most of these controversies developed, in 1957 the Sellafield-Windscale site suffered the world's worst nuclear reactor accident before Chernobyl, when a nuclear pile caught fire and burned for some days before being quenched (Arnold, 1992). It emitted a plume of radioactive isotopes, mainly iodine and cesium, over much the same area of the Lake District of Western Cumbria as that affected by the Chernobyl fallout. This fire and its environmental effects were surrounded by a great deal of secrecy. Although farmers in the vicinity were forced to pour away contaminated milk for several weeks afterward, at the time they reacted without any overt hostility or criticism of the industry. Even in 1977 when they had the opportunity during a public inquiry to join with an emergent coalition of various forces against a major expansion at Sellafield, the local farming population largely kept out of the argument (Wynne, 1982). Significantly however, it was only after 1987 that the fuller extent of the Windscale fire cover-up emerged into the public domain. In 1990 it was revealed in a television program that the ill-fated pile had in fact been allowed to operate with faults, which meant that highly irradiated spent fuel had been lying in the air streams emitted up the stack. Thus, it was exposed that the fire had been a blessing in disguise for the authorities, since any discoveries of local environmental contamination could be attributed to the one-off fire itself rather than to longer-standing irresponsible management that had allowed routine uncontrolled radioactive emissions to occur for some time before. The parallels with the Chernobyl issue nearly 30 years later are remarkable, as explained below.

The farming population in the Cumbrian hills is relatively stable, most farmers having lived through these controversies and events as near-neighbors. Indeed many of them have relations, neighbors, and casual employees who also work at the Sellafield-Windscale site. Not only is it close physically, it is also never far away from contemplation. Far from giving Sellafield-Windscale some welcome relief, the Chernobyl emergency ironically brought it even more to critical public attention.

16.3 SCIENTIFIC KNOWLEDGE AND SOCIAL IDENTITIES

At first, the scientific advice was that there would be no effects at all from the Chernobyl cesium fallout. After six weeks, these confident public reassurances were dramatically overturned when on 20 June 1986 the Minister for Agriculture announced the complete ban on sheep sales and movements in several affected areas, including Cumbria (see Figure 16.1). However, the shock waves from this reversal were contained by the strong reassurances accompanying the ban that it would be for three weeks only, by which time radioactivity levels in lamb would, it was confidently expected, have reduced beneath the level at which intervention was required. This short ban could be accommodated because very few if any hill lambs would be ready to sell before late August.

Yet after the three-week period, instead of lifting the ban the government announced an indefinite extension, albeit for a smaller area. This represented an altogether more serious situation in which the hill farmers faced ruin, because not only lamb crops but also breeding flocks faced starvation and wholesale slaughter due to inadequate grazing. The government introduced a scheme to remove this threat: it allowed farmers to sell lambs contaminated above the limit if they were marked, in which case the lambs could be moved to other areas but not slaughtered until their contamination had reduced. This blight factor collapsed the market price for marked sheep, and many lowland farmers bought them and then made handsome profits when they sold them after the sheep had decontaminated on their farms.

The hill farmers were left in a quandary. If they sold, they had to run the gauntlet of the threatening bureaucratic system, which had been established to manage the restrictions, and which consisted of prior notification, tests and controls, and paperwork, and offered only a possible and partial future compensation for catastrophically low prices. If they held on to their sheep they risked ruin from starvation, disease and knock-on effects, or from the costs for buying in extra feed. Yet even after the initial contradiction of their scientific beliefs, the scientists advised farmers to hang on because, as they persisted in believing, the contamination would fall soon—it was just taking a bit longer than expected. When farmers did follow expert advice and waited, they found once again that the advice was badly overoptimistic, and had led them into a blind alley in which many costly complications to farm management cycles had been introduced, and compensation was cut off because they had not sold within the prescribed period. In the circumstances, it was not surprising that our interviews found many farmers bitterly accusing the scientists of being involved in a conspiracy with a government that they saw as bent on undermining hill farming anyway.

Through the troubled and confused summer of 1986, in spite of mounting evidence and their own public embarrassment, the scientists persisted in their belief that the initially high cesium levels would fall soon. Only later did it emerge that these predictions were based upon a false scientific model of the behavior of cesium in the upland environment. The prevailing scientific model was drawn from empirical observation of alkaline clay soils, in which cesium is chemically adsorbed and immobilized and so is unable to pass into vegetation. But alkaline clay soils are not found in upland areas, which have acid peaty soil. This type of soil had been observed, but only for physical parameters such as depth-leaching and erosion, and not for chemical mobility.

Thus, the scientists unwittingly transferred knowledge of the clay soils to acid peaty soil, in which cesium remains chemically mobile and available to be taken up by plant roots. Whereas their model had cesium being deposited, washed into the soil and then locked up by chemical adsorption, thus only contaminating the lambs on a one-pass basis, in fact the cesium was recycling back from the soil into vegetation, and thence back into the lambs. This mistake became apparent only over the next several years, as contamination levels remained stubbornly high and the reasons were urgently sought. What was not lost on the farmers, however, was that the scientists had made unqualified reassuring assertions, then had been proven mistaken, and had not even admitted making such a serious mistake. Their exaggerated sense of certainty and arrogance was a major factor in undermining the scientists' credibility with the farmers on other issues such as the source of the contamination. In any case, the typical scientific idiom of certainty and control was culturally discordant with the farmers, whose whole cultural ethos routinely accepted uncertainty thus lack of control, and the need for flexible adaptation rather than prediction and control.

The structure of the scientific knowledge in play also embodied and, in effect, prescribed a particular social construct of the farmers (Wynne, 1989). To summarize this analysis, the degree of certainty expressed in scientific statements denied the ability of the farmers to cope with ignorance and lack of control; and the degree of standardization and aggregation of the scientific knowledge, for example the spatial units of variation of variables such as cesium contamination, denied the differences between farms, even in a single valley (and even within the same farm). At the same time, the scientists ignored farmers' own knowledge of their local environments, hill-sheep characteristics, and hill-farming management realities such as the impossibility of grazing flocks all on cleaner valley grass, and the difficulties of gathering sheep from open fells for tests.

As a result, the farmers felt their social identity as specialists within their own sphere, with its adaptive, informal cultural idiom, to be denigrated and threatened by this treatment. This was a reflection of the culture and institutional form of science, not of what specifically it claimed to know.

A graphic example of the scientists' denial of the specialist knowledge of the farmers was the scientists' decision to perform experiments on the value of the mineral bentonite to chemically adsorb cesium in the soil and vegetation, thus helping reduce recontamination of grazing sheep (Beresford et al., 1989). The bentonite was spread at different concentrations on the ground in different plots; the sheep from each plot were then tested at intervals, and compared with controls on zero-bentonite land. However, in order to do this the sheep were fenced in on contained fell-side plots. The farmers pointed out that the sheep were used to roaming over open tracts of fell land, without even fences between farms, and that if they were fenced in they would waste (lose condition), thus ruining the experiment. Their criticisms were ignored, but were vindicated later when the experiments were quietly abandoned for the reasons that the farmers had identified. The farmers had expressed valid and useful specialist knowledge for the conduct and development of science, but this was ignored. Similar experiences occurred over other aspects of hill-farming and the scientific knowledge relating to the management of the crisis.

In respect of both the "conspiracy theory" and the "arrogance theory" of science, the Cumbrian sheep farmers felt that their social identity as a specialist community with distinct traditions, skills, and social relations was under fundamental threat. These two models of science, which reinforced each other in the experienced threat to social identity, are mutually contradictory if taken literally. The former implies omniscience ("they knew all along that the high levels would last much longer than they admitted"); the latter implies unadmitted ignorance in science. This apparent anomaly exposes the futility of expecting consistent formal "beliefs" about science in research on public understanding; coherence lies at a deeper level, of the defense and negotiation of social identities. We examine this dimension next.

16.4 PUBLIC BELIEF AND PRIVATE DISSENT

Before the Chernobyl plume deposited its radioactive burden on the fells of Cumbria, there had been little or no controversy about radioactive cesium and related contamination of sheep on the high fells. Amongst several other issues concerning Sellafield's environmental discharges, contamination of pastures and grazing animals along the lowland coastal plain near the plant had been found and debated, for example in monitoring by Friends of the Earth, the Sellafield operators British Nuclear Fuels, and the Ministry of Agriculture, Fisheries and Food (Friends of the Earth, 1987). But little or no scientific monitoring or public attention had been paid to the high fells and their sheep; and no allegations of contamination of the fells and their sheep had been made. When the Chernobyl restrictions were announced however, and then almost immediately extended indefinitely, questions were very soon circulating locally about the real source of the contamination almost centering on Sellafield persisted (Figure 16.1) against scientific long-standing, than was being officially admitted. The fact that a crescent of high contamination almost centering on Sellafield persisted (Figure 16.1) against scientific reassurances that levels would decrease within a few weeks, was prima facie evidence of a hitherto hidden Sellafield dimension. The first national maps of cesium contamination measured after Chernobyl (in June and July 1986) had already shown remarkably high levels in West Cumbria, near Sellafield (The Guardian Newspaper, 1986).[1] The fact that these measurements, which were taken from vegetation samples (Figure 16.2), did not tally with the distribution as estimated from a combination of rainfall data during the crucial period while the radioactive cloud was over Britain, and models of rain-out of cesium from the atmosphere, also invited the question of whether a hidden factor, such as unacknowledged long-term Sellafield discharges, had created the differences (Smith & Clark, 1986). This factor would be picked up by the vegetation samples method, but not by the rainfall-data method (Figure 16.3).

In the manifest scientific confusion and inconsistency, it was as if the farmers had suddenly found an outlet for fears and suspicions that they had previously entertained, but felt unable to voice. Ironically, it was radioactive contamination which

1. These were produced by the Institute for Terrestrial Ecology at Merlewood in South Cumbria. The map was first published in The Guardian Newspaper, 25 July 1986, after pressure on the then Director of the Merlewood laboratory not to publish. See also House of Commons Agriculture Select Committee, 1988, Chernobyl: the Government's Response (London: HMSO). The Merlewood scientists rapidly gained a reputation amongst the farmers for being more open about uncertainties and better at listening to farmers' knowledge than government and industry scientists.

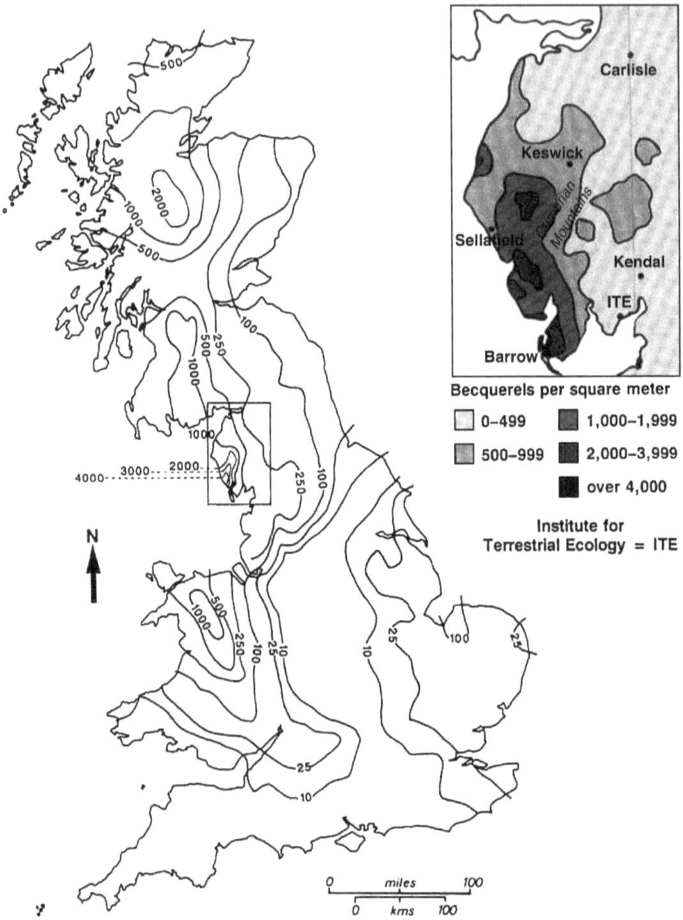

FIGURE 16.2 Contours of UK radioactive cesium contamination measured from vegetation June–July 1986. The data are in units of $Bq\,m^{-2}$.

scientists confidently proclaimed was nothing to do with Sellafield-Windscale, which gave the hill farmers their first embryonic voice about that local trouble-spot.

In our interviews, typical skepticism about the scientists' assertions of Sellafield's innocence was expressed as follows:

There's another thing about this as well. We don't live far enough away from Sellafield. If there's anything about we are much more likely to get it from there! Most people think that around here. It all comes out in years to come; it never comes out at the time. Just look at these clusters of leukaemia all around these places. It's no coincidence.

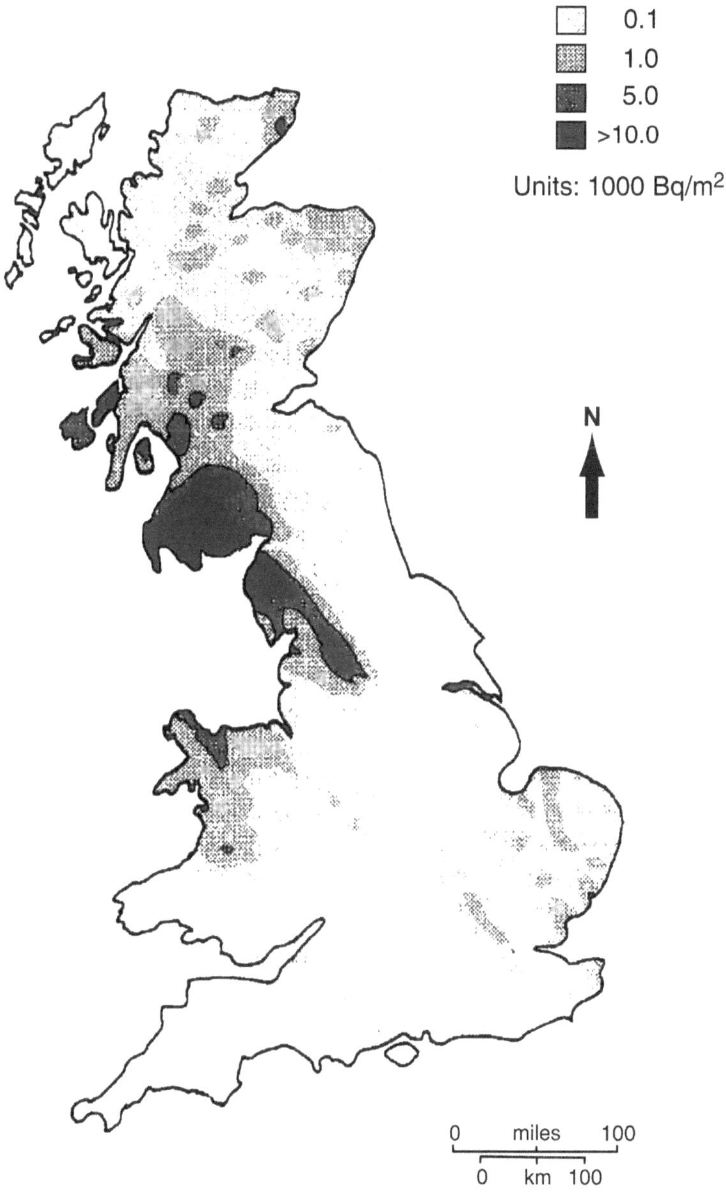

FIGURE 16.3 UK radioactive cesium levels estimated from rainfall data, Chernobyl cloud movement data, and models of cesium rain-out from the atmosphere.

They talk about these things coming from Russia, but it's surely no coincidence that it's gathered around Sellafield. They [MAFF, The Ministry of Agriculture Fisheries and Foods] must think everyone is completely stupid.[2]

These immediate local suspicions were only strengthened by the Ministry of Agriculture maps showing the restricted areas (Figure 16.1). Other farmers reinforced this logic, as did experience of the continuing secrecy surrounding the 1957 fire:

It still doesn't give anyone any confidence, the fact that they haven't released all the documents from Sellafield in 1957. I talk to people every week - they say this hasn't come from Russia! People say to me every week, "Still restricted eh - that didn't come from Russia lad! Not with that Lot on your doorstep".

The scientific view was that the Chernobyl cesium depositions could be distinguished from the cesium in routine Sellafield emissions, 1957 fire emissions, or 1950s weapons testing fallout, by the typical "signature", in gamma-radiation energy spectra, of the ratio of intensities of the isotopes cesium137 and cesium134 (each isotope has a characteristic gamma-ray frequency or energy). The half-life of the cesium137 isotope is about thirty years, while that of the cesium134 isotope is less than one year, and so the ratio of intensities of cesium137 to cesium134 increases with time. A typical Sellafield sample (from fully burnt-up fuel, usually stored for several years before reprocessing; or if from the 1957 fire, aged in the environment) would therefore show a greater ratio (about 10 to one) than a Chernobyl sample consisting of fresh fuel and fission products (about two to one). Thus, the deposits were said scientifically to show the so-called Chernobyl fingerprint, making an analogy with a form of evidence that is never questioned in law.

This scientific distinction, which exonerated Sellafield, was unequivocally asserted at public meetings and lectures with virtually complete consensus from scientists from the Ministry of Agriculture, Fisheries, and Food (MAFF) and the other scientific organizations involved in the issue, at least for the first year or more of the crisis. However, this too turned out to be less clear-cut than first claimed: it was later admitted that only about 50% of the observed radioactive cesium was from Chernobyl, the rest being from "other sources", including weapons testing fallout and the 1957 Windscale fire (House of Commons Agriculture Select Committee, 1988a). Nevertheless, at the time the difference in the fingerprints was represented as a very clear-cut scientific distinction, with Sellafield for once in the clear, and Chernobyl definitely to blame. Yet although

2. The quotes are from transcripts of interviews, which were taped and then transcribed in abridged form to record elements of relevance to this study. Over 50 interviews were conducted with affected farmers, farmers' wives, MAFF officials, scientists, farmers' representatives, and others. Each interview lasted between one and two hours: several repeat visits were made, allowing some observation of changing beliefs. The interviews were mostly conducted by Peter and Jean Williams. accompanied by the author on about 12 occasions. Public meetings and regular sheep markets were also attended and observed, by all of us. All the quotes in the text are verbatim quotes from interview tapes.

it was against their economic interests to entertain thoughts of a longer-standing but neglected (or covered-up) blight from Sellafield, and in the face of this solid scientific consensus, many hill farmers were ready, at least in private, to implicate Sellafield. Their reasoning tells us a lot about the deeper cultural and social structures of expert credibility.

It was striking that when we asked farmers skeptical about the scientists' exoneration of Sellafield to explain their reasoning, many of them talked about the 1957 fire and the secrecy surrounding it. At first we did not see this as an answer to the question, but then we realized that it was—they were explaining the lack of credibility of the present scientific claim about the Sellafield-Chernobyl distinction as due to the untrustworthy way in which the experts and authorities had mistreated them over the 1957 fire, and the longer history of perceived misinformation surrounding Sellafield:

Quite a lot of farmers around believe it's from Sellafield and not from Chernobyl at all. In 1957 it was a Ministry of Defence establishment—they kept things under wraps—and it was maybe much more serious than they gave out. Locals were drinking milk, which should probably never have been allowed—and memory lingers on.

The farmers thus embedded their reading of the present scientific claim about the isotope-ratio distinction firmly within the context of the unpersuasive and untrustworthy nuclear institutional body language that had denigrated them for thirty years or more. Their definition of risk was in terms of the social relationships they experienced, as a historical process.

They had a range of further reasons supporting this dissident logic. The empirical evidence of the maps (Figures 16.1 and 16.2) was powerful as far as they were concerned; and official disclaimers were ridiculed with a heavy irony only evident in a personal interview, such as (referring to a MAFF scientist) "she said she couldn't understand why the heaviest fallout from Chernobyl happened to fall around Sellafield."

Thus, the farmers gathered evidence that was drawn from science, including scientific inconsistencies, on which the scientists themselves did not focus. They entered the scientific arena in this sense, redrew its boundaries, and, operating with different presuppositions and inference rules, also redrew its logical structures, as well as its substantive conclusions.

Other direct empirical connections were drawn that may not have made scientific sense, but which served to make a consistent explanatory picture to people who saw the science to be either politically manipulated or naively overconfident in its own certitude.

Most farmers believe it's really from BNFL [Sellafield]. You'd have great difficulty convincing them otherwise. The area is a kind of crescent shape. If you're up on the tops [of the fells] on a winter's day you see the tops of the cooling towers, the steam rises up and hits the fells just below the tops. It might be sheer coincidence, but where the [radiation] hot spots are is just where that cloud of steam hits - anyone can see it if they look. You don't need to be a scientist or be very articulate to figure that one out. And a lot of the fanners may not be very articulate but they've figured it out all right. I think there's been low-level fallout ever since that placed opened, and Chernobyl has gone on top of it.

Interestingly, the apparently unfounded notion that high deposits occur where "the clouds" hit the fellsides is not unreasonable, because scientists themselves recognized the importance of intense "occult deposits" of radio-isotopes direct from low-lying clouds and mists that are typical of the Lake District climate.

Other farmers seemed to be exercising a strong penchant for irony when they put into skeptical perspective the experts' claims about the "coincidence" of Chernobyl deposition next to the local controversial nuclear site.

When you look at the stations around here, I said it was like a magnet, it just drew it in! [Then, relaxing the irony] I still think it was here before. They [the experts] won't have it ... We can't argue with them, but you can think your own ideas.

Often the justification for disbelieving the scientists on the Sellafield connection was simply that the same experts had very recently asserted, with similar confidence, first that there would be no effects of the Chernobyl cloud, and then that the restrictions, which were imposed after all, would be very short-lived. Since their self-confidence had been shown to be misplaced on those counts, why should they expect to be believed this time, especially when no admission of the earlier mistakes was forthcoming? The farmers scorned what they saw as the scientists' addiction to overconfidence and false certainty:

My theory, which is probably as good as anyone else's is this: we don't know... They keep rushing to conclusions before the conclusion has been reached - you understand what I'm saying? They'd have been far better to keep their traps shut and wait.

And a farmer's union representative put it:

We may be on the eve of a new age of enlightenment. When a scientist says he doesn't know, perhaps there's hope for the future!

It is important to note that scientific credibility was influenced not only by the evidence, which alternative logical presuppositions could select and render coherent, and not only by the prior intellectual mistakes, but by the way they were handled socially. This gave impetus to the alternative constructs.

The farmers also came into direct contact with the conduct of science on their farms, as hosts to a proliferation of monitoring, sampling, field analysis, and various other scientific activities. Again, they soon noted the inconsistency between the certainty pervading public scientific statements, and the uncertainties involved in actually attempting to create definitive scientific knowledge in such novel and open-ended circumstances. The experience of watching scientists decide where and how to take samples, of seeing the variability in readings over small distances, of noticing the difficulty of obtaining a consistent standard for background levels, and of gradually becoming aware of the sheer number and variety of less controlled assumptions and judgments that underpin scientific facts, corroded the wider credibility of official statements couched in a typical language of certainty and standardization. By accident, as it were, the farmers entered the "black-boxes" of constructed, "naturally-determined" science, and saw its indeterminacies for themselves (Latour, 1986). Referring to the live monitoring of sheep which was now obligatory, one farmer indicated how doubts set in:

Last year we did 500 [sheep] in one day. We started at 10.30 and finished at about six. Another day we monitored quite a lot and about 13 or 14 of them failed. And he [the monitor] said, "now we'll do them again" - and we got them down to three! It makes you wonder a bit ... it made a difference ... when you do a job like that you've got to hold it [the counter] on its backside, and sheep do jump about a bit.

These forms of reasoning were buttressed by further social evidence and judgment. There existed a widespread model of the capture of science by institutions with their own manipulative political agendas. Such judgments were supported by empirical observations, such as the refusal of MAFF officials to allow an American television team to film the lively debate with affected farmers at a public meeting in February 1987. The TV team was preparing a five-country documentary on the international response to Chernobyl. The producer's acid comment as he departed—that his team had received more open treatment in (pre-glasnot) Poland than in Britain—was widely quoted afterward among the farmers.

The farmers drew similar conclusions from MAFF's response to their requests for pre-Chernobyl cesium data on the fell-top vegetation, soils, and sheep; they asked for these in order to test MAFF's assertion that there had been no significant contamination before Chernobyl. However MAFF's reply was to refer first to an official document that contained only post-Chernobyl data (Ministry of Agriculture Fisheries and Food, 1987), and then to data that included pre-1986 monitoring only on the lowland coastal strip, but still no fell-top data. The farmers saw this as evidence that the authorities were trying to cover up—either that they did have data which showed high fell-top levels of cesium before Chernobyl, or that they had no data at all! If the former, they were guilty of straightforward lying and conspiracy. If the latter, they were guilty of at least gross complacency and incompetence, but possibly also conspiracy to remain deliberately ignorant of the levels before Chernobyl forced them to look. In addition, the 1957 fire had provided an ideal opportunity—apparently neglected—to have done the necessary research which would have avoided mistakes in the 1986 prediction:

Going back to the 1957 fire, nobody really knows what that did, what effect it had on the land and that, because they never tested it ... A lot of people have it in their minds that they [the UK authorities] were just waiting for something like this [Chernobyl] to blame.

This indicates a belief that the authorities had done secret research, had found high levels and had decided to cover up—waiting for the chance, which Chernobyl provided, to pass on the blame. It also encouraged the farmers to conclude that they and their families had been used as mere objects of scientific research. In fact the question of whether the authorities had done previous research in the Cumbrian fells, and thus knew that the radioactive cesium contamination would last much longer, is extremely complex. What counts as "previous research" is itself open to interpretive differences; some ecologists we interviewed said afterward that they knew, and told the government at the time, but that they were ignored by the "physicalist" ethos, which dominated the official scientific advisory mechanisms. This is the subject of further research. In evidence to the House of Commons Agriculture Select Committee (1988b), a local environmental group, Cumbrians Opposed to a Radioactive Environment, alleged cover-up, and also noted that the government's advisory body, the National Radiological Protection Board, had promulgated emergency reference levels for environmental radioactivity, only a month before Chernobyl, which completely overlooked the central environmental medium and food chain in the Chernobyl emergency, namely sheep meat.

These modes of reasoning interlocked with other judgments that the farmers made of the controlling institutions from which scientific claims were seen to emanate. Thus, another farmer related what he saw as deliberate official ignorance, in Sellafield's denial of claims that the site caused leukaemia:

> *The Department of Health could body monitor but they don't deliberately because if they did and found high readings then various ministries could one day be accused of irresponsibility in this regard. I think it self-evident that when BNFL [the Sellafield operator] were accused of being responsible for leukaemias they were quick to say "what evidence is there?" I have been told that if I make an accusation that my granddaughter has got leukaemia in the future and I suggest it was due to Sellafield they will say to me "what evidence have you?" It is a deliberate policy of government not to do this appropriate monitoring and testing so that they can protect themselves against an accusation of this nature. I would suggest we have another Christmas Island situation. The first such situation was at BNFL (it was then the Atomic Energy Authority) in 1957. Now we have Chernobyl Cumbria, Chernobyl Wales, South Scotland and Ireland... When you have bottomlesss financial pits like Sellafield sponsoring this, that and the other in order to blackmail local feeling, why could they not instead do something positive like supporting controlled experiments to answer all the questions that need to be answered?*

Of course, we can judge that these views were encouraged by probably unrealistic ideas about what can be expected of scientific knowledge in a situation such as the post-Chernobyl emergency. Even allowing for this factor however, the expressed attitudes reflect a rich supply of evidence to support a model of the subordination of science to untrustworthy institutional and political interests, and of a deep flaw in the very nature of science which drives it toward unrealism, insensitivity to uncertainty and variability, and incapability of admitting its own limits. (These can be seen as contradictory models of science, but are better treated as rhetorical stances which deconstruct and delimit the authority of the social control which the science represented in the experience of the farmers.) Analysis of the logical structure of the farmers' responses to the scientific expertise indicates both a far greater open-endedness about scientific logical structures and its institutional and cultural forms than is usually recognized; and a greater need to acknowledge and negotiate these as a condition of science's social legitimation.

16.5 CREDIBILITY: THE SOCIAL DIMENSION

The way in which the farmers' skepticism was expressed suggests that Chernobyl acted to release a large historical backlog of more private disbelief, mistrust, and alienation from the authorities, which related to Sellafield, and which had been quietly simmering over the years as one experience of official perfidion led into another. This would also explain the apparently abrupt change in their position from acceptance to hostility: it was probably not nearly as abrupt as it may have seemed, because there was already a finely balanced "private" ambivalence, and not by any means a complete uncritical acceptance of the site and its expert apologists (Wynne, 1987, chap. 11).

However, the dimension of this issue which drew in the farmers, and on which they had the most confidence to judge the outside experts and to criticize them, was the fact that this time, expert responses to the crisis constituted massive interventions, disruptions and denigrations of their normal practices and livelihood. The administrative restrictions introduced by the Government to prevent contaminated lamb from reaching the market were tantamount to large-scale social control and reorganization, and denial of essential aspects of the farmers' social identity, to an extent that the outside experts and bureaucrats did not remotely recognize. The interventions required not only scientific understanding of the radioecology of cesium in this particular physical environment; they also required this to be integrated with knowledge of hill sheep-farming methods and decision-making processes, in what is a highly specialist and particular kind of farming.

Whereas the hill farmers were quite reserved in their skepticism toward the scientists on scientific matters, they were abrupt and outspoken about them when they saw the extent of the scientists' ignorance of hill-farming environments, practices, and decision making. Even worse was the way that the outside experts did not recognize the value of the farmers' own expertise, nor see the need to integrate it with the science in order to manage the emergency properly. An example that ruined the experts' credibility with many farmers was the advice given to farmers to keep their lambs a little longer on cleaner valley pastures so as to allow high cesium levels gained on the fell tops to decrease. This ignored the locally taken for granted fact that hill farming in such areas is organized around a severe short supply of valley grass, which would as one farmer put it, "be reduced to a desert in days" (and with knock-on effects into future breeding) if it were not very carefully husbanded.

Naturally, the farmers felt that their whole identity was under threat from outside interventions based upon what they saw as ignorant but arrogant experts who did not recognize what was the central currency of the farmers' social identity, namely their specialist hill-farming expertise. This expertise was not codified anywhere: it was passed down orally and by apprenticeship from one generation to the next, as a craft tradition, reinforced in the culture of the area. The impact of the scientists' hegemonistic cultural orientation on their general credibility showed itself repeatedly:

There was the official who said he expected levels would go down when the sheep were being fed on imported foodstuffs, and he mentioned straw. I've never heard of a sheep that would even look at straw as fodder. When you hear things like that it makes your hair stand on end. You just wonder what the hell are these blokes talking about? When we hill men heard them say that we just said, what do this lot know about anything? If it wasn't so serious it would make you laugh.

Another derided the experts' ignorance of what were elementary facts of life to hill farmers:

If you start fattening lambs and sell twenty, the next twenty get fat quicker, because you've got more grazing. But if you keep them all… [gesticulation of disaster]. But that's the problem with the ministry—trying to tell them those sort of things. That's where the job has fallen down a lot. They couldn't understand that you were going to sacrifice next year's lamb crop as well. They just couldn't grasp that!

Scientists and Ministry officials were often seen as indistinguishable; indeed the most prominent officials explaining and defending official decisions were scientists. But there was also a deeper structural convergence between the forms of monopolistic scientific representation of the issues, and the forms of administrative intervention and reorganization of farming practices. The significant elements of scientific representation in this respect were:

- its standardization of local physical environmental variations, farming conditions, and practices (hence farmers); and
- its ethos of prediction and control, which engendered an exaggerated sense of certainty, and which conflicted sharply with the farmers' ethos of adaptation and acceptance of intrinsic lack of control.

These coincided with the centralized formal nature of the administrative interventions, which reduced the long-established individualism, informality and flexibility of farm management decision-making to an extension of bureaucracy. The farmers were quite familiar with uncertainty on several fronts and thus with adaptation to factors beyond their control. This deep cultural outlook—reflected in their intellectual frameworks as well as in their whole way of life—was simply incompatible with the scientific bureaucratic cultural idiom of standardization, formal and inflexible methods and procedures, and prediction and control. These dimensions of incompatibility and lack of mutual credibility existed at a structural level which was deeper than that of evidence and information. They lay at the level of moral, or cultural recognition. Each side only recognized even as possible evidence, claims expressed within its cultural style. Thus for example the scientists had an *a priori* credibility gap to overcome when they stated things so categorically and universally, before the substance of the statement was even reached. By the same token, the farmers' expertise was not recognized because it was not formally organized in documentary, standardized, and control-oriented ways recognizable to scientific culture; and their later claims for compensation encountered the inflexible bureaucratic demand for formal documentation, dates, details, proofs, and signatures in a way that was entirely alien to their own culture.

This sense of being ensnared by an alien and unrecognizing combination of science and bureaucracy was neatly captured in two typical comments:

They've been watching too much of "One Man and his Dog" [a popular national television programme where shepherds compete in driving and penning sheep, under artificially simple conditions]... They think you just stand at the bottom of the fell and wave a handkerchief and the sheep come running.

Another, after a detailed explanation of complex differences between farming practices even within his own small valley, reflecting different microconditions, lamented:

this is what they can't understand. They think a farm is a farm and a ewe is a ewe. They think we just stamp them off a production line or something.

Thus, underlying overt clashes of knowledge, information, evidence, and belief were incompatible social and cultural structures, prescribed modes of social interaction. The scientific knowledge, in the levels of aggregation and standardization of data and parameters by which it was organized, also expressed commitments about the levels of political standardization and control of the farmers.

Thus, the scientific perspective was just as socially grounded, conditional and value-laden as the other. Its credibility was influenced not so much by what it said directly and explicitly, as in the way it was institutionally and intellectually organized, including lack of recognition of its own cultural and institutional biases—its own tacit social body-language. As explained later, it suffered from its own lack of reflexivity.

Analysis of this credibility gap allows us to identify factors that affect the social credibility of science. These are summarized below, as criteria by which laypeople rationally judge the credibility and boundaries of authority of expert knowledge. It is easier to understand the persistence of disputes over the authority of scientific knowledge when these several layers of the social and cultural framing of expert and lay discourses are recognized. They are structurally identical to the factors shaping the logics of dispute and development within science; it is just that in public situations the prior mechanisms of social closure are by definition less powerful.

Lay criteria for judgment of science

1. Does the scientific knowledge work?
   For example: predictions fail.
2. Do scientific claims pay attention to other available knowledge?
   For example: scientists monitor sheep without paying attention to where they graze, whereas farmers know where on open fells they graze.
3. Does scientific practice pay attention to other available knowledge?
   For example: when scientists devise and conduct field experiments that the farmers know will not work.
4. Is the form of the knowledge as well as the content recognizable?
   For example: degrees of expressed certainty, standardization, and aggregation.
5. Are scientists open to criticism?
   For example: no recognition of other legitimate knowledges and expert actors; no admission of errors, omissions, or oversights.
6. What are the social/institutional affiliations of experts?
   For example: imputed social political biases and interests; historical track record of trustworthiness, openness.
7. What issue "overspill" exists in lay experience?
   For example: from Chernobyl to Windscale-Sellafield; lack of rational connection for scientists because institutional dimensions defined out a priori, but for laypeople continuity, resulting from social dependency, institutional models of agency and responsibility in decision and knowledge construction.

This analysis suggests that reflexive recognition of its own conditionality is a prerequisite for science's greater public legitimation and uptake; yet this requires more than an intellectual advance from science, but institutional reform of its modes of organization, control, and social relations. This would involve, inter alia, recognition of new, socially extended peer groups which offer criticism of scientific bodies of knowledge from beyond the confines of the immediate exclusive specialist scientific peer-group. The social definition of such extended peer-groups would relate to the context of use of the scientific specialties concerned; and criticism would include explicit negotiation of the social criteria or epistemology of knowledge for the situation. (For an analysis of institutional negotiation over what counts as 'good science' in the public arena of US regulatory policies, see Jasanoff, 1990.) This approach to public understanding of science therefore underlines the point reflected in other sociological analysis of scientific knowledge, that the boundaries of the scientific and the social are social conventions, predefining relative authority in ways which may be inappropriate, and which are open to renegotiation (Jasanoff, 1987). The practical process of developing that negotiation first requires recognition that existing approaches and discourses misrepresent this conventional character as if it were naturally determined.

16.6 CONCLUSIONS: LAY REFLEXIVITY AND SOCIAL IDENTITIES

A productive way of analyzing the interactions between hill-sheep farmers and scientists in this case is to see each social group attempting to express and defend its social identity (Burke, 1991). The farmers experienced the scientists as denying, and thus threatening, their social identity by ignoring the farmers' specialist knowledge and farming practices, including their adaptive decision-making idiom. They also experienced the scientists as engaged in a conspiracy with government against hill farmers, initially to deny any need for long-term restrictions and later to claim an innocent mistake in prediction—misinformation that caused many farmers to make unfortunate decisions and to lose heavily as a result. On top of the further hardships and external controls besetting the hill farmers in an area that is a tourist-dominated National Park, their mistreatment by the scientists and bureaucrats after Chernobyl was almost the final straw in a baleful succession of blows to their cultural and social identity.

The scientists on the other hand were expressing and reproducing their intellectual administrative frameworks of prediction, standardization, and control, in which uncertainties were "naturally" deleted, and contextual objects, such as the farmers and their farms, were standardized and "black-boxed" in ways consistent with this cultural idiom. Whatever private awareness they may or may not have had of the cultural limits and precommitments of their science, they successfully suppressed these.

These social identities were not completely predetermined and clear, nor were they immune to interactive experience and negotiation. My main point is that this dimension should be seen as the level from which explanation of lay responses to science is to be derived, and in which the factors and processes shaping credibility or "understanding" can be identified.

The laypeople in this case showed themselves to be more ready than the scientific experts to reflect upon the status of their own knowledge, and to relate it to that of others and to their own social identities. Thus, for example, the farmers implicitly recognized their social dependency upon the scientific experts as the certified public authorities on the issues, even if, as they indicated in interview, they held dissenting informal beliefs which they could defend along the lines described before. As one farmer put it: "*You can't argue with them because you don't know - if a doctor jabs you up the backside to cure your headache, you wouldn't argue with him, would you?,*" the suggestion being that when the expert tells you unbelievable things, you don't overtly argue, thereby inviting denigration. As another said: "*We can't argue with them, but you can think your own ideas. I still think it [the radioactive cesium] was here before.*"

These more private beliefs were rarely displayed in public, and the farmers refused to confess to such dissent in media interviews. It was made clear to us that one reason for this was that the farmers identified socially with family, friends, and neighbors who were part of the Sellafield industrial workforce. They recognized their own indirect and sometimes direct social dependency upon the plant—not only neighbors but also close relatives of the hill farmers work there. Thus, underlying and bounding their expressed mistrust of the authorities and experts, there was a countervailing deep sense of social solidarity and dependency—of social identification with material kinship, friendship, and community networks that needed to believe Sellafield was well-controlled and its surrounding experts credible.

Thus social alienation and identification coexist in the same persons and communities, leaving deep ambivalence and apparent inconsistency in relevant beliefs and structures of "understanding". These can only be understood by reference to the multiplex, not necessarily coherent, dimensions of social identities expressed in interleaved social networks, relations, and experiences.

All this could be interpreted as yet another example of the lamentable inconsistency and impossible fickleness of lay beliefs. The conventional model of rationality would include a principle of cognitive consistency as measured against some canons of abstract logic. However, what is revealed in this case is a deeper and more complex consistency in public reasoning than that recognized by such simplistic models. In the real world, people have to reconcile or adapt to living with contradictions that are not necessarily within their control to dissolve. Whereas the implicit moral imperative driving science is to reorganize and control the world so as to iron out contradiction and ambiguity, this is a moral prescription that may be legitimately rejected, or at least limited, by people. They may opt instead for a less dominatory, more flexible and adaptive relationship with their physical and social worlds. In this orientation, ambiguity and contradiction are not so much of a threat, because control and manipulation are not being sought or expected. This is no less legitimate a form of rationality than the scientists', and the "public understanding of science" research field, as well as science in public practice, should recognize this, and build upon it.

The advance from focusing on cognitive dimensions (often assumed public deficiencies) to trust and credibility is important. But closer examination in this case-study of the basis of trust and credibility falsifies the predominant analytical tendency to treat these as unambiguous, quasicognitive categories of belief or attitude that people supposedly choose to espouse or reject (Renn, 1991). My analysis suggests instead that "credibility" and "trust" are themselves analytical artifacts that represent underlying tacit processes of social identity negotiation, involving senses of involuntary dependency on some groups, and provisional or conditional identification with others, in an endemically fluid and incomplete historical process. Thus, what the hill farmers believed about the scientists and their assertions was rich in insights and refinement, on several levels beyond the one-dimensional reductionism of scientific logic alone. But this richness was also pervaded by an ambivalence reflecting their multiple and conflicting social networks and relations. It would have been easy to have marked them down as mere "don't knows" in a more efficient attitude survey, even though this would have been a grotesque distortion of the true position.

Recognition of this multidimensional, even internally contradictory character of belief, allows a more accurate perspective on the apparent fickleness of public responses to risks and scientific knowledge that is much lamented by authorities. If we assume that a widely observed lack of public dissent to expert reassurances equals (voluntarily espoused) public acceptance, then apparently sudden shifts to opposition and rejection seem capricious, irrational, and uncontrollably emotive. If on the other hand, we recognize the alienation and ambivalence often underlying surface quietude, we may see that what looks like a sudden shift of attitude, a "betrayal", was nothing of the kind—it may have been only a very small shift in the balance of components of social identity which people are holding in tension with one another. This intrinsic instability of actors' "loyalties" is something that is not fully recognized in Latour and Callon's theoretical vocabulary of enrollment and representation of actors by scientists, as they build intellectual-social empires by tying in those actors, appropriately defined, to their particular role in the edifice. Thus Callon's account of the "betrayal" of the marine biologists by scallop fishermen of St Brieuc Bay who had seemed to have internalized the identity which the scientists had articulated for them, does not recognize the possible ambivalence of the fishermen about their designated identity even before the "betrayal", which may thus have been much less of a shift than it appears in Callon's otherwise superb account (Callon, 1986).

Thus, the cognitivist presumption that risks, or scientific knowledge, exist independently as an object for measurable public attitudes or beliefs, is left at least two steps behind. The first step is the recognition that the trustworthiness and credibility of the social institutions concerned are basic to people's definition of risks, or uptake of knowledge, and that this is reasonable, indeed unavoidable. However, the second step is to recognize that trust and credibility are themselves analytically derivative of social relations and identity-negotiation; thus, like risk, they too should not be treated as if they have an objective existence that can be unambiguously measured and manipulated.

Having advanced the case for social identity as the more fundamental concept for explaining responses to science and risks, however, it should be accepted that this term is itself not unproblematic. To claim that it offers more explanatory depth is not to claim it is empirically pure, coherent and unambiguously identifiable.

The theoretical orientation of this paper coincides with postmodernist concepts that treat identities as intrinsically incomplete and open-ended, and as an endlessly revised narrative attempting to maintain provisional coherence across multiple social roles and reference groups. (For this process in historical perspective, see Hobsbawn & Ranger, 1984.) Beliefs and values are functions of social relationships and patterns of moral and social identification. This stands in sharp contrast to the taken-for-granted (and hence rarely articulated) commitment underlying conventional approaches, in which values and beliefs are taken to be coherent, self-sufficient, and discrete entities, and where social identities are simply the aggregate of individual beliefs and values. In this perspective, social interaction is recognized only as an instrumental device to maximize preferences and values, not as an activity with moral and social meaning in its own right.

The case shows the unacknowledged reflexive capability of laypeople in articulating responses to scientific expertise. They are able to reflect on and develop their own social position as part of a "dependent" response in which they are supposed to enjoy no powers of independent critical rationality autonomous from "proper" assimilation of scientific understanding. Indeed, it is interesting that those who would be regarded as the representatives of traditional society showed this reflexive capability, whilst the putative representatives of enlightened modernity, namely the scientists, did not The scientists showed no overt ability to reflect upon their own social positioning that is upon the latent social models, which their scientific interventions imposed on the farmers. Perhaps the distribution of reflexive capability (or impulse) is itself a contingent function of social relations of power.

It is not true to say that scientists are not reflexive, in that they do show a capacity to reflect upon the nature of their practice, its contingencies, and limits. However, this may (for all social groups) be brought about only by criticism and a related sense of insecurity, rather than by any intrinsic qualities of self-criticism. Thus, the extent of such reflexivity in science is open to question, both in how deep it goes into examination of scientific founding commitments (hence identities) and in how openly such critical self-examination is expressed to other social groups, for example in public or policy debate. Such articulated self-criticism would display the uncertainties in scientific knowledge, and at the same time expose as negotiable science's definition and role in relation to other social groups. As I have suggested in this paper, ambivalence is usually treated as intellectual feebleness—the antithesis of rationality and "clear thinking". But it may be a necessary corollary of a social commitment to disavowing control of others, in which case the remit of scientific rationality (as usually conceived) is seen in a radically different light. (For suggestive correspondence with feminist critiques of science, See Harding 1986.)

The intellectual properties of reflexivity or its lack (or to put it another way, of the epistemological principles of science) are not independent of the institutional forms of science. Thus, it becomes evident why the quality of its institutional forms—of organization, control, and social relations—is not just an optional embellishment of science in public life, but an essential subject of critical social and cultural evaluation. It is a crucial missing part of the contemporary nondebate of science's social purchase and legitimacy.

ACKNOWLEDGMENTS

Different versions of this paper were delivered at: the Society for Risk Analysis European Meeting, IIASA, Laxenburg, April 1990; The Science Museum Conference on Policies and Publics for Science. London. April 1990; the American Association for the Advancement of Science, Washington DC. February 1991; and the Oak Ridge and Los Alamos US National Laboratories, February and May 1991. I am grateful to Deborah Barnes for extraordinarily multidisciplinary support. The work was supported by the UK ESRC. I am also grateful to colleagues and friends under the Science Policy Support Group network of research on public understanding of science, for rewarding discussion and moral support, especially to Peter and Jean Williams, John Wakeford, Mike Michael, Alan Irwin, John Ziman, and Frances Price.

REFERENCES

Arnold, L. (1992). *Windscale 1957: The anatomy of a nuclear accident.* London, New York: Macmillan.

Beresford, N. A., Lamp, C. S., Mayes, R. W., Howard, B. J., & Colgrove, P. M. (1989). The effect of treating pastures with bentonite on the transfer of 137Cs from grazed herbage to sheep. *Journal of Environmental Radioactivity, 9*, 112–122. (This paper recognizes the loss of body weight suffered by the sheep in these tests, though it suggests this may be due to the direct effects of the bentonite, and ignores the possible effect of their being fenced in).

Burke, P. (1991). We the people: popular culture and popular identity in Modern Europe. In S. Lash & J. Friedman (Eds.), *Modernity and identity* (pp. 293–310). Oxford, Cambridge, Mass: Blackwell. (This capability could be equated with the informal resistance of people in solidary sub cultures to meanings, identities or rationalities imposed on them from 'outside', as in Bailey, F. G., 1968, A peasant view of the bad life. Peasants and Peasant Society, edited by T. Shanin (Harmondsworth: Penguin Books)).

Callon, M. (1986). Some elements of a sociology of translation: domestication of the scallops and the fishermen of St Brieuc Bay. In J. Law (Ed.), *Power, action and belief Sociological reviews monographs* (Vol. 32, pp. 196–233). Keele University Press.

Collins, H. (1987). Certainty and the public understanding of science. *Social Studies of Science, 17*(4), 689–713. (Wynne, B., 1991, Knowledges in context. Science, Technology and Human Values, *16*, 111–121).

Douglas, M. (1986). *Risk acceptability according to the social sciences.* New York: Russell Sage Foundation.

Friends of the Earth. (1987). Fallout over Chernobyl: a review of the official radiation monitoring programme in the UK. In P. Green & P. Daly (Eds.), London: Friends of the Earth. (British Nuclear Fuels plc, 1987, Annual Reports on Environmental Monitoring (London: BNF plc); Ministry of Agriculture, Fisheries and Food, 1987, Radionuclide Levels in Food. Animals and Agricultural Products (London: HMSO). See also, MAFF's Annual Terrestrial Radioactivity Monitoring Programme, TRAMP, No I, 1986 (London: MAFF), and House of Commons Agriculture Select Committee, 1988, Chernobyl: the Government's Response (London: HMSO)).

Gardner, M. J., Snee, M. P., Hall, A. P., Powell, C. A., Downes, S., and Terrell, J. D. (1990). Results of case-control study of leukaemia and lymphoma among young people near Sellafield nuclear plant in West Cumbria. *British Medical Journal, 300*, 423–425.

Harding, S. (1986). *The science question in feminism.* Milton Keynes: Open University Press.

Hobsbawn, E. J., & Ranger, T. (1984). *Invention of tradition*. Cambridge and New York: Cambridge University Press.

House of Commons Agriculture Select Committee. (1988a). Chernobyl: The Government's response. *Minutes of evidence*. (Vol. 2). London: HMSO. (See also Chernobyl I and II, Farmers Weekly (London), 4 and II March, 1988).

House of Commons Agriculture Select Committee. (1988b). *Chernobyl: The Government's response*. London: HMSO.

Howard, B. J., & Beresford, N. A. (1989). Chernobyl radiocaesium in an upland farm ecosystem. *British Veterinary Journal*, *145*, 212–224. (Livens, F. R., and Loveland, P. J., 1988, The influence of soil properties on the environmental mobility of caesium in Cumbria. Soil Use and Management, 4(3), 69–75).

Jasanoff, S. (1987). Contested boundaries in policy-relevant science. *Social Studies of Science*, *17*(2), 195–230. (Starr, S. L., and Griesemer, J., 1989, Institutional ecology; translations and boundary objects: amateurs and professionals in Berkeley's Museum of Vertebrate Zoology, 1907–1939. Social Studies of Science, 19(3), 387–420).

Jasanoff, S. (1990). *The fifth branch: Science advisers as policymakers*. Cambridge, Mass: Harvard University Press.

Knorr-Cetina, K. (1989). Epistemic cultures; forms of reason in science. *History of Political Economy*, *23*, 105–122. (Latour, B., and Woolgar, S., 1979, Laboratory Life: The Social Construction of Scientific Facts (London, Beverley Hills: Sage)).

Krimsky, S., & Golding, D. Eds., (1992). *Social theories of risk*. New York: Praeger.

Latour, B. (1986). *Science in action*. London: Open University Press.

Lave, J. (1986). The values of quantification. In J. Law (Ed.), *Power, action and belief. A new sociology of knowledge? Sociological reviews monograph* (vol. 32, pp. 88–111). Keele University Press. (Also Potter, J., and Wetherell, M., 1987, Discourse and social psychology: Beyond attitudes and behaviour (London: Sage). For a critique in the context of risk perception research, see Wynne, B., 1989, Building public concerns into risk management. Environmental Threats. Edited by J. Brown (London: Belhaven), pp. 118–132).

Luhmann, N. (1990). Technology, environment and social risk: a systems perspective. *Industrial Crisis Quarterly*, *4*, 223–231. (Gambetta, D., 1988, Trust: Making and breaking cooperative relations (Oxford: Blackwell)).

Ministry of Agriculture, Fisheries and Food. (1987). *Radionuclide levels in food, animals, and agricultural products*. London: HMSO. (This account is taken from the author's field notes from attendance at the meeting, in Broughton-in-Furness, Cumbria, February 1987).

Renn, O. (1991). Risk communication and the social amplification of risk. In R. E. Kasperson & P. J. M. Stallen (Eds.), *Communicating risks to the public: International perspectives* (pp. 287–324). Dordrecht, Lancaster: Kluwer.

The Royal Society of London. (1985). *The public understanding of science*. London: Royal Society.

Smith, F. B., & Clark, M. (1986). Deposition of radionuclides from the Chernobyl cloud. *Nature*, *322*, 690–691. (A more complete paper was published in 1988, which contained the map as shown in Figure 16.3: Clark, M. J., and Smith, F. B., 1988, Wet and dry deposition of Chernobyl releases. Nature. 332, 245–249).

Unger, R. M. (1976). *Law in modern society*. New York: Free Press.

Wynne, B. (1980). Technology, risk and participation: on the social treatment of uncertainty. In J. Conrad (Ed.), *Society, technology and risk* (pp. 167–202). London, New York: Academic Press.

Wynne, B. (1982). *Rationality and ritual: The Windscale inquiry and nuclear decisions in Britain.* Chalfont St Giles: British Society for the History of Science. London and New York: Earthscan, 2011. Republished as *Rationality and Ritual: participation and exclusion in nuclear decision-making.*

Wynne, B. (1987). *Risk perception, decision analysis and the public acceptance problem. chapter 11 of Wynne, B. Risk management and hazardous wastes: Implementation and the dialectics of credibility.* London, Berlin, New York: Springer.

Wynne, B. (1989). Sheepfarming after Chernobyl: a case study in communicating scientific information. *Environment, 31*(2), 10–15. 33–39. Reprinted in Bradby, H., 1991, Dirty Words: Writings on the History and Culture of Pollution (London: Earthscan), pp. 39–60.

Wynne, B. (1992). Public understanding of science research: new horizons or hall of mirrors? *Public Understanding of Science, I*, 37–43.

Chapter 17

Stakeholder Engagement in Regaining Decent Living Conditions after Chernobyl*

Jacques Lochard
Centre d'étude sur l'Evaluation de la Protection dans le domaine Nucléaire (CEPN), Fontenay-aux-Roses, France
E-mail: lochard@cepn.asso.fr

Chapter Outline

17.1 Introduction 311
17.2 A Key Feature of the Situation of the Population in the Contaminated Territories: The Loss of Control on Daily Living Conditions 313
17.3 The ETHOS Project 316
17.3.1 The Approach 316
17.3.2 Phase 1: 1996–1999 317
17.3.3 Phase 2: 1999–2001 320
17.4 Key Lessons 322
17.5 New Approach to Long-Term Rehabilitation 325
17.6 Conclusion 326

17.1 INTRODUCTION

The long-term rehabilitation issue emerged progressively during the years following the Chernobyl catastrophe when it became more and more evident that the countermeasures adopted after the emergency phase, basically aiming at moving away the inhabitants from the most contaminated areas and reducing and controlling the contamination in the environment whenever possible, were insufficient to durably protect the population still residing in less contaminated but large territories. The long-lasting contamination in these territories, even if the associated levels of exposure were low, was a permanent worry for the population as far as health was concerned because of the remaining uncertainty

*Parts reprinted, revised and adapted from Lochard, J. (2007). Rehabilitation of Living Conditions in Territories contaminated by the Chernobyl Accident: The ETHOS Project. *Health Physics*, 93, N° (5), pp. 522-526, with permission of Wolters Kluwer Health.

Radioactivity in the Environment, Volume 19
ISSN 1569-4860, http://dx.doi.org/10.1016/B978-0-08-045015-5.00017-4

concerning protracted exposure, particularly due to internal contamination. It was also a very serious handicap for the long-term preservation of the socioeconomic infrastructures and the quality of life of the inhabitants.

This led the governments of Belarus, Russia, and Ukraine to elaborate and adopt ambitious national laws in the early nineties in an attempt to organize the health, social, and economic living conditions of the population residing in the contaminated territories. The objective of these laws was mainly to address long-term issues through a series of national countermeasures and compensation mechanisms, designed mainly according to radiological protection criteria. Schematically, the adopted rehabilitation programs relied on restricting further the human presence in the contaminated territories (mandatory or voluntary relocation, free stays in sanatoriums outside of the territories for children...) and on strictly controlling the level of contamination in foodstuffs and the whole body contamination of individuals. Many countermeasures were focused on the control and improvement of the radiological quality of agricultural products in collective farms; private productions being restricted as much as possible because of the difficulty to control and monitor their qualities.

However, despite the diversity of the countermeasures and the huge amount of national resources dedicated to their implementation, the adopted strategy failed to take properly into account the complexity of the situation created by the contamination. In particular, they did not succeed to mobilize the local communities and the individuals who progressively felt completely powerless in the face of the radiological situation. The general loss of quality of products, commodities, and assets combined with the rising concern about the presence of the contamination and its potential health consequences continued to undermine the quality of life of the families, in fact the large majority of the population, which had made the choice to stay in the affected territories. Altogether these factors contributed to generate among the inhabitants a general feeling of loss of control on daily life, exclusion and abandonment.

During the mid-nineties, the continuous degradation of the economic situation due to both the collapse of the USSR and the financial burden of the rehabilitation programs pushed the inhabitants of the territories to restart private production and to rely ever more on the products of the forest to ensure their daily subsistence. In the absence of individual know-how and adequate means to control the radiological quality of the foodstuff at the local level, the effect of this change was inevitably a significant increase of the level of exposures within the population and particularly among children. The importance of diary products in their diet put a strong pressure on the authorities and experts and contributed to aggravate further the loss of confidence of the population in their ability to manage the situation. In this difficult context, the authorities of Belarus supported the proposal of a group of French experts who were involved since several years in the evaluation of the radiological, psychological, social, and economic consequences of the Chernobyl accident in the contaminated territories in Ukraine and Belarus to develop a Pilot Project with the objective to directly involve the population in the management of the radiological situation to improve durably, using the existing resources, both its protection and living conditions.

The project, named ETHOS, started in spring 1996 in the Stolyn District in the South of Belarus. From the French side, it involved an interdisciplinary team of 12 experts in radiation protection, agronomy and local development, social risk-management, communication, and cooperation in complex situations, supported financially by the European Commission. From the Belarus side the project was directly supported by the Chernobyl Committee with the technical assistance of 2 Belarus institutes dealing with radiation protection and agronomy. The first phase of the project (1996–1999) was implemented in only one village of the District. It allowed a few tens of voluntary villagers to participate together with the team of experts to a step-by-step expertise of the radiological situation of the village (coexpertise) and its surrounding to progressively regain control on their daily life and to noticeably improve their protection and living conditions. In a second phase (1999–2001), at the request of the local authorities, the Project was extended to 4 other villages of the District to evaluate the feasibility of the diffusion of the approach by local professionals and authorities. The Project went to an end in autumn 2001 with the running of an International Seminar that gave the opportunity to all those who participated actively in its implementation to express their views on its results and lessons.

The paper presents an overview of the ETHOS Project, which by several aspects brought a new insight into the long-term rehabilitation strategy after a nuclear accident or a radiological event. Section 17.2 describes briefly the key features of the situation of the population in the contaminated territories of the Community of Independent States (CIS) in the mid-90s that led to the implementation of the Project. Section 17.3 presents a short insight about the methodological and ethical basis of the approach and the two phases of the Projects with their main results. Section 17.4 and 17.5 present respectively the lessons learnt through the project and their impacts on the long-term rehabilitation strategy for contaminated territories.

17.2 A KEY FEATURE OF THE SITUATION OF THE POPULATION IN THE CONTAMINATED TERRITORIES: THE LOSS OF CONTROL ON DAILY LIVING CONDITIONS

Several studies performed in the context of the European Commission (EC)-Community of Independent States cooperation program (1991–1995) to evaluate the consequences of the Chernobyl accident had shown that a salient characteristic of the situation in the territories considered administratively as contaminated was the general and progressive loss of control of the population on its daily life (Girard & Hériard Dubreuil, 1994, 1995). This observation, shared by several Belarus authorities and experts, undoubtedly gave the impetus to launch the ETHOS project when it became evident that not only the national rehabilitation program encountered difficulties to improve the radiological situation, but on the contrary this last one was deteriorating in some areas, despite all the efforts of the authorities and experts to try to restore decent living conditions. It is beyond the scope of this paper to describe in details this process of loss of control but it is however worth to highlight some of the factors that can explain it and also contributed to its reenforcement over the years.

First, the invisible and intangible presence of radioactivity in the environment resulting from the Chernobyl accident created a radically new and very difficult to conceive situation for the millions of individuals who were directly confronted to it. It induced severe disturbances of their daily life at the health, social, economic, and cultural levels. In the absence of past experience and memory this complex situation was unspeakable by the population and gave rise to a multitude of interrogations that remained in most cases without answers and generated a lot of different concerns among the population (Hériard Dubreuil & Girard, 1996). Furthermore, the management of the contamination and its potential consequences introduced a set of scientific terms, measurement units and new technical procedures, which were meaningless for nonspecialists. People confronted to the contamination felt ignorant and progressively renounced to get involved in the day-to-day management of the situation. They finally left this task to a limited number of experts who in most cases were unable to express in a common language the meaning and usefulness of the proposed actions. As a result, this process generated among the general population a feeling of exclusion that prevented the development of personal know-how and direct involvement of the individuals. It is interesting to note that the implementation of a vast radiation monitoring system within the territories did not really help the population to understand and regain control on the situation. In fact, measurements, independent of their more or less sophisticated technical aspects difficult to understand and the fact that they were rarely shared and explained to the population, rather contributed to increase the interrogations and concerns of the population and generated anxiety.

Another aspect that greatly contributed to the loss-of-control process is the centralized and prescriptive management of the situation adopted by the authorities. A retrospective overview of the strategies, including those adopted after 2001 in the context of the new national laws on "the safe living conditions of the population", shows that these strategies were mainly driven by the scientific and technical features of the radiological situation and implemented throughout the affected territories by means of countermeasures without neither involving the inhabitants and the local professionals in the decision making process nor providing them with feedback information on the successes and difficulties. In most cases they were decided by experts and authorities on cost-benefit considerations, and applied at the collective level on a centralized and planning mode. Basically, they aimed at maintaining individuals at a certain distance from the contamination or reducing, as far as reasonable, the levels of contamination in the environment and particularly in agricultural products. However, for many inhabitants of the contaminated territories, the implementation of the countermeasures, particularly those that were affecting ancestral relationships with the environment and the traditional modes of production were experienced essentially as intrusive and generating interdictions before to be considered as a means of protection. This process kept the individuals aside the real problems and re-enforced their feeling of exclusion and confiscation by experts and authorities of the control of the situation (Lochard, 1996).

In this context in which the population was unable to cope with the complexity of the situation and felt excluded from its management, individuals developed strategies ranging from denial to anxiety. The first was the most frequent, characterized by the withdrawal into oneself and the ignorance of the risk as a means to support the situation. For example, many people in the contaminated territories of Belarus adopted the belief that after several years of exposure to the contamination they became immunized against radiation like Colorado beetles against pesticides. Other inhabitants just gave up. They adopted a fatalist attitude, felt abandoned and victims of a great injustice. Both attitudes led to neglect basic protection actions and resulted in an increase of exposure. Altogether, the management of the situation, focused on its radiological dimension, progressively re-enforced among the population the strange and hostile character of the environment and the concern about health. Neither the contamination and exposure limits nor the countermeasures were able to generate social confidence and to incite individuals to take control again on their future situation. Taking into account the uncertainty inherent to the long-term effects of radiation, the inevitable debate about the norms to manage the situation and the global depreciation of the economic value, the social confidence in authorities and experts deteriorated rapidly. This led generally the authorities to call for more technical expertise and to develop communication strategies to try to close the gap of perception of the situation with the population. These approaches did not bring any re-insurance and on the contrary even re-enforced distrust toward the authorities as far as they tended to substantiate the feeling among the people that they had not been properly protected so far. Based on general recommendations not linked to the local living conditions, they also failed to address the individual situations and the specific actions for their daily management.

Combined with the intrinsic difficulty for the individuals to cope with the new reality of the contamination, the top-down, centralized, prescriptive and normative approaches adopted by the authorities, which were necessary in the very emergency phase, have progressively become inefficient and even counterproductive in the long-term rehabilitation perspective. They did not adequately addressed the complexity, and also the specificity, of the situations created by the long-lasting contamination and failed to give the inhabitants the means to regain control on their day-to-day life and to protect themselves and their offspring. On the contrary, they contributed to generate several negative side effects, among which:

- A dependency culture within the affected population, which evidence was given by the continuous demand of the population to increase the State efforts throughout assistance programs to the "victims" of Chernobyl;

- Social distrust and loss of confidence in authorities and experts, corollary of the several reorientations of the national strategies and budgets reductions to respond to the increasing economic difficulties perceived by the local affected populations as a decline in their protection; and
- A general feeling of abandonment and fatalism and a phasing out of personal initiative, which excluded further the population from the rehabilitation strategies and acted as a strong blocking factor for self-help initiatives.

17.3 THE ETHOS PROJECT

17.3.1 The Approach

Based on the findings of the studies that preceded the launching of the Project, the intention was to develop an original approach addressing jointly the technical and societal dimensions of the situation created by the presence of the contamination, and involving actively all local, regional and national stakeholders in a decentralized management in order to improve concretely the protection and the day-to-day quality of life of the affected population. The aim was not to produce any new scientific knowledge about the radiological situation but to apply the existing ones in the development of a practical know-how for the population. Furthermore from the beginning, this approach was clearly presented to all stakeholders as complementary to the national rehabilitation program implemented by the Belarus authorities.

Beyond these general intentions, there was no preestablished methodology to implement the project but a shared willingness between the French teams to build, what became later on the "ETHOS approach", pragmatically and step by step with the local actors. However, before to start of the missions in the territories, the team carefully developed its ethical position as far as the meaning of its intervention was concerned. Indeed, working with inhabitants of contaminated territories to help them to live better in these territories raises very difficult ethical questions. This led the ETHOS team to adopt the clear position that the proposed approach to rehabilitation of living conditions could only be engaged with persons having decided to stay in the territories despite the risk, assuming that they were informed about this risk and had the means to evaluate it. As a consequence, the team went to the territories with the intention to help the inhabitants to understand the radiological situation they were living in and its associated risks but also with the position to refuse to decide for them if they should stay or leave their living places. This difficult decision was to remain the responsibility of each family.

Such a position was deliberately opposite to the one that is "traditionally" expected from the experts to take the decision for those who are facing a risky situation. It also led the ETHOS team to implement the Project in a so-called "voluntary relocation zone" i.e. a zone defined by the law as contaminated and opening the right for families to be relocated with the support of the State if they wish so, but where the risk was considered law enough to allow further living for those having made the choice to stay taking into account the countermeasures implemented within the national rehabilitation program and assuming that people were following the official recommendations for their protection.

17.3.2 Phase 1: 1996–1999

The Project started in April 1996 with a first mission of the French team in order to select a site for its implementation. In total, the team visited 6 contaminated districts in the Southern part of Belarus and met with local authorities, professionals and inhabitants to present the Project. An important criterion was the voluntary commitment of the local community. After discussions and negotiations, the candidate village of Olmany (1300 inhabitants) in the Stolyn district (Brest region) close to the Ukrainian boarder and about 250 km West from the Chernobyl power station, whose population clearly expressed its willingness to participate in the Project, was selected (Figure 17.1). According to the Belarus law of 1991 on the "Safe living conditions of the population", the village was located in a "voluntary relocation zone" where the estimated annual average exposure ranged between 1 and 5 mSv with a ground contamination ranging between 185 and 555 kBq/m^2. In fact, only a few families with very young children left the village after 1991 when it was officially recognized as contaminated. Despite an on-going debate among the authorities on the appropriateness of relocating the population because of the local living conditions

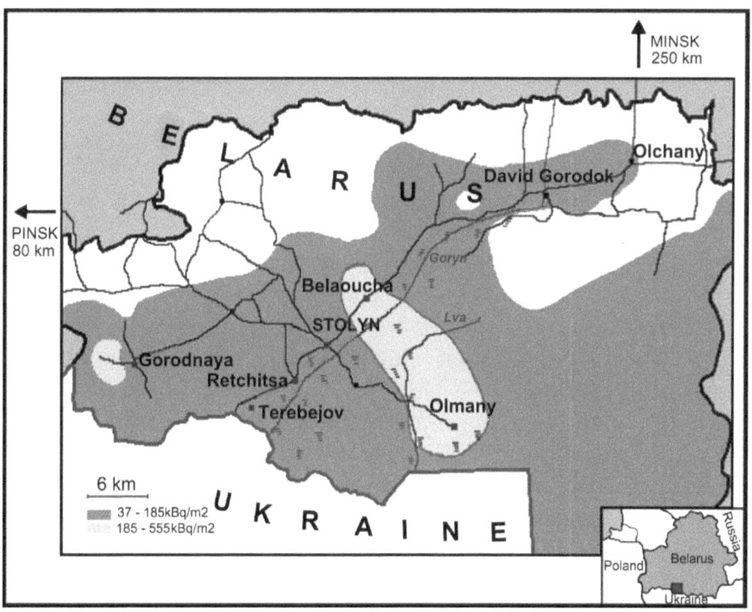

FIGURE 17.1 The district of Stolyn-Belarus with the villages in which the ETHOS Project was implemented. (For color version of this figure, the reader is referred to the online version of this book.)

(remoteness of the village, importance of forests and swamps...), there was a strong opposition from the villagers to leave the place.

A cooperation agreement was signed in July 1996 between the Ministry of Chernobyl, the authorities of the district of Stolyn, the village of Olmany, and the Ethos team setting up the basic principles of the cooperation and the respective commitments of the partners. As far as the ETHOS team was concerned, the members committed themselves to work with the voluntary villagers to evaluate their radiological situation and to come 4 times a year during 3 years to help them to improve their protection and their living conditions. Altogether the ETHOS team spent more than 60 days of presence in the village.

During the first two missions of the Project, the French team mainly listened to and learned from the villagers about their concerns. Then, after discussion and negotiation with the villagers, 6 working groups involving about 100 volunteers were created with the specific objective to try to respond to the concerns considered as priorities (Hériard Dubreuil et al., 1999):

- the "young-mothers" group aiming at the radiological protection of the children;
- the " milk-producers" group focused on the production of clean milk;
- the " meat-producers" group aiming at restoring the marketing of privately produced meat;
- the " teachers" group working on the development and diffusion of a practical radiation-protection culture at school;
- the "waste" group dealing particularly with the management of the contaminated ashes resulting from the burning of contaminated wood in stoves; and
- the "teenagers" group aiming at shooting a video presenting the life in the village.

Each group, with its own rhythm, went through a process of collective learning and assessment of the local situation with the help of the French team and, when necessary, with the support of local professionals and authorities. The local radiological situation was assessed by means of ambient dose rate and foodstuff measurements managed directly by volunteers. It is to note that, although, several monitoring systems had been put in place by the authorities, none of them were providing back information to the individuals. These measurements that were analyzed collectively by each group allowed passing progressively from a very fuzzy and negative appreciation of the radiological situation to a more contrasted and reliable picture with sometime good surprises and sometime bad ones. Altogether this common expertise (coexpertise) and requalification of the radiological situation allowed identifying many possible protection actions to be implemented locally with a minimum of additional resources to improve the protection of the inhabitants, and particularly the children, and the quality of food products. The concrete implementation of these actions sometimes required the population to make difficult choices between conflicting interests and wishes. For example, a typical dilemma for the villagers was to abandon the secular tradition of consuming products of the forest. It also often called for the involvement and support of local authorities and, for some actions of national ones. In the course of the process, new villagers, but also representatives of local professionals (medical doctors, nurses, teachers, technicians...), joined the working groups to bring their expertise but also to try to solve their own problems either as professionals or individuals.

It is beyond the objective of this paper to present in more details the coexpertise process and its results. The experience of the "young mothers" is described in Lochard, 2000 and the experience of the "milk producers" is described in Lepicard & Hériard Dubreuil, 2001. From the radiation protection point of view, one of the most interesting results, beyond the demonstration of the feasibility to directly involve villagers in the day-to-day management of the radiological situation, was the development by the group of "young mothers" of local criteria to help individuals to behave safely as far the ambient dose rates in the environment and the daily intake of contamination through food products were concerned. These criteria were established following long discussions on the interactions between the local radiological situation as resulting from the numerous measurements (in houses, gardens, recreational areas around the villages, food products growing in the village) and the local habits and customs of the villagers including their diet, private and professional activities, as well as considerations concerning the natural background and the official dose limits set up at the national level.

More concretely, the first phase of the ETHOS Project resulted in a significant improvement of the radiological quality of the milk and the whole body contamination of the children. The cooperation between the private farmers, the local collective farm, the District authorities, and the national administration allowed to increase the production of milk with contamination below 111 Bq/l (national limit at that time) from 25% to 55% in winter and from less than 10% to 80% in summer between autumn 1996 and autumn1998. As a concrete consequence, the milk from Olmany was again collected by the Stolyn's dairy from March 1999. This reaccess on the market, which was interrupted after the Chernobyl accident, became a source of income for private farmers that brought them new opportunities in terms of restoration of their living conditions. The quality of the meat was also improved with the possibility to sell part of the village production on the local market. In particular, a protocol was elaborated between the private farmers, the collective farm of Olmany and the slaughterhouse from the city of Pinsk, to ensure the private farmers with the provision of clean food for their animals as well as the follow-up of the radiological quality of the meat. These improvements largely favored a change in attitude among many families who decided to directly manage the radiological exposure of their children through a better control of the contamination of food products. A positive dialogue and cooperation between the mothers of the villages and the medical staff of the District hospital led to a reduction of the internal contamination of the children under school age by about 30% in average. With regard to education, 6 projects were performed by the teachers of the school allowing the pupils to get knowledge of the experience gained in the different projects implemented in the village and to develop the basis of a practical radiation protection culture for living day-to-day in a contaminated environment. Another positive result was the recovery of self-confidence and initiative among many villagers and a noticeable restoration of confidence in the authorities. Unexpected at the beginning of the Project, this change in the attitude of the population was certainly one of the strong incentives for the authorities to develop further the cooperation between all concerned stakeholders.

17.3.3 Phase 2: 1999–2001

Considering the achievements during three years in Olmany, the local and national authorities proposed the French team during the summer 1999 to extend the ETHOS Project for 2 more years with the objective to evaluate the conditions and means of the diffusion of the approach in other villages of the Stolyn district. A key point for this new phase was to test the ability of the local professionals and authorities to implement the coexpertise process developed in Olmany with the support of Belarus institutes. After discussions on the radiological situation in the district, the villages of Belaoucha, Gorodnaia, Retchissa, and Terebejov volunteered to participate (see Map above) and the villagers of Olmany decided to continue with the Project. A new cooperation protocol between the authorities of the 5 villages, the District, the National Chernobyl Committee, 2 Belarus scientific institutes, and the ETHOS team was signed in March 2000 validating the joint commitment of all partners and clarifying their means and resources. For this second phase, in addition to the support of the European Commission, the Project received complementary supports from French organizations (FERT, IPSN, EDF, and COGEMA) and from the Swiss Ministry of Foreign Affairs. In total, the French team participated in 8 missions representing more than 40 days of presence in the 5 villages.

In this new phase, the main role of the French team was to train and to help the local professionals to implement practical projects with voluntary inhabitants in the selected villages concerning:

- the improvement of the radiological protection and the health surveillance of children;
- the production by the private sector of good quality foodstuffs and their marketing; and
- the development of a practical radiological protection culture at school.

A seminar was organized in March 2000 in Stolyn to provide the dosimetrists, physicians and nurses, teachers, agronomists, local authorities, involved in the Project, as well as members of the Belarussian Research Institute for Soil Science and Agrochemistry (BRISSA) and the Brest Branch of the Research Institute of Radiology (BB-RIR), with the basic necessary skills to set up these practical projects in cooperation with the villagers.

The implementation of the projects in the various villages led to new interesting results. Based on the cooperation between the dosimetrists, medical doctors and nurses of the villages, the Stolyn hospital, BB-RIR, the French team and many families, the concept of an operational and "inclusive" monitoring system for the measurement of external dose rates, contamination of foodstuffs and internal contamination of inhabitants was developed and tested. The objective of this system was to establish direct links on the one side between the measurements and the daily activities and habits and, on the other side, between the radiological quality of the food products and internal exposure of persons at the level of each village. Another interesting result was the successful testing of a new technique for the production of good-quality potatoes by private farmers. This test demonstrated the potential for improving both the radiological and organic qualities but also the yield of the production (Bogdevitch, 2003). Building on the work performed during the first phase in Olmany, the project also allowed delineating further the key elements of a practical radiological protection culture for living in contaminated territories.

The main result was certainly the coexpertises of the radiological situation performed in each village and validated by all parties. These coexpertises comprised for each village a contamination map of the village and its surrounding, the range of contamination of the food products, the radiological quality of milk according to the pastures and the seasons, the distribution of the internal contamination of children, and the margin of maneuvre on the daily intake of contamination by children through the diet. The results largely confirmed the nonhomogeneity of the contamination in the environment and the very wide distribution of the individual internal doses that was already identified in Olmany. They invalidated the current approach by the administration to use average values by settlement for the management of the countermeasures and validated the large room for reducing the internal contamination of the most contaminated children. The table below, related to the situation in Olmany by the end of the project, shows for example that for a given daily diet the resulting contamination could range between 750 and 35 Bq/day, i.e. a factor 20, according to the level of contamination of the food products available in the village. This finding was without any doubt one of the driving factors for motivating families to better control the radiological quality of their daily diet through regular measurements of the food products from their village (Table 17.1).

The second phase of the ETHOS Project ended with the running of an International Seminar that took place in Stolyn in November 2001 with 150 participants representing the local, national, and international levels (ETHOS, 2001). Results from the Project were presented in detail directly by the villagers and the local professionals and authorities through a series of panels leaded by members of the French team. The discussions allowed to clearly delineating the lessons but also the limitations of the Project. The seminar was also a first opportunity to debate in the presence of national authorities (Ministries of health, agriculture, education…), Belarus institutes (BRISSA, BB-RIR, BELRAD,…), foreign institutes (IPSN-France, GSF-Germany, NRPB-UK,…), representatives of

TABLE 17.1 Example of the Possible Range of Contamination Associated with a Daily Diet of a Child in the Village of Olmany-Belarus according the Radiological Quality of the Food Products Available. The Minimum and Maximum Values are Those, Which Were Measured for Each Type of Product in Autumn 2001

		Minimum Contamination		Maximum Contamination	
Products	Grammes	Bq/kg	Bq ingested	Bq/kg	Bq ingested
Bread	250	10	2.5	60	15
Vegetable soup	300	10	3	100	30
Meat	200	10	2	300	60
Sauerkraut	300	15	4.5	50	15
Potatoes	100	10	1	80	8
Blueberries	200	100	20	2800	560
Milk	100	10	1	650	65
			Total: 34		Total: 753

numerous other districts, and representatives of the International Community (French Embassy in Belarus, Swiss Ministry of Foreign Affairs, European Commission, United Nations Development Programme (UNDP), World Bank) about the perspective opened by the ETHOS Project for the sustainable rehabilitation of living conditions within the contaminated territories in Belarus. At the end of the Seminar, a Declaration entitled: "The rehabilitation of living conditions in the territories contaminated by the Chernobyl accident: the contribution of the ETHOS approach in the Stolyn District" was adopted unanimously by the conference participants. This Declaration (see Annex) recognized the practical effectiveness of the ETHOS approach to complement the national rehabilitation program and called for an international cooperation in the area of the rehabilitation of living conditions taking into account the experience of the ETHOS Project and associating both radiological protection and the sustainable economic development in the affected territories of the Republic.

17.4 KEY LESSONS

The ETHOS Project demonstrated first that the direct involvement of the local stakeholders in the day-to-day management of the radiological situation is feasible and can significantly improve the individual situations as well as the effectiveness of the collective actions from the authorities. Because exposures are mainly driven by individual behavior and family modes of living, the most effective way to control the exposures is to rely on the vigilance and responsibility of each individual to assess and manage her or his own environment. In this perspective, the ETHOS Project evidenced the potential for implementing many protective actions in the day-to-day life in complement of the collective actions taken by the authorities.

The Project also demonstrated that the involvement of the population and the local professionals and authorities can contribute to overcome the vicious circle of exclusion and loss of control, which is in no way a fatality. It can also contribute to restore and maintain social trust and confidence in the experts and authorities. To be effective, this involvement must rely on the development of a "practical radiological protection culture" shared by all affected stakeholders and particularly the relevant professional's networks that are inherently committed in the rehabilitation process i.e. the healthcare professionals, the teachers, the professionals in charge of radiological monitoring, the responsible of agriculture, etc. The experience from the ETHOS project has shown that the development of such a culture should be based on 3 key pillars (Lochard, 2004):

- An inclusive radiation monitoring system. This system calls for a basic comprehensive practical knowledge about the mechanisms through which man is exposed and a direct access to monitoring equipment, by which the radiological quality of the environment can be evaluated and the levels of internal and external exposure of individuals and the whole population can be controlled.
- A health surveillance strategy responding to the precautionary principle as large uncertainties still remains on the potential health consequences of chronic exposures at low doses. It calls for a system based on regular clinical investigations as well as the development of registries to monitor important indices in public health, including the level of individual exposure. Such system should allow to identify any changes in the health status of the population that could occur and to investigate whether these changes could be related to radiation or other factors.
- The transmission of practical knowledge about the control of the radiological situation to future generations through the education system.

These above lessons have been progressively integrated into the international reflection on the role of radiation protection in the long-term rehabilitation of living conditions in contaminated territories, particularly in Europe in the framework of Research and Development Projects supported by the European Commission like the SAGE (Lepicard, Fiedler, Nesterenko, Nisbet, & Sudas, 2005) or EURANOS (Raskob et al., 2010) Projects. They also largely contributed to the methodological developments on stakeholder involvement in radiation protection that is now recognized by professionals as a key element for the success of radiological risk assessment and management in complex exposure situations of population living into a contaminated environment resulting either from a nuclear accident or a radiological event. (NEA/CRPPH, 2006). Finally, these lessons formed the basis of the recommendations of the International Commission on Radiological Protection to the protection of people living in contaminated areas after a nuclear accident or a radiation emergency published in 2009 (ICRP, 2009).

Secondly, the ETHOS experience revealed the need to develop dedicated local structures for the follow-up and the management of the local radiological situations by all affected stakeholders to overcome the difficulties encountered by the traditional administrative forms of management. The complexity resulting from the presence of the contamination, which is affecting all the dimensions structuring the living conditions (psychological, health, environmental, social, economic, cultural, etc.) induces de facto a close interaction between all concerned actors and the classical sector-related approach is rather ineffective.

Thirdly, the project demonstrated that to be sustainable the management of the radiological situation by the stakeholders must rely on a dynamic of economic development based primarily on individual initiatives of the local actors in partnership with the national and international institutions and organizations. The implementation of countermeasures is raising large resources coming from outside the affected territories, which are not guaranteed in the long term. Furthermore, the durability of the radiological protection actions is largely depending on the living conditions of the population. It is problematic to maintain a high level of commitment of the population for its protection in a context of economic difficulties.

An important lesson from the ETHOS Project has also been the need to go deeper in the public health problematic in a context of long-lasting contamination of the environment. Although the Project has not tackled the health dimension directly, but only indirectly through the clearly announced objective to reduce exposures in a precautionary perspective as much as possible, this dimension has been all along the years a matter of interrogation and concerns not only for the population but also for the local health care professionals and the authorities. It was obvious that, by the end of the project 15 years after the catastrophe, many questions concerning the long-term potential effect of the contamination were unresolved. Local medical doctors and nurses who participated actively to the project regularly reported a deterioration of the health situation, notably among children without being able to identify clearly the cause. Without a clear picture about the numerous factors that could have explained this trend, among which of course the radioactive contamination, an interrogation always persisted among the participants of the ETHOS Project as to the possibility for human beings to continue to reside for generations in the affected territories. This interrogation is still existing by more than 25 years after the accident and it is of utmost importance that a coherent and systematic scientific research effort be maintained in a cooperation embracing the population, the local healthcare professionals and the national and international experts. This effort should focus on the general health status of the population and its evolution over the years, a better understanding of the potential factors that may affect this status (radiological exposures, chemical pollution, nutrition, general living conditions...) and the deployment of new public health approaches fitting the complexity of the situation.

17.5 NEW APPROACH TO LONG-TERM REHABILITATION

The ETHOS Project undoubtedly was a "turning point" in the development of long-term rehabilitation strategies. Following an investigation of the situation of the population in the contaminated territories of the CIS countries during the summer 2001, the United Nations produced an important report entitled the "The Human Consequences of the Chernobyl Nuclear Accident - A Strategy for Recovery" (UNDP-UNICEF, January 2002), which not only confirmed the findings of the previous EC-CIS studies but also recognized the positive results of the ETHOS approach and largely endorsed its key lessons.

The report underlines that the vast national and international research efforts that took place in the early nineties spectacularly improved the scientific knowledge as far as the causes, the consequences, and the remediation means to deal with radioactive contamination of the environment is concerned. This allowed to implement technical, administrative, and economic measures that have effectively contributed to reduce the collective and individual radioactive burden on the population. However, the report also recognizes that these measures had failed in reversing the downward spiral of deterioration of the living conditions in the affected territories and that a large fraction of the population was still facing, 15 years after the catastrophe, a difficult economic, social, and health situation. The contamination of the environment acts as a blocking factor in the economic development for many communities and results in strong economic and social disturbances, which the most affect those people already vulnerable to poverty. This situation deteriorates the individual capacity to maintain a vigilant attitude toward radiological risk and, on the contrary, favors risky behaviors increasing radiological exposures. As a corollary, it reenforces within the population the reluctance and distrust toward the authorities, the increase of anxiety and the general feeling of loss of control described in Section 17.2. Finally, it excludes the population even more from a potential development process.

Building on the UNDP report and the declaration that concluded the Stolyn International Seminar on the ETHOS Project, the Chernobyl Committee of Belarus acted immediately after this Seminar with the support of several international organizations (EC, UNDP, World Bank), embassies in Belarus (French, German, Great-Britain, Italy) and various national and international experts, including the French team that implemented the ETHOS project, to develop a new strategic approach for the sustainable rehabilitation of the living conditions in the contaminated territories of Belarus. This resulted in the launching, in October 2003, of a 5 years international program named CORE (Cooperation for Rehabilitation) in 4 contaminated districts of Belarus: Bragin, Slavgorod, Stolyn, and Chechersk. The program was based on the active involvement of local actors in the rehabilitation process and the implementation of support mechanisms to accompany the initiatives of the local population and professionals. It was also structured with the objective to take into account all the key affected dimensions (health, social, economic and ethical) within a sustainable development perspective. The program relied on the direct access of the population to a pluralistic monitoring of the radiological quality and also included an education dimension to ensure the intergenerational transfer of the indispensable knowledge and know-how to live in a contaminated environment as well as the memory of the Chernobyl catastrophe (Trafimchik, 2005).

By paving the way to a renewed approach for the long-term rehabilitation of the living conditions in Belarus and potentially in the other CIS contaminated territories, the ETHOS project initiated a new "governance" approach for the long-term rehabilitation. Based on the close cooperation of the local, national, and international levels of actions to favor the direct and active involvement of the population and local professionals in the day-to-day management of the radiological situation, this approach allowed the inhabitants in the territories to regain control on their future and to inverse the exclusion process, which was at the source of individual resignation, social distrust, and "risky" behaviors among the population. In this perspective, two key factors, which are also encountered in other types of public participation approaches, were decisive (Beierle & Cayford, 2002): first the participation of a wide panel of stakeholders to avoid possible exclusion of persons or groups that can reveal to be in fact the key actors in the process and, secondly, the empowerment of inhabitants of the territories to favor their appropriation of the local situations and their autonomy in the implementation of their personal projects.

17.6 CONCLUSION

The sudden irruption of radioactivity in the close environment of a population, whether resulting from accidents, malevolent events, or discoveries of consequences from past activities, is creating a new reality that is altering all aspects of individuals' living conditions of those affected—psychological, health, social, economic, and cultural. The experience in the contaminated territories of the CIS countries by the Chernobyl catastrophe has revealed that the daily confrontation with such situation over years is progressively creating a general feeling of loss of control and insecurity that is difficult to solve by just introducing radiological norms and countermeasures. The prescriptive and normative traditional modes of administrative management are unable to properly address the complexity of the problems resulting from the contamination. The radiological protection dimension tends to become dominant in the daily choices concerning the living conditions of the inhabitants. Failing to be integrated into a qualitative approach of protection grounded on a shared knowledge of the radiological situation and its consequences, the prescriptions and norms gradually alienates these living conditions without the possibility to rebuild safe and responsible new ones.

How to reduce the complexity characterizing a post accident situation after the emergency phase? How to avoid perverse mechanisms introducing distrust, loss of confidence and the demobilization of the population? These are the main challenges associated with the long-term rehabilitation of living conditions in contaminated territories. The ETHOS project was an attempt to bring first elements of response. Its achievements paved the way to a renewed approach of rehabilitation strategies. It demonstrated the necessity and also the feasibility of the direct involvement of the inhabitants and the local authorities and professionals in the long-term rehabilitation process of living conditions to overcome the difficulties generated by the presence of the contamination. However, shifting from a centralized, top-down, prescriptive and normative approach to a more decentralized, bottom-up, and quality-driven approach is not straightforward. In particular, there is a need to combine rehabilitation with sustainable development and to adopt an integrated approach dealing with health, economic, and social development, environment, education, and culture at the same time.

It has often been argued that the CIS experience after the Chernobyl accident is hardly transferable elsewhere in the world because of the social, economic, and legal arrangements that prevailed in these countries. However, the current experience in Japan in the areas affected by the consequences after the Fukushima Daiichi nuclear power plan accident is demonstrating that beyond the differences in the administrative structures and in the state of technical and socioeconomic development, the administrative responses are quite similar and the populations are confronted to same concerns and problems affecting their daily life. How to find benchmarks faced with environmental contamination? What is safe and is not? How to adapt everyday activities? How to envisage the future? Like Chernobyl, the Fukushima experience reveals the prominent role of the human dimension of the consequences of a nuclear accident. By giving the affected individuals the possibility to regain control on the radiological situation and to restore decent living conditions for them and their children, the ETHOS project has demonstrated the possibility to respect at the same time the fundamental value of prudence in the management of the radiological risk and the dignity of the affected people.

ACKNOWLEDGMENTS

The Project ETHOS was the meeting and engagement in a real human adventure of personalities who shared the same values. Under the leadership of Gilles Hériard Dubreuil (Mutadis) the group formed by Jean-François Guyonnet and Gilles Le Cardinal (UTC), Henry Ollagnon et Vincent Pupin (INAPG), Jacques Lochard, Samuel Lepicard and Thierry Schneider (CEPN) accompanied with its experience but also its questions and doubts the population of the District of Stolyn in Belarus for 5 years. The author of this article has contracted an intellectual debt to each member of the group and to the people of the villages of Olmany, Belaoucha, Gorodnaia, Retchissa, and Terebejov.

REFERENCES

Beierle, T., & Cayford, J. (2002). *Democracy in practice – Public participation in environmental decisions. Resources for the future*. Washington, DC.

Bogdevitch, I. (2003). *Remediation strategy and practice on agricultural land contaminated with 137 Cs and 90 Sr in Belarus*. 25–26 November 2003, Environment and Radiation Protection. Seminar 4, pp. 83–92, Paris: Eurosafe.

ETHOS. (Novembre 2001). La rehabilitation des conditions de vie dans les territories contaminés par l'accident de Tchernobyl: la contribution de l'approche ETHOS, Actes du séminaire international (pp. 15–16). Collège Agro-économique, Stolyn, Biélorussie, Première partie: Verbatim, 163 pages. Deuxième partie: Observatoire de la qualité radiologique, 103 pages. (In French) http://www.cepn.asso.fr/fr/ethos.html.

Girard P., Hériard Dubreuil G. (1994). Conséquences sociales et psychiques de l'accident de Tchernobyl. Rapport Mutadis 93/JSP2/PG/GHD/003, (in French).

Girard P., Hériard Dubreuil G. (1995). Conditions de vie dans les territoires contaminés 8 ans après l'accident de Tchernobyl - evaluation de la situation dans le district de Tchetchersk. Rapport Mutadis 95/JSP2/PG/GHD/003, (in French).

Hériard Dubreuil, G., Lochard, J., Girard, P., Guyonnet, J.F, Le Cardinal, G., Lepicard, S., et al. (1999). Chernobyl post-accident management: the ETHOS project. *Health Physics*, *77*, 361–372.

Hériard Dubreuil, G., & Girard, P. (1996). Stress in accident and post-accident management at Chernobyl. *Journal of Radiological Protection*, *16*(3), 167–180.

ICRP. (2009). Application of the commission's recommendations to the protection of people living in long term contaminated areas after a nuclear accident or a radiation emergency. ICRP Publication 111. *Annals of the ICRP*, *39*(3).

Lepicard, S., Fiedler, I., Nesterenko, V., Nisbet, A., & Sudas, A. (2005). Strategies and guidance for establishing a practical radiation protection culture in Europe in case of long term radioactive contamination after a nuclear accident: the SAGE project. In: "The Scientific Basis for Environment Protection Against Radioactivity", ECORAD 2004, Proceedings of the International Congress, Aix-en-Provence, 6–10 September 2004, *Radioprotection*, *40*(1), 865–870.

Lepicard, S., & Hériard Dubreuil, G. (2001). Practical improvement of the radiological quality of milk produced by peasant farmers in the territories of Belarus contaminated by the Chernobyl accident – the ETHOS project. *Journal of Environmental Radioactivity*, *56*, 241–253.

Lochard, J. (1996). Psychological and social impacts of post-accident situations: lessons from the Chernobyl accident. *Proceedings of the ninth international congress of the international radiation protection association, 14–19 April 1996*. (Vol. 1) Vienna, Austria: IRPA, 105–111.

Lochard, J. (2000). Stakeholder involvement in the rehabilitation of living conditions in contaminated territories affected by the Chernobyl accident. In *Proceedings of the international symposium on 'restoration of environments with radioactive residues'* (pp. 495–506). Arlington, VA, USA: IAEA International Symposium. 29 November–3 December 1999, IAEA-SM-359/5.2, 2000.

Lochard, J. (2004). Living in contaminated territories: a lesson in stakeholder involvement. In H. Métivier, et al. (Ed.), *Current trends in radiation protection* (pp. 211–220).

NEA/CRPPH. (2006). *Stakeholders and radiological protection: Lessons from Chernobyl 20 years after*. NEA Report No. 6170, OECD.

Raskob, W., Gering, F., Lochard, J., Hugon, M., Nisbet, A., Starostova, V., et al. (2010). Overview and main achievements of the EURNAOS project: European approach to nuclear and radiological emergency management and rehabilitation strategies. In: Enhancing nuclear and radiological emergency management and rehabilitation – key results of the EURANOS European project, *Radioprotection*, *45*(5), 9–22.

Trafimchik, Z. (2005). *The CORE programme in Belarus: A new approach to the rehabilitation of living conditions in contaminated areas*. Vienna: Chernobyl Forum. http://www.docstoc.com/docs/44675417/The-CORE-Programme-in-Belarus-A-new-approach-to.

UNDP-UNICEF. (January, 2002). *The human consequences of the Chernobyl nuclear accident: A strategy for recovery*. UNDP.

Conclusions and Recommendations of the International Conference Held In Stolyn – Republic of Belarus, 15–16 November 2001 (Translation from The Russian Original Text)

"The Rehabilitation of living conditions in the territories contaminated by the Chernobyl accident: the contribution of the ETHOS approach in the Stolyn District"

The conference brought together representatives of the Chernobyl Committee at the Soviet of the Ministries of the Republic of Belarus, the Ministry of Education, the authorities of the district of Stolyn, and the oblast of Brest, directors and professionals of the kolkhozes, the inhabitants of the contaminated localities, scientists from the National Academy of Sciences and the Academy of Agrarian Sciences, representatives of the European Union, the European Commission, the UNDP, the World Bank, members of the European interdisciplinary group ETHOS and other representatives of international NGOs.

The conference participants came to the following conclusions:

1. Numerous health, ecological, economic, and social problems in Belarus, generated by the accident, have a durable character and remain the focus of attention of the administration of the Republic, researchers, and the world community. A series of factors, in particular the degradation of the economic situation, the disintegration of the USSR etc., have reinforced the consequences of this catastrophe. One of the decisive factors is also that the populations are lacking the knowledge, which would enable them to have a certain autonomy in the evaluation of the authenticity of the information on the consequences of the catastrophe—which is often contradictory—and to take the measures that would allow them to reduce the radiological risks resulting from living in the contaminated territories.
2. Wide-ranging approaches undertaken by the State have enabled to reduce appreciably the negative consequences of the catastrophe. The protection measures in the public agricultural sector assure a production that corresponds to the norms and all things considered, to reduce the expected exposition dose of the population. However, in the private sector, the level of production exceeding the norms still remains very important. This is true notably for forest products. The solution to these problems demands a

particular attention on the part of the local authorities. It is necessary also to think of the improvement of acknowledged approaches and the development of new methods in interaction with the population. For the protection of children, the measures adopted in the agricultural sector must be able to assure a production that corresponds to the most recent international radiological exposition norms.

3. The complex problem of the rehabilitation of the living conditions in the contaminated territories, which includes the restoration of economic and social activity, taking into account the necessity to secure the living conditions of the population, is becoming the foremost priority 15 years after the accident. This problem is without a parallel throughout history due to its complexity and its scope. The current time is characterized by an intense search for approaches to the solution of the problem of rehabilitation. Moreover, it is important to continue the research concerning the health of the inhabitants of the contaminated territories.
4. Taking into account the above, the ETHOS project approach, financed by the European Community, which has been conducted since 1996 in the Stolyn district merits to be studied, developed and disseminated. This approach is complimentary to the Program of the Belarus State concerning the consequences of the accident. It is based on the involvement of the population and the local specialists in the management of the radiological situation, which necessitates the development of a specific radiological culture concerning life in the contaminated territories. The efficiency of this approach has been confirmed in practice, as well as in the context of this conference following the example of the villages of Olmany, Gorodnaia, Belaoucha, Terebejov, and Retchissa.
5. The participants of the conference call upon the organisms of the Belarus administration concerned, as well as international organizations, to envisage a long-term cooperation in the area of the rehabilitation of living conditions in the territories of the Republic affected by the accident, particularly in relation to economic aspects. One overall objective would be the development of a new project that would take into account the experience of the ETHOS project and that would associate both sustainable economic development and radiological rehabilitation.

Chapter 18

Public Participation—Potential and Pitfalls

Sven Ove Hansson,[1,*] and Deborah Oughton,[2,*]
[1]*Royal Institute of Technology (KTH), Division of Philosophy, Stockholm, Sweden,*
[2]*Centre for Environmental Radioactivity, Norwegian University of Life Sciences, Ås, Norway*
**Corresponding authors: E-mails: soh@kth.se, deborah.oughton@umb.no*

Chapter Outline

18.1 Introduction **333**
18.2 What is Participation? **335**
18.3 The Role of Participation in Democracy **335**
18.4 Two Ways to Justify Participation **338**
18.5 Quality Criteria for Participative Procedures **339**
18.5.1 Representativeness 340
18.5.2 Transparency 340
18.5.3 Impact on the Decision 341
18.5.4 Early Involvement 341
18.5.5 Full Access to Expert Knowledge 342
18.6 Conclusion **343**

18.1 INTRODUCTION

In the last one or two decades we have seen an increased focus on the importance of stakeholder involvement and public participation in a number of policy areas, particularly those concerned with environmental issues and technology evaluation. Formally, the requirement for participation was clearly stated in Principle 10 of the Rio Declaration:

Environmental issues are best handled with the participation of all concerned citizens, at the relevant level. At the national level, each individual shall have appropriate access to information concerning the environment that is held by public authorities, including information on hazardous materials and activities in their communities, and the opportunity to participate in decision-making processes. States shall facilitate and encourage public awareness and participation by making information widely available. Effective access to judicial and administrative proceedings, including redress and remedy, shall be provided. (UNEP, 1992)

Radioactivity in the Environment, Volume 19
ISSN 1569-4860, http://dx.doi.org/10.1016/B978-0-08-045015-5.00018-6

Nearly 10 years later, The Aarhus Convention on Access to Information, Public Participation in Decision-making and Access to Justice in Environmental Matters was legally ratified. This was described by Kofi Annan as "the most ambitious venture in the area of environmental democracy so far undertaken under the auspices of the United Nations" (Aarhus convention, 1998). Since then, the requirement for stakeholder participation in decision making has been restated for instance in Article 26 of the Johannesburg World Summit on Sustainable Development, in a number of EC directives (e.g. EC Directive, 2001/18 2001), and in the ICRPs recommendations on nuclear accident and emergency management (ICRP, 2009).

In the area of radiation protection, various forms of stakeholder involvement are common, mostly in relation to nuclear energy and in particular the siting of nuclear waste repositories (Krütli, Stauffacher, Flüeler, & Scholz, 2010; Lidskog & Sundqvist, 2004). But involving stakeholders in decision making is not without controversy. Outcomes have been criticized both for taking too much notice of public opinion—"it is the job of those in power to take decisions necessary for the sound management of waste" (House of Lords Select Committee on Science and Technology, 1999)—and for paying too little attention to stakeholder advice (Dienel & Renn, 1995; Webler, 1995). Procedures have also been criticized for being a waste of resources and even for undermining democracy. For example, in Europe, where stakeholder participation has become particularly popular in nuclear issues, a number of Nongovernmental Organizations (NGOs) have expressed concern with the meaningfulness of these procedures (LLRC, 2003). One of the most common complaints is that those in charge of the process have misled participants on what impact it can have on the eventual decision. Many participation exercises have been accused of simply being a subtle form of public relations or propaganda. Despite the general consensus that public participation is important within policy making, the debate continues on exactly how that participation should be undertaken and by what criteria the various methods of participation should be evaluated (Oughton, 2008).

These questions are related to the underlying philosophical assumptions concerning the role of participation in a democratic society, and we will start by examining those assumptions. In order to establish robust evaluation criteria, we need to be explicit about *why* public and stakeholder participation is considered beneficial, *who* should participate, as well as exactly *what* kind of process the stakeholders are involved in. The answers will impact on how one evaluates the process of involvement. After explicating the notion of participation in Section 18.2 we discuss in Section 18.3 what role participative procedures can and should have in a democratic society. After some further discussion in Section 18.4 on how participation can be justified, we present in Section 18.5 a list of five quality criteria for participative procedures that are constructed to evaluate to what extent these procedures fulfill their democratic purpose.

18.2 WHAT IS PARTICIPATION?

It is often unclear what is meant by "participation" in a decision-making process. In democratic theory, "participative democracy" refers to a democratic procedure in which a large portion of the citizens are involved in politics in various ways in addition to voting, for instance by taking part in local meetings or in web-based political discussions (Musso, Weare, & Hale, 2000; Pateman, 1970). As it is usually interpreted, participative democracy consists in actual partaking; unused opportunities to partake do not make democracy participative. This is the sense in which we will use the term "participation" here; it will refer to actual involvement in activities influencing the decision to be taken. Those who participate may or may not be among those who will finally make the decision.

The term "decision-making process" can be taken in more or less inclusive senses. An insightful description of democratic decision-making processes was made by Condorcet in his justification or the French constitution of 1793. He divided decision processes into three phases. In the first phase, one "discusses the principles that will serve as the basis for decision in a general issue; one examines the various aspects of this issue and the consequences of different ways to make the decision." In this phase, the opinions are personal, and no attempts are made to form a majority. After this follows a second discussion in which "the question is clarified, opinions approach and combine with each other to a small number of more general opinions." In this way, the decision is reduced to a choice between a manageable set of alternatives. The third phase consists of the actual choice between these alternatives (Condorcet, [1793] 1847, pp. 342–343. Cf. Hansson, 2007).

It is not uncommon for participative procedures to take place only after the first of these phases has been completed, which means that the options open for choice have already been selected. There can be a large difference between taking part in the decision-making process as a whole and being asked about the choice between options that have been preselected by others (Hansson, 2013; Section 7.2). In this paper, we will take the term "decision-making process" in an inclusive sense that covers all three of Condorcet's phases. We will take "decision-making process" and "decision-making procedure" to be synonymous, and we will distinguish between participation in the different stages of the process.

18.3 THE ROLE OF PARTICIPATION IN DEMOCRACY

We take democracy to be a rule by the people, meaning that the people are the real decision-makers. This is a difficult ideal to realize, and one that has only been imperfectly realized. Some authors have tried to rid democracy of this ideal. The most prominent proponent of this approach was Joseph Schumpeter (1883–1950). In his view, democracy does not consist in the people taking political standpoints and electing officials to implement them. Instead, it consists in a market-like competition for leadership positions (Schumpeter, 1942. Cf.

Joseph Lawrence, 1981). In what follows we will stick to the classical definition of democracy as rule by the people.

When a decision is made in a democratic society, what role should those who are personally affected preferably have in the decision? There are three possible types of roles that they can have. The affected persons can

1. be decision makers,
2. be offered the opportunity to express their views, but not be decision makers, or
3. be entirely left out of the decision making, and not even be heard.

Democratic theory does not have a simple answer to our question what role the affected persons should have. The answer will have to depend on which the group of affected is and how it relates to the group(s) held to be sovereign in the issue at hand. Traditional democratic theory primarily assigns sovereignty to geographical units, namely nations and in some issues (by delegation) established political subunits of nations. If the group of affected persons (roughly) coincides with a country or with a smaller political unit such as a municipality, then democratic theory and practice will support option (1), i.e. the affected persons should be decision makers through the ordinary decision-making process. But in risk-related issues, the affected persons are often not delimited in this way. They may form an "ad hoc" geographical unit that does not coincide with any local political unit, or they may form a nongeographical unit including for instance the workers, neighbors, and customers of an industry. (For examples, see chapters 10, 11, 14, 16 and 17.) This creates problems for democracy, since democracy, as we know it, is based on stable communities that are mostly geographical. Decision making in ad hoc units, created for each particular issue, would be exceptionally difficult to implement, and for two reasons: First, the delimitation of these units (presumably by higher-level decision makers) would create opportunities for gerrymandering of hitherto unknown proportions. Secondly, the functionality of democratic decision-making depends to a large extent on the persistence of the decision-making units. Decision makers who meet to make one single decision cannot be expected to be as willing to listen to each other and find a workable compromise as decision makers who expect to make many future decisions in the same constellation. For these reasons, the creation of ad hoc decision-making bodies for a particular decision is seldom a practicable option. Therefore, answer (1) to our question has to be rejected in such cases due to its impracticability, although it would, democratically, be the ideal answer. Instead, we have to resort to answer (2), i.e. the affected persons should be offered the opportunity to express their views, but the actual decision has to be made by others.

In many risk-related issues, "those affected" form a group that is heterogeneous both in terms of the nature and the strength of their involvement in the issue. Often there is both a relatively small group of rather heavily affected persons, such as the neighbors of a projected new railroad, and a much larger group of persons who are affected to a smaller degree, such as all those whose travel time will be reduced a few minutes by the new railroad. There are strong reasons why the interests of each heavily affected person should have more weight than those of each slightly affected person. However, it would be difficult to construct legitimate and practicable decision procedures in which influence is apportioned according to degree of interest. Therefore, the distribution of impact over the decision will have to be dealt with through judicious politics rather than through the application of formal procedures that divide power unequally among the participants.

In conclusion, the practical implementation of the democratic ideal that all decisions should be made jointly by those concerned requires considerable adjustment and compromise. Decisions have to be made in stable units (such as nations and regions), and often the group of persons affected by a particular decision does not coincide with any of these units. However, these practical limitations in the implementation of the democratic ideal do not make it less urgent that those affected should be involved in the decision process and have an influence on its outcome. For this reason, a democratic government does not (as some seem to believe) replace or reduce the need for participative procedures through which those most affected can influence a decision. On the contrary, such participative procedures are a necessary component of a well-functioning democracy, in order to make up for the unavoidable misfit between decision-making units and the groups of persons affected by many decisions.

But the shifting and somewhat ephemeral nature of the group of affected persons is not only part of the justification for participative procedures as additions to the regular structures of representative democracy. It is also the source of one of the major difficulties in the construction of such procedures, namely the so-called stakeholder identification problem: The persons affected for instance by an engineering project cannot in typical cases all be identified (Long, 1983). Since there are many degrees of affectedness, it is not possible to draw a sharp line between affected and unaffected persons. In many cases, such as that of a nuclear power plant, those affected are distributed over a large geographical area. This is also a major reason why the notion of informed consent cannot be transferred from medicine to the management of engineering projects. In the latter case, the group of affected persons has ill-defined boundaries, whereas for instance a surgical procedure or an hour with a psychiatrist "is for a specific, known individual" (Long, 1983, 60–61. Cf. Hansson, 2006). (Schinzinger & Martin, 1983 claimed that this problem can be solved with vicarious decision-making, but that is a way to bypass the problem rather than to solve it.)

The view on participation procedures that we have presented here is based on the classical view of democracy as rule by the people. In the discussion on participative procedures, other views are being heard that are closely related to the Schumpeterian view of participation as an unnecessary and potentially counterproductive component of democracy. According to such views, communication between authorities and laypersons should be primarily one way, namely provision of information from the former to the latter. This has been described as the "Decide-Announce-Defend" (DAD) model. One of its consequences is that the public is invited to public relations activities rather than public participation activities. It has been criticized for being more technocratic than democratic and for leaving the field open for lobbying from groups with vested interests while closing it for less vociferous and organized laypersons (Oughton & Forsberg, 2005).

Our approach is much more compatible with another approach to stakeholder and public participation that has gained much in approval in the last two decades, namely that which requires a two-way communication and dialogue. This has often been called a bottom-up or deliberative democracy approach (Nielsen, Lassen, Sandoe, 2004). Its general aim is to involve all stakeholders as early as possible in an open-ended decision-making process, and to solicit as broad a range of opinions and knowledge as possible. As a contrast to the DAD model, this style was rather succinctly described as a "Meet-Understand-Modify" (MUM) approach by a former president of the ICRP, Roger Clarke (Clarke, 2002).

18.4 TWO WAYS TO JUSTIFY PARTICIPATION

There are two major classes of justifications for methods of decision making, namely justifications based on the expected outcomes of a method and justifications that refer to properties that the method has independently of what the outcome will be. In short, we can express this as a difference between outcome-oriented and process-oriented justifications. This distinction has often been explicitly referred to in the literature on democracy. Some authors have justified democracy exclusively as a means to improve decision outcomes. Hence, John Stuart Mill distanced himself from what he called "the metaphysical radicals, who hold the principles of democracy not as means to good government, but as corollaries from some unreal abstraction—from 'natural liberty,' or 'natural rights'." Instead, he counted himself as one of the "philosophic radicals", namely "those who in politics observe the common practice of philosophers—that is, who, when they are discussing means, begin by considering the end, and when they desire to produce effects, think of causes" (Mill, [1837] 1982, p. 353). In his utilitarian view, outcome-oriented arguments alone could justify democracy.

Others have justified democracy as an intrinsically just way of making decisions, in particular with reference to how it satisfies individuals' rights to have influence on the society in which they live. Such views have grown in importance in the last half-century. In an interesting article, the Canadian political scientist H.B. Mayo (1911–2009) argued against outcome-oriented justifications of democracy (Mayo, 1962). Different persons may have different views on what the desired outcomes are, he said, and therefore democracy cannot be adequately justified in outcome-oriented terms. Democracy is "an organizing principle, the principle of order and priorities and harmonization; what it is about, what gives it content, is morality, economics, religion, education and so on" (p. 559). Instead, democracy can be justified by process-oriented principles such as the intrinsic value of popular control and political equality.

In the discussion of stakeholder participation both outcome-oriented and process-oriented arguments have been appealed to. In practice there are two major classes of outcome-oriented arguments. First, there are arguments that refer to the general usefulness of incorporating many inputs into an open-ended decision-making process. These inputs can improve the outcome in otherwise unforeseeable ways.

Secondly, there are the outcome-oriented arguments of those who see participatory procedures as means to achieve acceptance of a preselected decision option. The following is a clear statement of that way of thinking:

"Community groups have in recent years successfully used zoning and other local regulations, as well as physical opposition (e.g., in the form of sitdowns or sabotage), to stall or defeat locally unacceptable land uses. In the face of such resistance, it is desirable (and sometimes even necessary) to draw forth the consent of such groups to proposed land uses."

(Simmons, 1987, p. 6)

We consider the second type of outcome-oriented argumentation to be counter productive since it only promotes decision processes that do not fulfill the functions of participative procedures in a democracy, as outlined in Section 18.3.

Many authors have promoted stakeholder participation in societal decision-making with process-oriented arguments such as principles of fairness and due process. Such principles imply for instance that everyone's interests should be taken into account and that all affected persons should have an opportunity to express their views on the imposition of any risk associated with a particular policy or technology (Shrader-Frechette, 1991).

As we see it, there is no need to choose between outcome-oriented and process-oriented justifications of participation. Strong arguments of both types support the idea of participation. (We refer of course only to the first of the two types of outcome-oriented justifications that we distinguished between above.) People have a right to influence on decisions that affect them, and when they exercise that right decision outcomes tend to improve.

18.5 QUALITY CRITERIA FOR PARTICIPATIVE PROCEDURES

We are now going to present five quality criteria for participative procedures. They all refer to how well the procedure fulfils the function of such procedures in a democratic system, as outlined above in Section 18.3. We will consider both outcome-oriented and person-oriented aspects of their function.

Criteria for the quality of participative procedures have been proposed before (e.g. Rowe & Frewer, 2000). Our approach differs from others primarily in the way in which they are based on democratic theory.

18.5.1 Representativeness

Representativeness is important for both outcome- and process-related reasons. Under the assumption that including the perspectives of all concerned improves the quality of the decision outcome, the absence of some of these perspectives will have a negative impact. Under the assumption that the procedure has democratic legitimacy that is independent of its outcome, such legitimacy will have to depend at least in part on the representativeness of the participants.

However, due to the stakeholder identification problem discussed in Section 18.3, the criterion of representativeness is notoriously difficult to specify in a precise manner for a particular procedure. In a procedure concerning the possible environmental effects of a temporary storage facility for nuclear waste, how distant neighbors should be invited to participate? The workers at the plant should be included, but what about those of subcontractors, or plants sending waste to the facility?

In addition, once we have found an answer the question who should be represented, it remains to determine by whom they should be represented. On what grounds can an individual legitimately represent a group? Does (s)he have to be elected, or is it sufficient just to be a member who is willing to take part in the procedure? The former answer gives priority to representativeness, but the second makes for more open procedures. Inclusiveness has sometimes been interpreted as meaning that all individual stakeholders should have the opportunity to be involved (Hunt, 2003).

The actual forms of representation will in many cases have to depend on practical constraints and limitations: Are representative organizations such as trade unions and community-based grassroots organizations available? How many persons in different categories are willing to participate? The difficulties in combining representativeness with openness can sometimes be solved by having parallel forms of participation, some of which are open so that all who wish to participate can do so in some way.

18.5.2 Transparency

Transparency can refer both to the procedure itself and to the discussions performed within it. The process should be transparent to enable the public to see what is going on and how decisions are being made. In addition, the reasoning behind actions, deliberations, and eventual decisions should be made publicly available.

Transparency is particularly important for participative procedures that are affected by the stakeholder identification problems, i.e. procedures with vaguely delimited groups of affected persons. If groups or persons with some potential interest in the matter have been excluded, then it should be possible for them to react against their exclusion and demand a right to participate. This requires that they have insight into the procedure so that they can judge for themselves whether their exclusion was justified or not. This is important for both outcome-oriented and process-oriented reasons. To the extent that they can contribute new information to the process, the outcome will be less well informed if they are not given the opportunity to do so. To the extent that they are legitimate participants in the procedure, the legitimacy of the procedure as a whole will be lessened by their involuntary exclusion.

It should be added that the transparency criterion is a general democratic requirement that is applicable also to nonparticipative procedures. The delegation of a decision for instance to a group of experts is much more easily defended if the public has the insight needed for an informed judgment on whether the delegation should continue or be revoked. Obviously, other considerations can on occasions outweigh the arguments in favor of transparency, but in such cases, the lack of transparency can often be partly compensated for, for instance by posterior disclosure or by disclosure to a representative oversight committee.

18.5.3 Impact on the Decision

For a participative procedure to fulfill its function, its output should at least potentially have a genuine impact on policy. If that is not the case, then the procedure can obviously not be justified with outcome-oriented arguments. And if participation has no impact on the decision, then it cannot fulfill the function of participation in a democracy, and therefore it has no process-oriented justification either.

As mentioned above, one of the most common complaints against participative procedures is their lack of influence on policy. According to Wallentinus and Paivo (2001), there have been instances where bodies have sat down to listen to stakeholders in a succession of meetings, but no effort was made to adjust proposals to their suggestions. Beder (1999) cites the lack of influence as evidence that decisions have already been made and that policy makers are just going through the motions for the sake of public relations. Participants in some of the earlier nuclear power consultations concluded that they were there "to legitimize agency decisions, defuse opposition, warn the agency of possible political obstacles and satisfy procedural and legal requirements" (Kraft & Clary, 1993). To date, there is relatively good evidence that stakeholder participation can provide a valuable input of information to the decision-making process, but much less evidence of procedures having a direct impact on policy. We see this as a major problem area for research and for the development of new participative procedures and methodologies.

18.5.4 Early Involvement

As we mentioned above, participative procedures can affect either the whole decision-making procedure or only parts of it. It is in the early stages (Condorcet's first phase) that issues and tasks are defined and delimited. Therefore, to the extent that stakeholders have valuable information to contribute, that information should preferably be entered into the decision process at an early stage when it is most useful. This makes early involvement important for outcome-related reasons. It is also important for process-related reasons, since limitations in the influence of stakeholders reduce the legitimacy of the procedures. As we have already mentioned, it is a common criticism that stakeholders have only been involved at a late stage, after the decisions have de facto already been made. However, the nature of stakeholder involvement may have to change as the process develops. For example, if there is strong expert disagreement over factual or technical information, then the appropriate role of laypersons will usually be to insure that expert discussions take place in the proper way and with the right instructions, rather than to participate themselves in detailed discussions on difficult technical issues (Cf: Kaiser & Forsberg, 2002; Oughton & Strand, 2003). In other parts of the early discussions, such as those concerning the criteria by which alternative policies should be evaluated, direct layperson participation is essential.

Often the early stages of participation focus on information gathering that may include surveys, polls, in-depth interviews or focus groups. Polls, surveys, and interviews have the advantage of being cheap, and at least potentially representative. However, they have the preponderant disadvantage of being one-way communication, not offering the participants the opportunity to inform themselves and listen to the viewpoints of others before forming an opinion. From that point of view, focus groups and other groups where participants actually meet and discuss are much to be preferred.

18.5.5 Full Access to Expert Knowledge

Efficient participation in a decision-making process requires full access to relevant expert knowledge. Participants with and without access to experts will not be equal in terms of their ability to promote the ends they prefer the decision to serve. Access to expertise is important primarily for outcome-related reasons; it is not difficult to find examples verifying that more informed persons can provide more useful inputs to decision-making procedures.

But in many participative procedures, access to experts is remarkably limited. The public hearing is a typical example of this. Whereas the elite have opportunity to meet and discuss with experts on equal terms, most citizens only have access to them at meetings with experts and authorities on stage and the public in a mass audience. There are, however, other types of procedures, such as the citizen's panels of Peter Dienel (1923–2006), in which the participants are provided with the same type of access to experts as the decision makers (Dienel, 2002). In particular in technically complex issues such as those relating to nuclear energy, we believe it to be of utmost importance that the public has full access to experts. This means that experts have to be available for group discussions and for hearings in small formats, not only for meetings with large audiences.

Information about uncertainties is important for decision makers, since their decisions may depend on it. Our requirement that the public should have full access to expert knowledge includes the information that experts have about uncertainties and lacunae in science. It was previously thought that the general public has a hard time understanding uncertainties, but recent research has shown that this need not be so and that the public wants that type of information (Frewer et al., 2002; chapters 16 and 17 in this book). There is also considerable experience showing that failure to report uncertainties can result in mistrust with ensuing communicative difficulties (Jensen & Sandøe, 2002; Miles & Frewer, 2003).

18.6 CONCLUSION

Judging by the development in the last two decades, we are likely to see an increase in stakeholder involvement and public participation in environmental and other policy areas, not least in nuclear and other radiation-related issues. However, it is much less clear whether such participation will be performed in ways that lead to better and more legitimate decisions. For decision makers who have vested interest or who have already made up their minds, it is always convenient to arrange participation procedures in forms that minimize the risk of their preconceived plans being thwarted. But that is not what we need.

What we need is participative procedures that complement representative democracy by making it possible for those affected by a decision to form and voice opinions and to have an influence on the decision outcome. The five quality criteria that we have proposed—representativeness, transparency, impact on the decision, early involvement, and full access to expert knowledge—can be used to evaluate whether a participative procedure has the capacity to fulfill its democratic purpose.

REFERENCES

Aarhus Convention. (1998). *The Aarhus convention on access to information, Public Participation in Decision-making and Access to Justice in Environmental Matters* (adopted June 1998, ratified October 2001) (accessible online: www.unece.org/env/pp/welcome.html).

Beder, S. (1999). Public participation or public relations? University of Wollongong, In B. Martin (Ed.), *Technology and public participation*. Australia: University of Wollongong Press.

Clarke, R. (2002). *Radiological protection of the environment: The path forward to a new policy*. Paris: NEA/OECD.

Condorcet. (1847). Plan de constitution, presenté a la convention nationale les 15 et 16 février 1793. [1793], In A. Condorcet O'Connor & M. F. Arago (Eds.), *Oeuvres de Condorcet* (Vol. 12, pp. 333–415). Paris: Frimin Didot Frères.

Dienel, P. C. (2002). *Die Planungszelle. Der Bürger als Chance* (5th ed.). Wiesbaden: Westdeutscher Verlag.

Dienel, P. C., & Renn, O. (1995). Planning cells: a gate to 'fractal' mediation. In O. Renn, et al. (Ed.), *Fairness and competence in citizen participation, technology, risk and society* (Vol. 10, pp. 130). Kluwer Academic Publishers.

EC (2001) European Commission directive 2001/18 of 12 March 2001 on the deliberate release into the environment of GMOs (revision of directive 90/220) (available online: www.europa.eu.int).

Frewer, L. J., Miles, S., Brennan, M., Kuznesof, S., Ness, M., & Ritson, C. (2002). Public preferences for informed choice under conditions of risk uncertainty. *Public Understanding of Science*, *11*, 363–372.

Hansson, S. O. (2006). Informed consent out of context. *Journal of Business Ethics*, *63*, 149–154.

Hansson, S. O. (2007). Social decisions about risk and risk-taking. *Social Choice and Welfare*, *29*, 649–663.

Hansson, S. O. (in press). *The Ethics of Risk*. Palgrave MacMillan.

House of Lords Select Committee on Science and Technology. (1999). *Management of nuclear waste*. London: The Stationery Office.

Hunt, J. (2003). Consultation and participation. In K. Anderson (Ed.), *Proceedings of Valdor 2003, values in decisions on risk* (pp. 9–13). Stockholm: Sweden.

ICRP. (2009). Application of the commission's recommendations to the protection of people living in long-term contaminated areas after a nuclear accident or a radiation emergency. ICRP publication 111. *Annals of the ICRP*, *39*(3).

Jensen, K. K., & Sandøe, P. (2002). Food safety and ethics: the interplay between science and values. *Journal of Agricultural and Environmental Ethics*, *15*, 245–253.

Joseph, L. B. (1981). Democratic revisionism revisited. *American Journal of Political Science*, *25*, 160–187.

Kaiser, M., & Forsberg, E.-M. (2002). Consensus conference on environmental values in radiation protection: a report on building consensus among experts. *Science and Engineering Ethics*, *8*, 593–602.

Kraft, M. E., & Clary, B. B. (1993). Public testimony in nuclear waste repository hearings: a content analysis. In R. E. Dunlap, et al. (Ed.), *Public reactions to nuclear waste*. London: Duke University Press.

Krütli, P., Stauffacher, M., Flüeler, T., & Scholz, R. W. (2010). Functional-dynamic public participation in technological decision-making: site selection processes of nuclear waste repositories. *Journal of Risk Research*, *13*, 861–875.

Lidskog, R., & Sundqvist, G. (2004). On the right track? Technology, geology and society in Swedish nuclear waste management. *Journal of Risk Research*, *7*, 251–268.

LLRC. (2003). Low Level Radiation Campaign. Dad is dead; long live UNCLE (Unlimited Nuclear Consultations Leading to Exhaustion) Radioactive Times. Vol. 4 No 2, www.llrc.org/rat/subrat/rat426.htm.

Long, T. A. (1983). Informed consent and engineering: an essay review. *Business and Professional Ethics Journal*, *3*, 59–66.

Mayo, H. B. (1962). How can we justify democracy? *American Political Science Review*, *56*, 555–566.

Miles, S., & Frewer, L. J. (2003). Public perception of scientific uncertainty in relation to food hazards. *Journal of Risk Research*, *6*, 267–283.

Mill, J. S. (1982). Fonblanque's England under seven administrations. [1837]. In John Stuart Mill (Ed.), *Collected works* (Vol. 6, pp. 351–380). Toronto: University of Toronto Press.

Musso, J., Weare, C., & Hale, M. (2000). Designing web technologies for local governance reform: good management or good democracy? *Political Communication*, *17*, 1–19.

Nielsen, A. P., Lassen, J., & Sandoe, P. (2004). Involving the public–participatory methods and democratic ideals. *Global Bioethics*, *17*, 96–102.

Oughton, D. H. (2008). Public participation – potential and pitfalls. *Energy & Environment*, *19*, 485–496.

Oughton, D. H., & Forsberg, E.-M. (2005). Ethical advice to policy in its problematic contexts. In *Biotechnology-ethics: An introduction* (pp. 295–304). Firenze: Angelo Pontecorboli Editore. (ISBN 88-88461-49-3).

Oughton, D. H., & Strand, P. (2003). The Oslo Consensus conference on protection of the environment. *Journal of Environmental Radioactivity*, *74*, 7–17.

Pateman, C. (1970). *Participation and democratic theory*. Cambridge: Cambridge University Press.

Rowe, G., & Frewer, L. J. (2000). Public participation methods: a framework for evaluation. *Science, Technology and Human Values*, *29*, 3–29.

Schinzinger, R., & Martin, M. W. (1983). Commentary: informed consent in engineering and medicine. *Business and Professional Ethics Journal*, *3*, 67–78.

Schumpeter, J. A. (1942). *Capitalism, socialism and democracy*. New York and London: Harper & brothers.

Shrader-Frechette, K. S. (1991). *Risk and rationality*. Berkeley: Univ. California Press.

Simmons, J. (1987). Consent and fairness in planning land use. *Business and Professional Ethics Journal*, *6*(2), 5–20.

UNEP. (1992). *Rio declaration*. United Nations Environment Programme. (UNEP). available online at: www.un.org/documents/ga/conf151/aconf15126-1annex1.htm.

Wallentinus, H.-G., & Paivo, J. (2001). The Hallandsas railway tunnel project in Sweden. In T. Hilding-Rydevik (Ed.), *The role of environmental impact assessment in the planning and decision process of large development projects in Nordic countries*. Stockholm: Nordregio Report.

Webler, T. (1995). "Right" discourse in citizen participation: an evaluative yardstick. In O. Renn, et al. (Ed.), *Fairness and competence in citizen participation, technology, risk and society* (Vol. 10, pp. 35–86). Kluwer Academic Publishers.

Chapter 19

Compensation or Bribery? Ethical Issues in Relation to Radwaste Host Communities

Mike Hannis,[1,*] and Kate Rawles,[2]
[1]*Environmental Politics, University of Keele, Keele, Staffordshire UK*
[2]*Outdoor Studies, University of Cumbria, Ambleside, Cumbria, UK*
**Corresponding author: Email: mikehannis@gmail.com*

Chapter Outline

19.1	Introduction	347
19.2	Host Communities	350
19.3	Ethical Acceptability and Public Acceptability	350
19.4	Site Selection	351
19.5	Effects on the Host Community—Negative Impacts	352
19.6	Positive Impacts	354
19.7	Appropriate Compensation: Process	355
19.8	Appropriate Compensation: Outcome	356
19.9	The Inadequacy of Existing Guidelines	357
19.10	Compensation, Bribery, and Incentives	358
19.11	Is Bribery Unavoidable in This Context?	359
19.12	Mitigation and New Build	361
19.13	Bribery, Benefits, and Independence	363
19.14	Financial Compensation	364
19.15	Ethical Decision Making and Commensurability: Against Algorithms	366
19.16	Responses	368
19.17	Problems with Cost-Benefit Analysis	369
19.18	Conclusions	371
19.19	Afterword	372

19.1 INTRODUCTION

How should we deal with radioactive waste? Any attempt to answer this question will raise ethical issues. Despite the gravity of these issues—inevitably associated with decisions about substances that remain toxic for thousands of years and that have the potential to harm people and other forms of life for

Radioactivity in the Environment, Volume 19
ISSN 1569-4860, http://dx.doi.org/10.1016/B978-0-08-045015-5.00019-8

generations to come—the ethical aspects of radioactive waste management have been relatively neglected. Both within the industry and the literature there has been a strong focus on the scientific and technical dimensions of questions about how such waste should be managed, and on the scientific and technical dimensions of assessing risk and safety. Critical though these dimensions are, questions of risk and safety clearly also imply a raft of ethical issues and value judgments. These range from big picture issues such as whether it is legitimate to create energy that produces this kind of waste, and the fairest way to deal with radioactive waste now it exists, to more subtle analyzes of the value judgments implicit in the concepts of risk and safety and the ethical questions arising, for example, in relation to whose criteria of risk should be consulted and by what process. Overlooking these issues does not make them go away: it just means that ethical and evaluative positions are adopted implicitly, without due consideration (For a general review of ethical and evaluative issues in the context of radioactive waste management, see Rawles, 2000).

Arguably, the view that the ethical and evaluative dimensions of radioactive waste management are inextricably bound up with the technical ones, and hence unavoidable, is now more widely accepted. Recent years have seen research into these questions not only in the academic literature but also within the field: the UK government-sponsored Committee on Radioactive Waste Management (CoRWM) included an explicit focus on ethical issues as part of its research and subsequent report (CoRWM, 2006).

In the UK, it is likely that higher activity radioactive waste will eventually be put into a geological disposal facility. Whatever decisions are made about the best way of dealing with radioactive waste in the UK and elsewhere—whether this be long-term management underground or short-term surface storage, in one central location or many, with or without options of retrieval, and so on—the waste is bound to have a physical location. This means that one or more communities of people—and other beings—may be especially impacted by their proximity to a radioactive waste-management facility (RWMF).

This chapter is based on research that initially set out to explore the ethical issues associated with the question of how these "host" communities should be compensated. These relate, not only to the amount or nature of such compensation, but also to who should decide, and how. Ethical judgments are inevitably made in selecting the process by which compensation decisions are reached as well as in the decisions themselves. In pursuing these questions, however, it rapidly became apparent that they are only one aspect of a general and more fundamental concern with the relationship between the host community and society as a whole. What does it take to ensure that this relationship is ethically sound rather than, for example, exploitative, misleading or manipulative, or in other ways unfair? Questions for consideration here include the nature of the relationships already in place between "society as a whole" and the host community, the sorts of decision-making processes already in place for making decisions about compensation, and who takes part in them. Who, for example, identifies impacts as negative (or positive): and who decides whether, or at what point, it is reasonable to expect a host community to accept them? A further critical set of questions fall under the heading of environmental justice. Is the host community significantly poorer than other parts of society? Does it already play a role in bearing the negative impacts of the nuclear industry? Is the host community in any way disenfranchised in terms of political or economic power? In short, questions about fair compensation shade into a wider set of ethical issues relating to the social context within which compensation is to be assigned.

The purpose of this chapter is to introduce a discussion of ethically appropriate compensation for host communities, understood as located within these wider socioethical concerns. It has two main aims. The first is to draw attention to key ethical issues, and areas where value judgments and/or judgments about ethics are unavoidable. Once identified, these ethical and evaluative judgments can be opened out for critical discussion and debate. A guiding assumption of this chapter is that ethical judgments are more likely to be made well if made explicitly and on the basis of reasoned debate, rather than, as may often have been the case, endorsed implicitly and hence without any attention to their underlying rationale. The second main aim is to identify a range of key themes and questions that call for further attention.

The chapter begins with some preliminary points about the difficulties of defining the boundaries of "the host community", the troublesome relationship between public acceptability and ethical acceptability, and the question of what constitutes a legitimate process for the initial selection of the host community. Although trying to define criteria for these processes is outside the scope of this chapter, some outline comments will be offered, as an attempt to establish appropriate compensation for a host community cannot in our view be cleanly separated from questions about the nature of the process used to select that community in the first place. Conversely, questions about how the host community is to be treated will clearly inform the site-selection process. These issues will therefore (at least sometimes) need to be considered together.

The main focus of the chapter, however, is on the issue of compensation, and in particular the question of how compensation is to be distinguished from bribery. At what point does the first become the second? How do we ensure that recompense for the disbenefits of hosting an RWMF is, and is understood to be, compensation rather than bribery? The nature of the compensation in question rapidly emerges as an important factor here. Whether compensation is financial (as opposed to, for example, improved educational facilities or infrastructure) appears to have psychological as well as ethical implications, with research revealing some initially counterintuitive results about the way in which financial compensation can act as a disincentive (Frey & Jegen 2000; Renn, Webler & Kastenholz 1998). This research is considered in some detail. Finally, it is argued that while many discussions of compensation are implicitly located within an essentially utilitarian ethical framework, this framework rests on a range of assumptions that are deeply problematic. A key recommendation is that discussions about host community compensation should acknowledge the ethical frameworks they are located within and should avoid taking a utilitarian approach.

19.2 HOST COMMUNITIES

An important qualificatory point is that there are real problems implied by any attempt to discuss "the host community" as if it were a homogeneous group. Age, gender, class, income, and many other variables may all, for example, affect attitudes to risk. Insofar as proximity to an RWMF is understood as risky, these variables are likely in turn to affect people's views about whether a proposed RWMF should be accepted or rejected. There are thus important issues about how different interests, attitudes, beliefs, and values within the host community are to be fairly represented in decision-making processes. Whose views should be consulted? Which groups within a given community should be represented, and how should this be decided? What counts as "representation"? A further set of issues arises with regard to the physical and nonphysical boundaries of the host community. Are they contiguous with local administrative boundaries, or is the extent of the community defined by other criteria altogether? Are there differing views on this? Does the host community include those who are not resident but regularly visit the area, or have other strong connections to it? Who should decide?

The aim here is not to try to answer these questions but simply to draw attention to them. The key point is that the notion of a "host community" can obscure important and relevant differences within a group of people, and this needs to be borne in mind whenever the term is used.

19.3 ETHICAL ACCEPTABILITY AND PUBLIC ACCEPTABILITY

Widespread recognition that decisions about radioactive waste management need to be acceptable to the public (see e.g. CoRWM, 2006; NEA, 2004; Nilsson, 2001; Sjöberg, 2004) has led to increasing public involvement in decision making about radioactive waste. Clearly public involvement and public acceptability must be key components of a long-term strategy for radioactive waste management, for tactical and pragmatic reasons as well as ethical ones. However, an important point, often overlooked, is that *public acceptability and ethical acceptability are not the same thing.*

While public involvement in decision making may be necessary for ethical legitimacy, it is not sufficient. Failure to involve the public in decision making about RWMFs will arguably render the decisions unethical. But the opposite does not apply. Even a high degree of public involvement in RWMF decision making will not guarantee that the decisions made are ethically sound. A decision could be accepted by the public (typically meaning the majority) and still be ethically unacceptable. It is claimed, for example, that a majority of people in the UK would be in favor of reintroducing capital punishment. This, even if true, would not in itself make the reintroduction of hanging an ethically sound policy.

A related point is that the requirement to find solutions that are acceptable to the public does not legitimize any possible way of getting the public to come on board. Certain kinds of approaches to securing public acceptance, such as withholding relevant information, deception, brain-washing, bribery, or threat are implicitly ruled out. Such approaches are ruled out on ethical grounds, not according to whether or not they would work. In other words, an ethical position is already implicit in the concern to make RWMF decisions acceptable to the public, and it is this concern that is in fact the primary one. To ensure ethical legitimacy, both concerns—ethical and tactical—should be explicit. There should be a clear and stated aim to make decisions that are both ethically and publicly acceptable: if the focus is only on how to gain public approval, there is a real danger of the exercise becoming a public relations one, and ethical legitimacy will not be guaranteed in this case.

Ethical decision making requires reflection and critical deliberation on the rationale behind different points of view. Ethical judgments can and should be critically evaluated. Some are held on the basis of better grounds than others, or are based on more humane principles. Ethical and value judgments, to be sound, need to be informed and developed as a process of critical reflection and dialogue. Two main conclusions follow from this. First, we need to know *why* a majority or "the public" think something should be done before we can tell whether doing it would actually be the right thing. Second, the process of engaging people with critical ethical deliberation may be an important part of a legitimate process. Some, but by no means all, methods of public consultation involve dialogue and critical discussion. Others simply involve eliciting currently held opinions and then entering these into a decision-making process that is indifferent to the reasons why these opinions are held. Treating host communities ethically will therefore require the creation of forums where constructive and rigorous ethical deliberations can take place, rather than simply eliciting people's views in a snapshot manner.

19.4 SITE SELECTION

In any given case, it will not, arguably, be appropriate even to consider the question of fair compensation unless the community that is to host the radioactive waste management facility has been, and has been seen to have been, chosen by a fair process. In an outline, a fair process will be one that is genuinely open (rather than intended to reach a prearranged outcome) and guided by criteria that are clear, explained, and reasonable. It will be informed by debate and dialogue between well-informed representatives of all significant stakeholders (both local and distant), and will be consistently sensitive to questions of unrepresented interests and unheard voices.

This generates the need for imaginative approaches to deliberative decision-making procedures, which will often constitute an effective extension of current democratic process. Establishing the exact nature of these procedures is beyond the scope of this chapter. Clearly, however, it is necessary for both ethical and tactical reasons to continue to find ways of moving beyond the so-called "decide-announce-defend" approach, and any other decision-making procedures which are not open to public scrutiny or in which public consultation plays a minimal role.

In short, a due process will be one that can generate robust and genuine answers to the question, "why us, and was it fair?" (or, even better, one in which this question would never arise); which pays serious attention to deliberation carried out between a broad base of representatives; and which can demonstrate, and show the validity of, the chains of reasoning involved. Throughout this chapter, it will be assumed that the site or sites will have been chosen by a process that meets these and any other conditions held to constitute an ethically legitimate process for selecting host communities.

19.5 EFFECTS ON THE HOST COMMUNITY—NEGATIVE IMPACTS

It is often argued (or assumed) that host communities, even having been selected as a result of a fair process, should receive compensation. This raises a range of questions about what would count as appropriate compensation, who should decide, by what process, and so on. These questions in turn rest on the assumption that there is something to be compensated for. The verb to compensate derives from the Latin *compensare* meaning to weigh one thing against another. In ethical and legal contexts, compensation is typically understood as the attempt to make amends for perceived or actual loss, injury, harm or damage sustained to persons or property, or for the imposition of certain levels of risk. Legal definitions of compensation are often in narrower monetary terms. In other words, the loss, injury, harm, damage, or risk is weighed against a sum of money. Here, however, the notion of compensation is used in the wider sense of simply weighing one thing against another. What, then, should be placed in the scales?

In this context, the scales are seeking to balance the compensation to a particular community, on one side, against the negative impacts considered to follow from agreeing to host an RWMF on the other. A first question then is what these negative impacts are likely to be. The main candidates can, arguably, be divided into three categories: actual harms, risk of harms, and fear of harm. Fear of harm is clearly, on one level, different from risk of harm, which is different from actual harm; and it is tempting to arrange these hierarchically as more or less significant. Moreover, fears can arguably be held in ways that are more or less rational or justified. However, a neat hierarchy here should be resisted. Fear in itself is a real phenomena, and some forms of fear can be as debilitating as some actual harms, or indeed even more so. Moreover, while difficult questions would arise in relation to compensation for fears held (by some parties in the debate at least) to be completely groundless, even fears accepted by those experiencing them as not well grounded can have a detrimental impact on people's lives and well-being. Focus group participants, for example, have expressed awareness that their fears might not be justified, but indicated that these fears still adversely affected their lives (see e.g. Hunt & Simmons, 2001). All of the above categories of harm, held for a wide range of reasons, should therefore be considered as reasonable candidates for compensation.

Relevant harms, whether actual, risked, or feared, are likely to include harms to:

- human health, physical and/or mental, now and in the future;
- human well-being, including peace of mind and happiness, now and in the future;
- the environment, including ecological systems and processes, and individuals of other species, now and in the future;
- the community, including harm to local economic and social development resulting from loss of tourism income, from "blight", or from the reluctance of companies to invest in the area;
- the aesthetic value of the landscape;[1]
- the character/quality/nature of the place;[2]
- reputation and self-esteem.[3]

Clearly the nature of the actual or likely negative impacts in a given case will need to be established empirically in order to identify appropriate compensation. But a preliminary set of ethical issues can already be identified. Describing something as a negative impact involves making a value judgment: in describing the thing as undesirable in some respect it goes beyond giving factual information. Value judgments will thus already have been made, implicitly or explicitly, before the issue of compensation is even raised, simply in virtue of talking about negative (or positive) impacts. Decisions as to whose views about what counts as negative are to be acknowledged, and whose disregarded, involve further value judgments. A key question here is that of what should be accepted as a

1. This consideration featured strongly, for instance, in Cumbria County Council's case at the Nirex Rock Characterisation Facility (RCF) Inquiry (1995–96). Aesthetic impact (as well as safety considerations) also featured in the Secretary of State's reasons for rejecting the RCF: "the adverse impact on visual amenity...and the natural beauty of the National Park caused by the surface elements of the proposed RCF are serious and...warrant refusal of the application." (DoE, 1997)

2. In focus groups conducted by Lancaster University, a recurring theme was people's conviction that the waste-management facility would change the place they live into a place they want to leave. This could lead to a desire to move away and start a new life somewhere else—in uncomfortable tension with the recognition that family ties, economic constraints, and so forth may make this unfeasible. (Hunt & Simmons, 2001)

3. Fears about social stigma and self esteem are very real in this context, as illustrated by the opinion voiced at a Cumbria County Council meeting in December 2001 that the West Coast of Cumbria is perceived by outsiders as a dumping ground for radioactive waste and dead cattle.

negative impact, whose views on this should be consulted and how different opinions should be prioritized or resolved. Answering these questions involves making judgments that are ethical and evaluative rather than empirical.

A further point is that there will be levels of negative impact considered inappropriate for any community to bear, no matter what compensation might be offered. Examples would be if the risk to human health were very high, or the "blight" effect so great that quality of life were radically undermined (see e.g. Renn, Webler, & Kastenholz, 1998). In many jurisdictions, certain kinds of negative impacts (particularly significant harm to human health and particular kinds of environmental damage) have already been judged as unacceptable, and ethical judgments to this effect are embedded in existing regulatory mechanisms. Such impacts are dealt with, not via a mechanism for compensation, but by the existence of a regulatory system and framework designed to ensure that these impacts do not occur. These regulatory mechanisms need to be working, and seen to be working. However, the fact that such mechanisms exist does not mean that no further ethical issues are raised. Regulatory mechanisms embody implicit ethical judgments and these implicit judgments can, and should, be made explicit and critically reviewed.

19.6 POSITIVE IMPACTS

There may be positive as well as negative implications for a community that agrees to host an RWMF. Positive impacts could include, for example, job creation or improved infrastructure. Such benefits may be immediate, or may only become significant over longer timescales (though it should be noted that these are still likely to be short when compared to the timescale of the risks involved). Questions follow about the relationship between positive impacts and compensation. Do they count *as* compensation? The kinds of positive impacts referred to here would not have been intended as compensation, and would have arisen independently of the compensatory process. They could, nevertheless, be considered to constitute part of the compensation package. Alternatively, it could be argued that compensation should be considered and established only in relation to the negative impacts of the facility.

If positive impacts were considered part of compensation, issues about commensurability would be raised. In what sense does an increase in jobs (for example) compensate for an increase in fear of harm to health? This question of commensurability, of course, arises with any form of compensation, and will be considered further below. Particular to this context, however, are questions about fairness that arise if positive impacts are included as part of the compensation package. It might be held that it is unfair to include positive impacts in the *process* of calculating compensation; that, as a point of principle, the amount and nature of compensation should not be affected by positive impacts that were not intended as compensation and that arose independently of the compensation process.

On the other hand, if compensation were established independently of any positive impacts, this could be held to result in an unfair *outcome* in terms of the distribution of benefits. Communities in which the presence of a waste-management facility did not lead to positive impacts but only negative ones would end up with less overall benefits after compensation than communities in which the presence of the facility did lead to positive impacts in addition to compensation. This illustrates a common tension between procedural justice and distributional justice. The classic case is Robin Hood: robbing from the rich to help the poor can be seen as violating (some notions of) procedural justice in order to further (some notions of) distributional justice.

In this case, (some versions of) procedural fairness would suggest that compensation should be established independently of positive impacts, while (some versions of) distributional justice would suggest that positive impacts be considered part of compensation, at least if there is to be more than one host community and if positive impacts are likely to differ. An additional point is that inclusion of positive impacts could effectively reduce the amount of compensation to be provided. This could be considered an unmerited advantage to the provider of the compensation. These dilemmas require further consideration, and generate the familiar second order questions about who should be involved in their resolution and by what process.

19.7 APPROPRIATE COMPENSATION: PROCESS

Suppose, then, that a fair process for identifying negative and positive impacts, and the relationship between positive impacts and compensation, has been identified; and conclusions reached about the relevant negative and positive impacts in a given case. Judgments then need to be made about how compensation that is appropriate, given these impacts, is to be decided. First, what process should be used, and who should be involved in it?

A key issue will be the extent to which members of "the public" are to be involved. Arguably, questions about what counts as appropriate compensation should be open to public deliberation, with dialogue and debate at both national and local levels. Constructing a fair process in this context will thus involve considerations similar to those outlined in relation to host community selection. A fair process will be one that is genuinely open, rather than intended to reach a prearranged outcome; that is guided by criteria that are clear, well explained, and reasonable; and that is informed by debate and dialogue held by well-informed representatives of all significant stakeholders, carried out at local and national levels. In addition, as with any ethical process, it should be possible to demonstrate the validity of the reasoning involved. The process should be informed by relevant ethical expertise (in addition to scientific and technical expertise) that can reveal the presence of value judgments and offer some analysis of these. In addition, it will be argued below that the identification of appropriate compensation should not be made exclusively or primarily by utilizing a cost-benefit analysis comparison of negative impacts with the benefits of compensation.

Some of those owed-compensation will not, for a variety of reasons, be able to take part in the process of deciding what this compensation should be. Obvious examples include young children, future generations as yet unborn, and other species. While there are many other contexts in which proxies more or less successfully represent the interests of other existing people (such as children) who are affected by decisions they cannot be party to, compensation owed to future generations of people likely to be affected by the negative impacts of the facility raises more complex issues, as do impacts on nonhuman species, ecological processes, and the environment in general. Ways of effectively and fairly representing the interests of future generations and nonhumans need to be further developed, and adopted (see e.g. Dobson, 1996). This is clearly a key challenge in securing ethical legitimacy in this context.

19.8 APPROPRIATE COMPENSATION: OUTCOME

Let us assume that a process for identifying compensation is developed that responds to these challenges and that meets the criteria outlined above, as well as any other criteria that might be identified at a later date. One approach would be to argue that appropriate compensation would then simply be whatever is decided upon as a result of this process. In our view, however, some independent criteria constrain what will count as a fair outcome of this process; and these criteria can be identified in advance of the process. Key here will be that the compensation is proportionate in quantity, and of the right kind. These criteria are briefly discussed below.

Compensation can clearly be inadequate in terms of quantity—for example, a small amount of money for a very serious injury. This raises difficult questions about how the seriousness of different harms, risks and fears is to be quantified, and how this scale of seriousness is to be related to the quantity of compensation. However, these questions are eventually answered: it is clear that, for example, losing a leg is a more serious harm than losing a toe, or that the risk of contracting leukemia is more serious than the risk of flu. Compensation should be proportionate to the seriousness or weightiness of the risk or harm or fear in question. Compensation can be disproportionate by being excessive as well as inadequate. For example, if the negative impacts on the host-community turnout, in fact, to be relatively minor, huge amounts of compensation could be both intrinsically disproportionate and amount to unfairness toward the rest of society, which is footing the bill.

It may well be the case that different kinds of compensation will be appropriate for different kinds of negative impacts incurred by the host community. Compensation needs to be appropriate in the sense of effective and fitting; in the sense of being a suitable kind of response to the negative impact in question. For example, counseling and the provision of information could be an appropriate response to certain sorts of fear; while actual harms, for example, to health, would call for other forms of response—including treatment—which in turn would not make sense if the impact is a risk. Those who see themselves as stigmatized by the presence of radioactive waste may argue that appropriate compensation should include improved infrastructure, the development of high-status jobs, and so on.

Detailed answers to the questions of what counts as appropriate compensation will clearly need to be developed in the context of particular cases and is beyond the scope of this chapter. However, the general points that both the quantity and the nature of the compensation need to be considered, and that compensation can be excessive as well as inadequate, will be valid in relation to all particular examples.

19.9 THE INADEQUACY OF EXISTING GUIDELINES

There are of course a range of existing contexts in which various forms of compensation may be offered in recompense for harm, risk of harm, or fear of harm, and a range of legal guidelines for identifying the sort and quantity of compensation held to be appropriate in particular cases. Clearly existing guidelines can be drawn on as a useful starting point for discussion in this case. In our view, however, there are three main reasons why they cannot be used as an "off the shelf" way of resolving questions about appropriate compensation for RWMF host communities.

First, it can be argued that there is something unique about radioactive waste management, the kinds of harms it may entail, the length of time for which these harms persist and the way radioactivity is perceived and understood—the sorts of social meanings it has for people. Second, even supposing that guidelines exist for the compensation of harms that are relevantly similar to those under discussion here, the existence of legal guidelines is no guarantee of their ethical adequacy. Current levels of compensation for industrial accidents in British law, for example, are often held to be unacceptably low and significantly out of proportion with what losing a leg actually means in terms of a person's subsequent need to rethink their life plans, as well as dealing with pain, disability, and the overall impact on their quality of life. The compensation recommended in existing guidelines *could* be adequate of course; but it will not necessarily be so.

Third, in existing legislation and literature on compensation, the identity of the person(s) to be compensated is usually clear. In this case, however, while those to be compensated will include current members of the host community, they will also include people who are to live in this community or area in generations to come, as well as members of other species. Key questions here will be how far into the future compensation is required to reach and, as already indicated, how to identify appropriate compensation for people who are not yet alive, and others who cannot take part in debate. In short, while current compensation legislation is certainly worth examining with a view to what light it may shed on the compensation of host communities, it will not be possible simply to deduce what counts as appropriate compensation for host communities from existing legislation.

So far, then, we have outlined a range of areas in which ethical and evaluative judgments are implicit and unavoidable, and in which ethical issues and questions are raised in relation to the context in question—identifying appropriate compensation for communities who host an RWMF. The remainder of the chapter will focus in more detail on two issues: the delicate relationship between compensation and bribery, and some of the ethical problems associated with using a utilitarian framework to try to balance the compensation versus negative impact scales.

19.10 COMPENSATION, BRIBERY, AND INCENTIVES

If compensation is perceived as bribery, it may act as a disincentive to accept whatever is being compensated for. Moreover, bribery is considered unacceptable for ethical and not just tactical reasons. But the question of what distinguishes bribery from compensation—and also from incentives—is harder to answer than it first appears. The fact that money is offered, for example, is neither necessary nor sufficient to constitute a bribe: not all forms of financial reward constitute bribery. Furthermore, although a bribe must constitute a "reward" of some sort for an act or omission, the reward is not limited to money. A bribe may consist of material goods or opportunities.

One obvious thought is that while compensation is normally offered retrospectively, bribery is done in advance. But this is not an adequate way of distinguishing between the two. If it were known that compensation would follow the installation of an RWMF, knowledge of the compensation could act as a bribe, even if the actual hand-over of the money or goods were after the event. Moreover, incentives are offered in advance, but do not normally constitute bribery.

The purpose of an incentive is to give someone a reason (the reward) to act in a way they would not do otherwise, such as work harder, or take risks. In this respect it is similar to a bribe. The difference is that the cost involved—the extra work or the risk (for example, when divers are paid "danger money")—is borne by the person accepting the incentive, rather than by third parties. Where there is a cost or negative impact that will be borne by someone other than the person accepting the incentive, the reward is open to being construed as a bribe. It would definitely constitute a bribe in cases where agreeing to negative impacts that will affect others amounts to unethical behavior. Similarly, if the action encouraged by the reward is unethical, illegal or unprincipled in other ways, accepting or offering the reward is a case of bribery rather than an incentive.

A further feature of a bribe is that it is offered in order to influence actions or decisions in a way that is beneficial to, or in the interests of, the briber. Again, this may also be true in the case of an incentive, for example, when an employer offers an incentive to employees to work harder. For an offer of reward to constitute a bribe, then, the following must all be true. The offer is intended to persuade someone to perform an action or make a decision they would not otherwise perform or make; this action or decision is illegal or unethical, either in that it entails unacceptable levels of risk or other negative impacts for others, or in other ways; and the change in the bribee's behavior serves the interests of the briber in some way.

By contrast, compensation is *not* offered with the intention of changing behavior, and does *not* have the result of persuading the persons compensated to perform unethical or illegal actions. An incentive, meanwhile, *is* intended to change behavior, but not to procure unethical or illegal actions. While the action or decision sought may be in the interests of those offering the incentive, and may entail risks to the person accepting the reward, it will not entail risks to others.

19.11 IS BRIBERY UNAVOIDABLE IN THIS CONTEXT?

If this analysis is correct, it suggests two conditions that would need to be met if offers of reward to communities considering whether or not to host an RWMF are to be construed as compensation rather than as bribery. These conditions are:

1. The offer of reward must not be made with the intention of persuading the community to, agree to be a host community.
2. The reward must not be offered as compensation for an action or decision that puts others at risk in a way that would constitute unethical behavior, or that is illegal, or that is unethical in other ways.

At first sight, both conditions suggest that keeping on the side of compensation rather than bribery will, in this case, be extremely difficult. Indeed, the first condition seems to imply that bribery is inevitable. The availability of compensation is one of the factors that will be included in discussion and it will almost certainly influence the decision. Moreover, one of the reasons for offering it is likely to be precisely in order to persuade a community to agree to host an RWMF.

One response would be to suggest that the question of compensation should not be discussed until after a decision about where the facility has been sited has been reached. But this goes against earlier conclusions about the importance of involving host communities in discussions from a very early stage. A second response would be to concede that offers of reward in this context *do* in fact inevitably constitute bribery. A third response, which is favored here, would be to argue for a modified condition (1) as follows.

For the offer of a reward to constitute compensation rather than bribery:

3. There must be reasons for selecting a potential site or sites that make sense independently of the compensation offer. These reasons will, presumably, take the form of a set of criteria referring to geological, geographical, and other features

of potential sites. These criteria should not simply be chosen and announced but should be informed by a process of public debate and deliberation, so that members of potential host communities will have been involved in their selection. The rationale behind the criteria should be transparent, well explained and understood to be robust. In short, the judgment that this is an appropriate site for a waste-management facility must be made on the basis that this is a good site according to a range of appropriate and appropriately chosen criteria.

Why might it be legitimate to modify condition (1) to (3) in this way? It was argued that intention to influence judgment is a feature of bribery. To distinguish compensation from bribery, then, whether or not there was an intention to influence judgment by offering a reward, and/or whether the judgment was in fact influenced by this offer, needs to be established. But it could be argued that the more important concern is with the nature and grounds of the judgment that this is an appropriate site for hosting an RWMF. If there are good independent grounds for the choice of potential sites, then there is a sense in which the intentions of those offering compensation for agreeing to accept an RWMF on such a site are not all that relevant. If there are good grounds for the choice of site that are independent of any offer of reward then (assuming this argument is valid) the offer could be understood as compensation rather than bribery—and this can be the case even if there is evidence to suggest that intention to influence the judgment was a feature of the situation. This argument is offered tentatively and, needless to say, it would be a clearer case of compensation rather than bribery if the intention to influence judgment were not present. Nevertheless, the presence of independent grounds that justify a particular site-selection decision does appear to be a significant factor.

The second condition claims that one cannot compensate someone for agreeing to act unethically. If a person is persuaded by a reward to act unethically, the reward is a bribe, and cannot be construed as compensation. The question here, then, will be whether agreeing to host an RWMF will cause or risk causing harm in a way that is unethical. Agreeing to something that will or might harm oneself would not normally be considered unethical: risks borne by parties to the agreement will probably not be ethically problematic, provided that the agreement is genuinely reached without coercion and with a full understanding of the implications. But agreeing to something that will or might harm those not party to the agreement (such as children, future people and nonhumans) is much more problematic.

To a certain extent, whether agreeing to something that risks harm to such groups is unethical will depend on how serious the risks and harms are. The actual, or likely, nature of the negative impacts of the facility will thus be important in distinguishing compensation from bribery. If the negative impacts of the RWMF are, say, more traffic and some planning blight then the offer of reward could arguably constitute compensation rather than bribery, even if the negative impacts will affect those not party to the agreement, because imposing these less serious impacts on others would not amount to unethical behavior.

The problem is that given the nature of radioactive waste and the length of time for which it remains potentially extremely hazardous, the negative impacts will inevitably include some degree of risk of serious harm to future generations of human beings, and to present and future generations of nonhumans, none of whom can give their consent. Given that exposing others to significant risks without their consent is normally considered unethical, the second principle also seems to suggest that *any* attempt to compensate for agreeing to host radioactive waste-management facility will in fact amount to bribery.

One important response is to recall that the radioactive waste that constitutes the ethically problematic risk in question already exists. The community is not agreeing to accept compensation in exchange for the creation of the waste, but in exchange for cooperation in the management of it, by agreeing to let this happen in a particular place. So in a sense the community is not making a decision that jeopardizes future generations—this decision has already been made by others. On the other hand, community members could be said to be agreeing to jeopardize particular groups of future people, including ones with whom they have special and ethically significant relations—for example, their children and grand children. And the situation is importantly different if the community, in agreeing to accept an RWMF, inadvertently contributes to the claim that radioactive waste has finally been dealt with. There is a real risk that appearing to "solve" the problem of radioactive waste by locating a disposal or management site will be used to legitimate the creation of more such waste through new build. This issue is considered further below.

19.12 MITIGATION AND NEW BUILD

It seems that the following at least is clear. There is a serious question about the ethics of agreeing to something that will pose significant risks to those not party to the agreement, and agreeing to the production of radioactive waste must be seen as falling into this category. There is a serious question, therefore, about the ethics of deciding to generate radioactive waste. It follows that the compensation case under discussion will inevitably be carried out in a context that is ethically problematic—there is a sense in which we are bound to be dealing with a set of options all of which are undesirable in some way. This does not necessarily mean that the decision to produce energy in this way should not have been made, all things considered. However, it does strongly suggest that the decision to produce energy in a way that produces radioactive waste can never be judged "the right decision"; and that the conditions under which the production of radioactive waste has been judged to be a reasonable thing to do, and the alleged need to produce energy in this way, should be kept under constant review.

How might this situation be mitigated? If the serious negative implications for future generations of people and for current and future generations of nonhumans are a key part of the reason why offering or accepting a reward for hosting an RWMF could be deemed unethical, then there is an absolute priority on keeping risks to these, and other groups who cannot take part in negotiations, to a minimum. But this presumably goes without saying. Another priority, already mentioned, is to continue with work aimed at devising ways in which future generations and nonhumans *can* be meaningfully represented in negotiations about issues that significantly affect their interests.

Further, it has already been argued that the deliberate production of substances known to be potentially lethal to various forms of life over thousands of years, is inevitably problematic from an ethical point of view. This strongly suggests that if there is any possibility that agreeing to host an RWMF will amount to a partial or full legitimization of continued radioactive waste production, then the compensation/bribery distinction will be seriously eroded. There should therefore be an explicit and genuine attempt to keep the question of where and how an RWMF is to be hosted separate from the question of whether radioactive material should continue to be produced. This was a recommendation of the CoRWM report (CoRWM, 2006, p.115), though unfortunately one that much recent UK government policy-making appears not to have heeded (see e.g. BERR, 2008; DECC, 2011c).

None of this is to prejudge the question of whether we should continue to generate energy using nuclear technology, but to note that there are unavoidable ethical problems with the production of energy in this way (as there are with other forms of energy), and to claim that there are very good reasons for ensuring that agreement to host a radioactive waste-management facility does not amount to an endorsement of the nuclear industry. Whether or not new power stations should be built (or indeed existing ones kept open) is a question that should be properly considered on its own merits, not closed by default. This is especially important in the current context, in which the undoubted need to mitigate climate change is cited as a justification for new nuclear power stations in a way that arguably hastens the decision-making process and short circuits the necessary wider debate.[4]

An important part of this wider debate should, of course, be the treatment of actual and potential host communities. Though the ethical issues involved will be broadly similar to those discussed here, there will be important differences and additional challenges. While further discussion of new build issues is

---

4. Discussion of changes made to regulatory frameworks in order to facilitate swift development of nuclear power is beyond the scope of this paper. See however National Policy Statements EN-1 and EN-6 (DECC, 2011a,b), which have effectively precluded first-principles discussion:

"For the UK to meet its energy and climate change objectives, the Government believes that there is an urgent need for new electricity generation plant, including new nuclear power. Nuclear power generation is a low carbon, proven technology, which is anticipated to play an increasingly important role as we move to diversify and decarbonise our sources of electricity." (EN-1, para. 3.5.1)

"[G]iven the urgent need to decarbonise our electricity supply and enhance the UK's energy security and diversity of supply, the Government believes that new nuclear power stations need to be developed significantly earlier than the end of 2025." (EN-6, para. 2.2.2)

beyond the scope of this chapter, it is worth noting the statement by one member of CoRWM that, currently,

> *Sites for new build nuclear power stations are identified simply on the basis that they were chosen in the past, though in quite different social circumstances. In the case of radioactive waste, the government has accepted CoRWM's proposal that communities must be willing to participate in a process of site selection and have a right to withdraw if they wish. Surely the same approach should be applied to the siting of new-build stations, which present great risks to safety and security and produce a legacy of waste that will last for several generations.*
>
> Blowers, 2008

19.13 BRIBERY, BENEFITS, AND INDEPENDENCE

In cases of bribery, the offer of reward is for an action or decision that benefits the briber. This feature alone is not enough for a reward to constitute bribery. As discussed above, incentives also manifest it. But all cases of bribery will have this feature. It is a necessary but not sufficient condition of bribery that the offer of reward, if successful in procuring the outcome it intends, will benefit the person or body offering it. From a logical point of view, that the person or body offering the reward stands to benefit does not mean that the offer is necessarily a bribe. However, the offer is more likely to be *perceived* as a bribe if the person making it stands to gain.

The psychological links between the issues of trust, perception and benefit strongly suggest that the body offering the compensation should not have, or be perceived to have, anything to gain by it. This will not be the case if the body offering the compensation is closely associated with the nuclear industry. It follows that there is a much greater chance of compensation being perceived as compensation rather than as bribery if this body is, and is seen to be, genuinely independent from the nuclear industry. In the UK context, it is noteworthy that the relatively independent Nirex UK has now merged with the Nuclear Decommissioning Authority, a government body.

The distinction between a logical analysis of the bribery/compensation distinction, (as offered above) and a psychological one, draws attention to the empirical question of the conditions in which an offer of reward is likely to be *perceived* as compensation rather than as a bribe. Edmundo Claro (2001) argues that, for this to be the case, three categories of conditions need to be met. Contextual conditions require that affected communities understand and believe that the proposed facility meets a genuine need. Procedural conditions require sufficient genuine public participation in siting decisions to allay suspicions of injustice and build trust between the host community and the relevant authorities. Lastly, consequential conditions require not only prioritization of health and safety concerns, but also confidence in the competence and integrity of operators and regulators. An important factor here is the historic record of previous management.

In addition to the logical arguments and distinctions offered above, there are thus a further set of psychological issues about the conditions under which an offer of reward is likely to be perceived as compensation rather than as bribery.

An outline summary of all of the above would read as follows:

1. potential host communities should understand the social context and need for the facility;
2. potential host communities should be active participants in discussions about siting;
3. there should be conditions of trust between communities, those offering compensation, managers of the facility and regulators;
4. health and safety must be, and must be seen to be, a priority; and
5. the body offering the compensation should not have anything to gain.

This seems to reinforce the conclusion that the body making an offer of compensation should be, and be known to be, genuinely independent of the nuclear industry. Whether this is achievable in practice by *any* public body in a country whose national government has declared itself to be in favor of nuclear generation is an important question—but one beyond the scope of this chapter.

19.14 FINANCIAL COMPENSATION

In a case in Switzerland (Renn et al. 1996), respondents were asked if they were willing to permit the construction of a nuclear waste repository within the geographical boundaries of their community. They were then asked the same question, but with the additional information that parliament had agreed to financial compensation for all residents of the host community. In the first instance, 50.8% agreed to accept the repository. This dropped to 24.6% when compensation was offered. The authors of the study suggest that one possible explanation is that the compensation was perceived as a bribe, and the facility was rejected for this reason.

A different explanation is put forward by Frey and Jegen (2000) in their discussion of this case. They claim that most attempts to compensate members of a community financially for accepting a so-called NIMBY (not in my back yard) facility—such as a nuclear waste-management facility—are unsuccessful, in the sense that the facility is rejected, even when the compensation is substantial. Their thesis is that where there is some "intrinsic" reason to accept the facility, such as willingness to contribute to the public good, sense of civic duty, or sense of ethical responsibility to the society as a whole, the offer of financial compensation in fact acts as a disincentive. The offer of financial compensation transforms an ethical relationship to a monetary one, in a way that undermines the initial motivation. The external motivation "crowds out" the intrinsic motivation—i.e. the motivation coming from within the person.

Frey and Jegen offer two suggestions about how this might work psychologically. The first is de-motivation through loss of self-control: the control shifts from motivation internal to oneself, to motivation that is external. The second is impaired self-esteem: offering money to do something is perceived as a failure to acknowledge, or a rejection of, the motivation that already exists:

> *[T]he person feels that his or her involvement and competence is not appreciated which debases its value. An intrinsically motivated person is denied the chance of displaying his or her own interest and involvement in or approval of an activity or project, when someone else offers a reward for undertaking it.*
>
> Frey & Jegen, 2000, p. 4

In the present context, there are two main issues here. One is a tactical one: if these authors are correct, offering financial compensation in the attempt to secure acceptance from a host community for an RWMF may be counterproductive. It may undermine a preexisting willingness to do something perceived as in the overall public interest. The second point is an ethical one, to the effect that that there is something intrinsically undesirable about transforming a willingness to contribute to public good into a financial relationship.

This connects with the argument explored below that certain kinds of relationships are degraded if financial compensation is offered, or betrayed if it is accepted. If a transformation from a nonmarket to market values contributed to the erosion of self-esteem and self-determination, this would be a further ethical argument against such a transformation. Clearly a key question here will be the extent to which the ethical or civic motivation actually exists, as the considerations offered above may be irrelevant if no such motivation is there in the first place. A further (large) question is whether and under what circumstances it would be ethical to encourage the development of civic motivation, and how this might be done. Leaving this aside, however, and assuming for the moment that civic motivation is already present, one response suggested by Frey and Jegen is to offer socially beneficial forms of nonmonetary compensation such as schools or fire stations. The thought is that ethical motivation would not be eroded by this kind of in-kind compensation, and that it would thus be more ethically acceptable to offer it.

Frey and Jegen's position is supported by Claro (2001, 2007), though Claro claims that while the bribe effect and/or the crowding out effect may be relevant in some circumstances, it is often the case instead that "an agent rejects financial payments and agrees to public good compensation...because this seems the correct or appropriate way of behaving" (Claro, 2001, p. 2). In other words, while people's choices are indeed influenced by reference to self-interest and to outcomes, procedural as well as consequential motives are nonetheless important determinants.

Claro's key claim is that how people think they should behave in specific circumstances will depend on the kind of social relations that exist between the parties involved, and on the norms of behavior associated with these kinds of relations. Following other writers in this area (e.g. Fiske & Tetlock, 1997) he suggests (Claro, 2007, pp.193–194) that there are three fundamental and distinct forms of exchange relations: communal sharing, equality matching, and market pricing. Different norms of exchange are held to be appropriate in each case.

Under communal sharing relations that correspond to the interpersonal value of care, people are often prepared to contribute a great deal without expecting much if anything in return. Under equality matching relations that correspond to the public value of justice, a contribution or favor would be expected to be matched in kind. Under market pricing relations that correspond to the private value of freedom, a contribution would typically be expected to be paid for.

Claro (2001) illustrates the point with three scenarios in which one person helps another to move house. What varies between the scenarios is the relationship between helper and helped. In the first, a son helps his father. This is taken to be a case of community-sharing. In the second, the helper and helped are close friends, a case of equality matching; and in the third, the helper is a commercial removal person. When subjects were asked to read these scenarios and choose between no compensation, nonmonetary compensation such as a gift or a meal and financial compensation, the response was as predicted: no compensation for the son, a pizza for the friend and monetary compensation for the professional remover. In other words, the nature of the compensation seen as appropriate in different cases varied according to the nature of the relationships involved. Claro accordingly suggests that, in the case of siting a waste-management facility,

> *[I]f someone who frames the siting problem as a communal sharing one is offered money as compensation, she will react with anger and indignation not only because the tradeoff is bizarre, but because it threatens fundamental aspects of her understanding of social relationships and society.*
>
> Claro, 2001, p. 7

In a more recent paper (2007), Claro offers detailed empirical evidence that in-kind compensation for communities hosting an RWMF is indeed often seen as more appropriate and thus more acceptable than monetary compensation. The critical factor is how members of a potential host community understand the relationship between themselves, the waste-management facility and those offering compensation.

19.15 ETHICAL DECISION MAKING AND COMMENSURABILITY: AGAINST ALGORITHMS

Compensation thus needs to be appropriate, or conversely, it needs to not be *in*appropriate, in an ethical sense. The danger of inappropriateness is clearly seen in the context of offering financial compensation for injury or loss of life. "No amount of money could compensate for the death of X, and it is offensive to offer it", would be a characteristic expression of this kind of inappropriateness. One way of understanding this sense of ethical inappropriateness is with reference to the idea of incommensurability.

Some views about compensation and how it should be established rest on the assumption that it must be possible to perform a particular kind of comparison between the compensation and the negative impacts being compensated for.

Where the negative impacts and the compensation are both financial, this is relatively straightforward. But some negative impacts are fundamentally different in kind from the sorts of compensation that might be offered. Hence, according to those who hold this view, a common scale is required, so that impacts and compensatory benefits that are of different kinds—risk of harm to health and better infrastructure, say—can be made commensurable in order to be weighed against each other.

The view that commensurability is required for certain sorts of decision making is inherent in the utilitarian ethical framework that underpins (often without explicit acknowledgment) many contemporary approaches to decision making. On a utilitarian view, the ethically correct action in any given context is the one that produces the best overall state of affairs, or best maximizes good outcomes. In order to establish which action will produce the best state of affairs, and hence to decide what ought to be done, it has to be possible to compare different options and to see which option leads to the best overall balance of negative and positive impacts. In order to do this, a common measure of value in relation to different kinds of negative and positive impacts is required.

One way of understanding what is happening with compensation is precisely this: that negative impacts are being compared with positive benefits (compensation). A tradeoff is being made between the two, with the aim of producing a state of affairs which overall has a maximized balance of benefits over costs. In this cost-benefit analysis process, monetary value is taken as the common measure into which all costs and benefits must be translated. On this approach, attempting to compensate for risk to health, say, will involve putting a financial figure on health, so that the figure offered is appropriate. As suggested above, such attempts to put a financial figure on negative impacts such as loss of life or harm to health often cause offence. Correspondingly, where compensation is to take a nonmonetary form, then a figure will need to be established for that too, so that some kind of equivalence or proportionality can be achieved.

John O'Neill argues, however, that certain kinds of social relationships are such that they are incompatible with market relations:

[T]here are certain social relations and evaluative commitments that are constituted by a refusal to put a price on them—there exist what are sometimes called constitutive incommensurabilities. [...] Given what love and friendship are, and given what market exchanges are, one cannot buy love or friendship. To believe one could, would be to misunderstand those very relationships. To accept a price is an act of betrayal, to offer a price is an act of bribery.

O'Neill, 2007, p. 23, 25

A commitment to an environment, or to a place as "home", can clearly be of this form, as O'Neill illustrates by quoting from a letter from an inhabitant of the Narmada Valley in Western India, threatened with displacement as a result of the Sardar Sarovar Dam.

You tell us to take compensation. What is the state compensating us for? For our land, for our fields, for the trees along our fields. But we don't live only by this. Are you going to compensate us for our forest?...Or are you going to compensate us for our great river—for her fish, her water, for vegetables that grow along her banks, for the joy of living beside her? What is the price of this?...We didn't buy this land; our forefathers cleared it and settled here. What price this land? Our gods, the support of those who are our kin—what price do you have for these? Our adivasi (tribal) life—what price do you put on it?

quoted in O'Neill, 2007, p. 23

O'Neill argues that in such cases, where social relations and evaluative commitments are constituted by a refusal to put a price on them, "the use of the term 'compensation' disguises what is going on" (O'Neill, 2007, p. 25). What is going on is an erosion of these relationships. This in turn can be seen as an erosion of human well-being, since relationships of this kind are, arguably, part of what it is to lead a full human life.

It is worth asking in the present context whether there are circumstances in which agreeing to be a host community, and to accept compensation for this, could reasonably be construed as a betrayal of the kinds of relationships or value commitments that are "constituted by a refusal to put a price on them". It seems likely that at least some of the various sets of relationships between current and future members of the community, members of the community and those outside the community, and community members and their environment will be of these kinds. Examples would include the relationships between those agreeing to host the management facility, their children, grand children and great grand-children; and the relationships between these people and valued features of their local environment. These relationships might reasonably be seen to be betrayed if, for example, the negative impacts include significant risk to the health or well-being of current or future members of the host community; or significant damage to the environment, including perhaps damage of a kind that would render it no longer "home".

19.16 RESPONSES

One response here is to return to the point that significant risks to life and health should be ruled out in advance by the regulatory system. On the other hand, given the immensely long time-scales involved, it is not feasible to claim that all risk of serious harm can be regulated out of the picture for tens of thousands of years. Some risks are bound to remain in virtue of the inevitable uncertainties that accompany this time-scale. Furthermore, even if the risk of serious harms to health and environment are taken to be negligible, it still makes sense to ask what other kinds of negative impacts will be involved in a particular case, and whether an offer of financial compensation in this case will erode the relationships and values involved.

If such erosion seems likely in a given case, what other responses might be appropriate? One would be to suggest that the compensation is not financial, but of a form that continues to embody the relationship commitments in question. So, for instance, if it is future generations of human community members that are in some sense threatened, then while accepting financial compensation might be conceived as a betrayal of this relationship, accepting a commitment to build and maintain schools and libraries might not. On the other hand, it could be argued that this is just a more subtle form of betrayal, and moreover, that it deprives people of autonomy in deciding how financial compensation should best be spent.

A second response would be to argue that in such a case the community would be quite correct to refuse to be a host community, and it would be inappropriate to offer compensation of any form. If agreeing to the negative impact amounts to a betrayal of significant relationships to current or future generations of humans or nonhumans, then the ethically correct response could well be to argue that these negative impacts are unacceptable. This strongly suggests a responsibility to be alert to these kinds of circumstances on the part of those offering the compensation, so that such situations can be avoided.

It can be argued that one cannot compensate for an ethical wrong-doing; or that compensation does not make the ethical wrong-doing right. It does not necessarily follow from this that it could never be acceptable to perform an ethically undesirable action. It might be, for example, that this is an unavoidable feature of the least worst option. What does follow is that there may be circumstances in which there is not a clear right answer, in which the interests and values involved conflict in a way that does not have a single clear resolution. For example, it could be that, given the existence of radioactive waste, there is no solution to the question of what to do with it that does not involve exposing people and environments to long-term risks that would normally be judged unethical. In this case, it is important to keep the circumstances in which all options are ethically compromised under constant critical review, and to seek to change these circumstances. This argument will be developed further below.

19.17 PROBLEMS WITH COST-BENEFIT ANALYSIS

The possibility of genuine ethical dilemmas is, however, effectively denied by cost-benefit analysis approaches. Unless there are two options where the various costs and benefits come out exactly equal, there will always be a right answer: the option with the best balance of benefits over costs. Further, because benefits and costs are reduced to a common unit, the benefits are in effect taken to erase the costs. As will already be apparent, the view taken here is that this approach is profoundly mistaken, and that it is important to recognize this for a variety of practical reasons.

It is sometimes argued that rational decision making as such must involve this kind of trading-off of costs against benefits by reducing them to a common unit of comparison, such as money, and that where a unit of calculation is absent rational choice is not possible. On this view, rational choice *consists* in trading off costs and benefits through the use of a common unit of comparison. In our view, however, not only is the possibility of trading-off in this way *not* a necessary condition of rational decision making but, in the kind of context in question, (where the decisions to be made have unavoidable ethical and evaluative dimensions), it is inappropriate to try to reach a resolution by this kind of algorithmic procedure.

The appeal to cost-benefit analysis rests on the assumption that ethical values are radically subjective. Values are effectively seen as equivalent to personal preferences or tastes, such as for thick or thin-cut marmalade. If values are understood as preferences in this way, then where they conflict the only rational procedure for decision making appears to be to identify the preferences and how strongly they are held, and choose the option that satisfies many or most of them. The assumption that ethical values are radically subjective and can be treated like preferences is, however, deeply flawed. Values, unlike preferences, are (or at least can be) held for reasons, and these reasons can be critically assessed. Some reasons are better than others, and hence a rational response to a conflict of values involves ascertaining and assessing the reasons why the values are held. Resolution of the kinds of issues under discussion thus requires dialogue, debate, and rational deliberation procedures.

This is a very different account of what rational decision-making involves from the mathematical model outlined above. Procedural accounts of rationality stress the ineliminable role of judgment, and allow room for the development and scrutiny of reasons for holding values. This contrasts strongly with the assumption that where a calculation is not available, rational choice is not possible. Indeed it can be argued not only that cost-benefit analysis-type procedures are not a necessary feature of rational decision-making, but also that, in this kind of context, the cost-benefit type approach should be resisted.

A first group of reasons for this conclusion, centering on incommensurability, have already been discussed above. Secondly, cost-benefit procedures are "reason blind". They ascertain the strength of people's preferences or values, and perform a calculation based on these, but not the reasons why these preferences are held. Deliberative and procedural conceptions of rationality, by contrast, emphasize the importance of reasons and of dialogue so that a reasoned position can be both explored and developed. A third sort of problem with cost-benefit type approaches (as with utilitarianism in general) is that achieving the greatest balance of benefits over costs is perfectly compatible with the imposition of grave injustices on some people, or with the violation of their rights.

Finally, it can be argued that cost-benefit approaches lead to ethical complacency; that the attitude arising from a conviction that "there is always a right answer" can itself be ethically problematic. Compare this attitude with that which follows from the view that, although the option that is probably the least worst has been chosen, there is still nevertheless an ineliminable element of ethical undesirability about it. This keeps alive the understanding that it would be better if the circumstances were such that this dilemma did not arise, and hence the possibility of exploring ways of preventing it from continuing to arise in the future. It tends toward an ongoing attempt to change the background circumstances and context within which the decision has to be made, and an ongoing alertness to the regrettable aspects of the decisions made in the context as it stands. It keeps open the question, what can be done so that in the future we have a different range of options to choose from?

Cost-benefit analysis in this regard is profoundly conservative, in the sense that it has no tendency to question the desirability of the circumstances in which the analysis is made, or to suggest that another set of options would be preferable. Furthermore, once the decision has been made, and characterized as the right decision, there is no reason or incentive to keep the situation under review. The right answer is the right answer, costs are outweighed by benefits, and that is that. This is clearly of relevance in the context under discussion. Given that radioactive waste exists, it has to be managed, and the facility for this has to be located somewhere. Where it should go and how those who receive it in their community are to be treated are genuinely difficult questions. Given that radioactive waste remains potentially dangerous to human and nonhuman life for many thousands of years, there can, arguably, be no clear right answer.

An awareness that all options have ethical drawbacks suggests the need for a continuous review of the circumstances that have led to our being compelled to choose from a set of options all of which are ethically problematic in some way. For example, it suggests that the question of whether radioactive waste should continue to be generated should be continuously under review, and that claims to have "solved" the issue of radioactive waste management by cost-benefit analysis must not be allowed to contribute to the shelving of this kind of review process.

The discussions above point, finally, to a set of deeper questions about what the underlying purpose or aim of compensation is taken to be. This can be understood in different ways. For example, it could be held that the aim of compensation is to trade-off costs and benefits in order to achieve an overall outcome in which negative impacts are outweighed by compensation (a consequentialist position); or to fulfill a set of obligations (a deontological position); or to meet the goal of acting virtuously. Such positions are likely to underpin compensatory projects, though they may be held implicitly rather than explicitly. It is well worth asking, therefore, what the underlying aim of compensation is taken to be, as different conceptions of this are likely to lead to different recommendations.

19.18 CONCLUSIONS

Questions about how communities who agree to host radioactive waste-management facilities should be compensated cannot be answered without making a wide range of value judgments and without raising a range of related ethical issues. One group of issues centers on the nature of compensation itself: what counts as compensation in relation to the negative and positive impacts involved, and how compensation is to be distinguished from bribery. Another set of issues centers on who should be making the range of unavoidable value judgments involved, such as what counts as negative and positive impacts, what counts as appropriate, what counts as significant fear and risk, and so on. A related third set involves the openness and inclusiveness of the process by which these judgments, and a range of other decisions, are made; more generally, it involves the criteria required if these processes are to be ethically legitimate.

Attention to these issues will be an essential part of ensuring that the wider relationships between host communities and the rest of society are also ethically legitimate. But three further essential points need to be made. First, it is important to note that, at a meta level, assumptions will inevitably be made about how decisions that have evaluative and ethical dimensions should be reached, and about what counts as reasonable or rational decision-making in this context. These assumptions are not neutral in terms either of the values embedded in them or the results they are likely to generate. They should therefore be revealed and opened out for critical scrutiny. Can we, for example, make such decisions using cost-benefit analysis techniques, even relatively sophisticated ones? What sorts of values and what sort of views about the nature of ethical judgments are implicit in this, and in contrasting approaches to decision making? A further area of assumptions concerns the underlying aim of compensation itself. Again, these assumptions will not be neutral, and should be examined.

Secondly, and much more practically, all the attention to openness, transparency, volunteerism, and hidden assumptions in the world will not lead to an ethical relationship between society and a host community if the overriding reason that a community agrees to accept radioactive waste turns out to be preexisting poverty. This has been strikingly documented in a key case concerning RWMF siting and Native American communities (Gowda, Rajeev, & Easterling, 1998), which underlined the importance of understanding, exposing and clarifying the relationships between poverty, social justice, and ethical decision making. The issue of environmental justice in the context of RWMF siting is an important and challenging one, raising hard questions that nevertheless require answers. It has been discussed more fully elsewhere (see e.g. Bullard, 2000; Schrader-Frechette, 2002).

Finally, the relationship between the ethical issues raised by radioactive waste management and those raised by other forms of energy production and consumption is significant and worth exploring further. Not consuming energy is not an option, though consuming less certainly is; and while all forms of energy production have some implications for the well-being of present and future generations of humans and other species, the generation of energy by nuclear means does raise a particular set of ethical issues that are both unavoidable and extremely difficult.

19.19 AFTERWORD

This chapter's journey to press has been a long one, though the issues discussed have not become any less pertinent along the way. It was largely written before the events at Fukushima, which in several countries have catalyzed a decisive turn away from nuclear power. In the UK however, nuclear generation still plays a key role in both energy and climate policy (see e.g. DECC, 2011a,b). At the time of writing (January 2013) a ministerial decision is awaited on a planning application for a new nuclear power station at Hinkley Point in Somerset, intended to be the first of a "new generation". The proposed eventual destination for waste from that station is the UK's as yet nonexistent Geological Disposal Facility (GDF).

Simultaneously, though not coincidentally, attempts to identify a site for a UK GDF have been reinvigorated. These are once again focused on an area of West Cumbria close to Sellafield, where the majority of the current UK inventory of radioactive waste is presently housed in above ground interim storage. Claims about the geological suitability of this area are hotly contested, as are claims about the levels of local support for hosting a GDF, or even for proceeding with further invasive geological testing in what is a highly protected and valued landscape. However, the issue of compensation and "benefit packages" for host communities is, perhaps inevitably, already prominent in both formal and informal public discourse on the proposals (see e.g. Copeland Borough Council, 2012). We offer these reflections as a contribution to this process, as well as to broader discussions.

ACKNOWLEDGMENTS

Parts of this chapter are based on a report originally written for UK Nirex Limited (Rawles, 2002). Nirex has now been integrated into the Nuclear Decommissioning Agency (NDA), and the authors thank the NDA for the use of this material in its current form. The views expressed are those of the authors and do not necessarily represent those of the NDA.

REFERENCES

BERR (Department for Business, Enterprise & Regulatory Reform). (2008). *Meeting the energy challenge: A white paper on nuclear power.* www.official-documents.gov.uk/document/cm72/7296/7296.asp.

Blowers, A. (2008). 'Back to the future' *The Guardian*, March 12, 2008.

Bullard, R. D. (2000). *Dumping in dixie. Race, class and environmental quality* (3rd ed.). Colorado: Westview Press.

Claro, E. (July, 2001). Counterbalancing environmental losses: Different compensation mechanisms and their performance in siting a sanitary landfill. *Paper presented at ESEE conference "Frontiers1: Fundamental issues of ecological economics"*, Cambridge, 4–7.

Claro, E. (2007). Exchange relationships and the environment: the acceptability of compensation in the siting of waste disposal facilities. *Environmental Values*, *16*(2), 187–208.

Copeland Borough Council. (*Final report of the West Cumbria managing radioactive waste safely partnership*. http://www.westcumbriamrws.org.uk/images/final-report.pdf2012.

CoRWM (Committee on Radioactive Waste Management). (2006). *Managing our radioactive waste safely: CoRWM's recommendations to government.* CoRWM. document no. 700 – available at: www.corwm.org.uk.

DECC (Department of Energy and Climate Change). (2011a). *Overarching national policy statement for energy (EN-1).* https://www.gov.uk/government/uploads/system/uploads/attachment_data/file/47854/1938-overarching-nps-for-energy-en1.pdf.

DECC (Department of Energy and Climate Change). (2011b). *National policy statement for nuclear power generation (EN-6).* https://www.gov.uk/government/uploads/system/uploads/attachment_data/file/47859/2009-nps-for-nuclear-volumeI.pdf.

DECC (Department of Energy and Climate Change). (2011c). *Waste transfer pricing methodology for the disposal of higher activity waste from new nuclear power stations.* https://www.gov.uk/government/uploads/system/uploads/attachment_data/file/42629/3798-waste-transfer-pricing-methodology.pdf.

DoE (Department of the Environment). (1997). *Secretary of state's decision letter on planning appeal relating to UK Nirex's application for a RCF facility at Sellafield, Cumbria.* http://www.westcumbriamrws.org.uk/documents/GONW_letter_to_Nirex_17_March_1997.pdf.

Dobson, A. (1996). Representative democracy and the environment. In Lafferty & Meadowcroft (Eds.), *Democracy and the environment, problems and prospects*. Cheltenham: Edward Elgar Publishing.

Fiske, A., & Tetlock, P. (1997). Taboo trade-offs: reactions to transactions that transgress the spheres of justice. *Political Psychology, 18*(2), 255–297.

Frey, B., & Jegen, R. (2000). *Motivation crowding theory: A survey of empirical evidence (CESifo Working Paper Series)*(Vol. 245). Zurich: CESifo.

Gowda, M., Rajeev, V., & Easterling, D. (1998). Nuclear waste and native America: the MRS siting exercise. *Risk: Health, Safety and Environment, 9*, 229–258.

Hunt, J., & Simmons, P. (2001). *The front of the front end: Mapping public concerns about radioactive waste management issues*. Report to UK Nirex (now part of the Nuclear Decommissioning Authority, Moor Row, Cumbria). http://www.csec.lancs.ac.uk/docs/the%20front%20of%20the%20front%20end.doc.

NEA (Nuclear Energy Agency). (2004). *Stepwise approach to decision making for long-term radioactive waste management experience: Issues and guiding principles*. Paris: OECD.

Nilsson, A. (2001). *Responsibility, equity and credibility: Ethical dilemmas relating to nuclear waste*. Stockhom: Kommentus Förlag.

O'Neill, J. (2007). *Markets, deliberation and environment*. London: Routledge.

Rawles, K. (2000). *Ethical issues in the disposal of radioactive waste*. Report to UK Nirex (now part of the Nuclear Decommissioning Authority, Moor Row, Cumbria) http://www.nda.gov.uk/documents/upload/Ethical-issues-in-the-disposal-of-radioactive-waste-December-2000.pdf.

Rawles, K. (2002). *Compensation in radioactive waste management: Ethical issues in the treatment of host communities*. Report to UK Nirex (now part of the Nuclear Decommissioning Authority, Moor Row, Cumbria) http://www.nda.gov.uk/documents/upload/Compensation-in-Radioactive-Waste-Management-Ethical-issues-in-the-treatment-of-host-communities-May-2002.pdf.

Renn, O., Webler, T., & Kastenholz, H. (1998). Procedural and substantive fairness in landfill siting: a Swiss case study. In R. E. Löfstedt & L. Frewer (Eds.), *The earthscan reader in risk and modern society*. London: Earthscan.

Shrader-Frechette, K. (2002). *Environmental justice: Creating equality, reclaiming democracy*. Oxford University Press.

Sjöberg, L. (2004). Local acceptance of a high-level nuclear waste repository. *Risk Analysis, 24*(3), 737–749.

Index

A

AAPM, 117
Aarhus Convention, 334
Abaiskii, 248–249
Abralinskii, 248–249
Academy of Agrarian Sciences, 330
Acceptability, public, 350–351
Accountability, 110, 115
Achievement, 220
ACR. *See* American College of Radiology
Act, 112
Active Practices, 224
Ad hoc units, 336
Adaptive control, 222
Adaptive responses, 139
Affected group involvement, 171
Affected people monitoring, 165–166
African customary law, 63–64
Agents, 111
Ahimsa, 60–61
ALAP. *See* As low as practicable
ALARA. *See* As low as reasonably achievable
ALARP. *See* As low as reasonably practical
Alma-Ata, 245
Almaty, 245
Alvarez, Robert, 199
Ambition, systemic, 230–231
American College of Radiology (ACR), 116
Annan, Kofi, 334
Anthropocentricism, 76–77, 79
  common principles of, 78
Anticonvulsive drugs, 95
Aplastic anemia, 133–135
Applied ethics, 34
  as multiculture, 186–188
  for nuclear science, 188–194
  safety culture and, 186
Apsalikov, Kazhek, 10
Arrogance theory, 290–291
Arrow, Kenneth, 48
As low as practicable (ALAP), 144, 152–153
As low as reasonably achievable (ALARA), 8, 39, 54–55, 143–145
  ALARP and, 144–145
  cost-benefit analysis, 151–152
  de minimis issue and, 147
  defining, 148–149
  economic compromises and, 149–151
  House of Lords on, 149
  HSE on, 151
  interpretive issues, 154
  legal status of, 146
  optimization and, 133–135, 145–146
  Precautionary Principle and, 153
  related principles, 152–154
  substitution principle and, 153
  three-levelled model, 146–148
As low as reasonably practical (ALARP)
  ALARA and, 144–145
  optimization and, 133–135
Australian Aboriginals, 65–66
Authority, 4
Autism, 99
Automated registries, 253
Autonomy, 59–60
  Beauchamp on, 59
  Childress on, 59
  cross-cultural ethics and, 59–60
Awareness, 119–120
Azamas, 16, 242

B

Backe, S., 160f
BACT. *See* Best available control technology
Bahá'is, 57
Bahá'u'lláh, 57, 59, 62–64
Balanced value profile, 224–225
Basic Safety Principles for Nuclear Power Plants, 184–185
Basis, value, 230–231
BAT. *See* Best available technology
BATNEEC. *See* Best available technology not entailing excessive costs
Bauer, Susanne, 10
BB-RIR. *See* Best Branch of the Research Institute of Radiology
Beauchamp, T. L., 56–59, 109, 252
  on autonomy, 59
  on justice, 62
Beder, S., 341

Bedford, T., 226–227
Belaoucha, 320, 331
Belarus, 161, 166, 171, 312–313, 326
  conference recommendations, 330–331
  contamination in, 315
  ETHOS project and, 318f
Belarussian Research Institute for Soil Science and Agrochemistry (BRISSA), 320
Belief, public, 291–298
Belikhina, Tatyana, 10
Beneficence, 109
  cross-cultural ethics and, 60–61
Benefits, 363–364
Benevolence, 220
Beninson, Dan J., 147
  on de minimis issue, 147–148
Bentham, Jeremy, 73–74
  methods of, 36f–37f
  on utilitarian theories, 35–36
Beria, Lavrentii, 244
Best available control technology (BACT), 153
Best available technology (BAT), 152–153
Best available technology not entailing excessive costs (BATNEEC), 153
Best Branch of the Research Institute of Radiology (BB-RIR), 320–321
Best practicable environmental option (BPEO), 153
Best practicable means (BPM), 153
Bhagavadgita, 57, 59, 62
Biocentricism, 73–74
  common principles of, 78
  deontological theories and, 73–74
  utilitarianism and, 73–74
Bioethics, 186–187
Biomedical ethics, 56–57
Bjerke, H., 160f
Black, Douglas, 286–287
Blight effect, 354
Bone sarcomas, 135
Bounded rationality, 224–226
BPEO. *See* Best practicable environmental option
BPM. *See* Best practicable means
Bribery, 358–359, 363–364
  avoidability of, 359–361
BRISSA. *See* Belarussian Research Institute for Soil Science and Agrochemistry
British Medical Journal, 139
British Nuclear Fuels, 291
Brucellosis, 245–246
Buddha, 59, 62
Buddhism, 60–61, 63–66

C

Cadmium, 95
Callicott, J. B., 75–77
Cancer, 94–95, 201–204. *See also specific types*
  in children, 247
  dose threshold, 99
  esophagus, 249–250
  lung, 44–45, 249–250
  primary bone, 135f
  risk, 242–243
  in Semipalatinsk, 247
  thyroid, 95, 251–252
Cataracts, 135–136
CDC. *See* Centers for Disease Control
Centers for Disease Control (CDC), 90
Central Physical Technical Institute, 248
Cerebral palsy, 95
Cesium, 180, 204
  from Chernobyl, 292–294
  deposition of, 289–290
  in Norway, 160f
  radiocesium, 165, 169–170
  reindeer husbandry and, 164
  in Sellafield, 292–294
  in UK, 292f–293f
Chelyabinsk, 241
Chemical hazards, 90
Chemical risk assessment, 87–88
  adverse health effects and, 93–98
  substance thresholds, 98–99
Chernobyl, 3, 7–8, 11, 72, 78–79, 193, 199, 253–254, 283–285, 327
  cesium from, 292–294
  daily living conditions and, 313–316
  early lessons from, 162–167
    affected people monitoring, 165–166
    information needs, 162
    local monitoring stations, 162
    permissible levels settings, 163–164
    psychosocial effects, 166–167
  liquidators, 136
  long-term challenges, 167–172, 311–313
    affected group involvement, 171
    compensation, 167–168
    health concerns, 172–174
    information needs, 167
    social countermeasures, 167–168
    technical countermeasures, 167–168
    zoning, 167–168
  Norway and
    contamination, 161–162

Cs-137 deposition map, 160f
safety culture and, 223
Chernobyl Committee, 313
Chernobyl Forum, 171
Children, 47–48
cancer in, 247
contamination of, 92–93
as FDNPS victims, 203–204
nuclear regulations and, 204–205
Childress, J. F., 56–59, 109
on autonomy, 59
on justice, 62
CHOOSE WISELY, 116
Christianity, 57, 63–66
CIS. *See* Community of Independent States
Clarke, R. H., 19, 27, 338
Claro, Edmundo, 363–366
Classified worker status, 133
Clearance, 21
Clinical audits, 121
Clinical equipoise, 38
Clinical medicine, 38
Clinical radiology, 107–109
awareness and, 119–120
clinical audits and, 121
communication and, 116–118
deterministic effects, 135–136
dose control, 133–135
dose limits, 108–109
ethics in, 109–114
act in, 112
agent in, 111
basic analysis, 111–114
consent in, 112–114
core principles, 109–110, 110t
recipient in, 112–114
hand damage, 134f
health economics and, 116
justification, 108–109, 118–121
failures of, 121
legal context for, 114–116
medical context for, 114–116
nonmedical exposures in, 121–122
optimization, 108–109
overutilization of, 116
pregnancy and, 123–124
referral guidelines and, 120–121
regulatory framework, 122–123
risk and, 116–118
social context for, 114–116
uncertainty and, 116–118
Close limitation, 54
Closed fuel cycle, uranium, 267–273. *See also* Nuclear fuel cycle
environmental friendliness and, 269–270
resource durability and, 269–270
safety and, 268
Co-expertise, 313
COGEMA, 320
Cold War, 242
Collective approaches, 35–39
Collective dose minimization, 42–43
Colley, S. K., 224–225
Commensurability, 354
compensation and, 366–368
Committee on Radioactive Waste Management (CoRWM), 348, 362
Common morality, 56, 66
Communication, 139
clinical radiology and, 116–118
Community of Independent States (CIS), 313
Compensation, 347–350, 354
bribery and, 358–359, 363–364
avoidability of, 359–361
commensurability and, 366–368
cost-benefit analysis and, 369–371
decision-making and, 366–368
financial, 364–366
inadequate guidelines, 357–358
mitigation, 359–361
in Norway, 167–168
outcomes, 356–357
process, 355–356
responses to, 368–369
Competence, 218, 218f
Competing values framework (CVF), 221–228
Computerized tomography (CT), 7, 55, 108
Condorcet, 335
Conformity, 220–221
Confucius, 58–59, 63, 65–66
Consent. *See also* Informed consent
in clinical radiology, 112–114
in NRRW, 138
worker, 138–139
Consistency, 136
Conspiracy theory, 290–291
Contamination, 90–91
in Belarus, 315
from Chernobyl in Norway, 161–162
of children, 92–93
foodstuffs, 322t
long-term, 170
properties of, 172–173
Contingency plans, 136–137

CORE, 325–326
Corten, O., 148
CoRWM. *See* Committee on Radioactive Waste Management
Cost-benefit analysis
ALARA and, 151–152
compensation and, 369–371
ethics and, 370–371
HSE on, 151
problems with, 369–371
Cotinine, 92–93
Coughlin, S. S., 252
Cranor, Carl, 6
Credibility, 298–302
Cross-cultural ethics, 53–55, 57–58
autonomy and, 59–60
beneficence and, 60–61
four principles of, 59–62
intergenerational equity and, 63–64
justice and, 61–62
need for, 58–59
non-maleficence and, 60–61
precaution and, 64–66
radiation protection and, 62–66
underprivileged and, 63
Crowding out, 365
CT. *See* Computerized tomography
Cumbria, 286, 288
restricted area map, 287f
Cumbrians Opposed to a Radioactive Environment, 297
Customer focus, 131
CVF. *See* Competing values framework

D

DAD. *See* Decide-Announce-Defend
Daigo Fukuryu Maru, 193
Daily living conditions, 313–316
Dalai Lama, 63–64
Data protection, 139
Davoudian, K., 226–227
DDT, 91, 95
De minimis issue, 146
ALARA and, 147
Beninson on, 147–148
Lindell on, 147–148
in radiation protection, 43–46
Rossi on, 147
Decide-Announce-Defend (DAD), 337–338
Decision-making
in ad hoc units, 336
compensation and, 366–368
impacts on, 341
participation and, 341
pluralistic, 187
process, 335
Declaration of Helsinki, 186–187
"Declaration Toward a Global Ethic", 57, 61–62
Declarative Practices, 224
Democracy
justifications of, 338–339
participation in, 335–338
stakeholder engagement and, 339
transparency and, 340–341
Democratic theory, 335
Deontological theories, 33, 41, 54, 65–66
biocentricism and, 73–74
on FDNPS, 191–194
safety discourses, 192–194
spatial justice, 191–192
Dependency culture, 315
DES. *See* Diethylstilbestrol
Desirable energy mix, 260
Deterministic effects, 19
from FDNPS, 181
ICRP on, 135–136
radiology and, 135–136
Developmental basis of disease, 92
Dhammapada, 60–61
Dienel, Peter, 342
Diet Report, 181–183, 185
on evacuation, 183
Dietary surveys, 166
Diethylstilbestrol (DES), 95
Disability legislation, 131–132
Discounting, 46
Disease, developmental basis, 92. *See also specific types*
Dispanser No. 4, 245–249
Disproportion factors, 152
Dissent, private, 291–298
Distributive justice, 62, 355
Dose control, clinical radiology, 133–135
Dose limits, 21–23, 201–202. *See also* Exposure limits
in clinical radiology, 108–109
ICRP, 145
occupational, 29
standards, 201–203
in worker protection, 133
Dose-response assessment
cancer and, 99
curves, 206
unified approach to, 99–100

Dose-response relationship, linear, 97–98
Dosimetry systems, 251–252, 320–321
Duffy, S. W., 247
Duty ethics, 41, 59

E

Earth Summit, 65
Earthquakes, 179–180
  Kobe, 200–201
Ecocentricism, 75–76
  common principles of, 78
Ecological risks, 71–72
Economic compromises
  ALARA and, 149–151
  all exposures, 150
  general exposure, 150
  individual workplace, 149
  in industry process, 149–150
  scopes, 150f
  in work process, 149–150
Economic inequality, 200
Economic viability, 266–267
Economics, health, 116
Edano, Yukio, 199
EDF, 320
Embrey, D. E., 226–227
Emergency Response Support System, FDNPS, 182
Enriched uranium, 261–262
  highly, 264–265, 273
Enrichment, 261–262
Environmental Defense, 90
Environmental friendliness
  closed fuel cycle and, 269–270
  of uranium, 265–266
Environmental injustice, 197–200
  children and, 203–204
  nuclear workers, 206–207
  poor people in, 200–201
Environmental protection
  ICRP on, 19–20
  monetary value of, 80–81
  philosophical theories on, 73–77
  theories on, 73–77
Environmental Protection Agency, 99, 201–202, 210
Environmental tobacco smoke (ETS), 92–93
Equality, 36, 40
Equity, intergenerational, 63–64
Esophagus cancer, 249–250
Essentialism, 284–285
Ethics
  acceptability, 350–351
  applied, 34
    as multiculture, 186–188
    for nuclear science, 188–194
    safety culture and, 186
  bioethics, 186–187
  biomedical, 56–57
  in clinical radiology, 109–114
    act in, 112
    agent in, 111
    basic analysis, 111–114
    core principles, 109–110, 110t
    recipient in, 112–114
  collective approaches, 35–39
  cost-benefit analysis and, 370–371
  cross-cultural, 53–55, 57–58
    autonomy and, 59–60
    beneficence and, 60–61
    four principles of, 59–62
    intergenerational equity and, 63–64
    justice and, 61–62
    need for, 58–59
    non-maleficence and, 60–61
    precaution and, 64–66
    radiation protection and, 62–66
    underprivileged and, 63
  duty, 41, 59
  ethical acceptability, 350–351
  fundamental, 33
  global, 58
  globalization and, 55–56
  holistic, 76
  ICRP terminology, 30
  individual approaches, 35–39
  informed consent and, 132
  land, 76
  of radiation protection, 4–6, 34–35
  rights-based, 41
  terminology of, 30
  Western, 55–56
  worker protection and, 137–139
ETHOS project, 11–12, 313
  approach, 316–317
  Belarus, 318f
  key lessons from, 322–325
  long-term rehabilitation and, 325–326
  phase 1 of, 317–320
  phase 2 of, 320–322
  recommendations, 330–331
  Stolyn District, 318f
ETS. *See* Environmental tobacco smoke
EURANOS, 323–324
European Commission, 17–18

Evacuation
autonomous approaches to, 183
Diet Report on, 183
from FDNPS, 181–183
paternalistic approaches to, 183
voluntary, 182–183
Expert knowledge, 342–343
Exposure limits, 40–41, 48. *See also* Dose limits
FDNPS and, 206
ICRP, 150–151
non-occupational, 150–151
occupational, 150–151
of temporary workers, 208

F

Fallout, 284
effects from, 288
weapons testing, 292–294
FDNPS. *See* Fukushima Daiichi Nuclear Power Station
FERT, 320
Financial compensation, 364–366
Flexibility, 222
Foodstuffs
in Norway, 322t
permissible levels in, 163–164
France, 320
Frankena, William, 76–77
Franklin, Benjamin, 35
French, S., 226–227
Frey, B., 364–365
Friends of the Earth, 291
Fuel cycles. *See* Closed fuel cycle, uranium; Nuclear fuel cycle
Fukushima Daiichi Nuclear Power Station (FDNPS), 3, 72, 161, 197–198, 260, 327
accident causes, 178–180, 198–199
deontological approaches to, 191–194
safety discourses, 192–194
spatial justice, 191–192
deterministic effects from, 181
Emergency Response Support System, 182
environmental injustice victims of, 197–200
children as, 203–204
nuclear workers, 206–207
poor people as, 200–201
evacuation, 181–183
events at, 177–184
exposure limits and, 206
future effects of, 208–210
human costs of, 180–181
IAEA on, 198–199
information disclosure, 181–183
intergenerational equity and, 209
moral hazards, 183–184
radiation levels after, 198–199
safety culture and, 184–188
stochastic effects from, 181
utilitarian approach to, 188–191
considered values of things, 189–190
local residents in, 190–191
national benefits and, 190–191
Fukushima Daini Nuclear Power Station, 191–192
Fukushima Genpatsu Dokuritsu Jiko Chosa Iinkai, 177–178
Fukushima Prefecture, 202
Fundamental ethics, 33
Future effects, 46–47, 356
nuclear regulations and, 208–210
safety and, 263

G

Gardner Study, 138–139
GDF. *See* Geological Disposal Facility
General exposure, economic compromises and, 150
Genetic mutations, 78–79
Genetic variability, 94
Genomic instability, 139, 209
Geological Disposal Facility (GDF), 372–373
Germany, 260
Global ethic, 58
Global Nuclear Energy Partnership (GNEP), 271–272
Globalization, 55–56
GNEP. *See* Global Nuclear Energy Partnership
Golden Rule, 61–62
Gorodnaia, 320, 331
Gospels, 57, 60–61
Government Report, 182–183, 185
Grounds, value, 230–231
Guldenmund, F. W., 229–230
Gusev, Boris, 10

H

La Hague, 268
Hale, A. R., 229–230
Hannis, Mike, 12
Hansson, Sven Ove, 5, 8, 12, 27, 29
Hazard and Operability study approach, 227
Hazards

moral, 183–184
worker protection and, 132, 136–137
Health and Safety Executive (HSE), 146–147
on ALARA, 151
on cost-benefit analysis, 151
on safety bias, 152
Health economics, 116
Health Physics Society, 44
Health surveillance, 323
Health surveys, 174
Health technology assessment (HTA), 116
Hedonism, 220
Helminthosis, 245–246
Helsinki declaration, 38
Heterogeneous populations, 97–98, 100
HEU. *See* Highly Enriched Uranium
Higby, D. P., 151
High-level waste (HLW), 264f, 267
Highly Enriched Uranium (HEU), 264–265, 273
Hinduism, 57, 60–61, 63–64
Hippocratic Oath, 58, 60
Hiroshima, 25, 193, 209, 242–243
HLW. *See* High-level waste
Holistic ethic, 76
Host community, 349–350
negative impacts, 352–354
positive impacts, 354–355
site selection in, 351–352
House of Commons Agricultural Select Committee, 297
House of Lords, 149
HRA. *See* Human reliability analysis
HSE. *See* Health and Safety Executive
HTA. *See* Health technology assessment
Human dignity, 59–60
Human Relation-Rational Goal Model, 225
Human reliability analysis (HRA), 219, 226–228
Human rights, 74
Hurricane Katrina, 200–201
Hydrogen bombs, 193

I

IAEA. *See* International Atomic Energy Authority
IARC, 206
ICRP. *See* International Commission on Radiological Protection
ICRU. *See* International Commission on Radiation Units and Measurements
Ikata Nuclear Power Plant, 191–192
IMAGE GENTLY, 116
IMAGE WISELY, 116
Incentives, 358–359
Inclusiveness, 340
Independence, 363–364
Indetectable effects, 44–45
Individual approaches, 35–39
Individual limit principle, 40, 59
Individual rights, 40
Individual workplaces, 149
Industry process, 149–150
Information needs
early, 162
long-term, 167–172
Informed consent, 59. *See also* Consent
ethics and, 132
NRRW and, 138
Institute of Biophysics, 250–251
Institute of Radiation Medicine and Ecology, 243
Interest rates, 46
Intergenerational equity
cross-cultural ethics and, 63–64
FDNPS and, 209
uranium closed fuel cycle and, 267–273
Internal Process Model, 225
International Atomic Energy Authority (IAEA), 28–29, 63, 72, 118
on FDNPS, 198–199
on safety, 262
on safety culture, 184–185
Safety Glossary, 264–265
International Commission on Radiation Units and Measurements (ICRU), 17–18
International Commission on Radiological Protection (ICRP), 4, 64–65, 72, 114, 181, 205
aims of, 18–21
defining, 17–18
on deterministic effects, 135–136
on dose limits, 145
occupational, 29
environmental protection, 19–20
formation of, 17–18
history of, 18
on justification, 145
moot points, 29–30
non-occupational exposure limits
occupational exposure limits
on optimization, 145
radiation protection
principles, 145
system for, 24

International Commission on Radiological Protection (ICRP) (*Continued*)
recommendations, 18–21
1951, 22
1954, 22–23
1957, 23
1958, 23, 26–27
1959, 143
1966, 23–24, 26
1973, 23
1977, 24, 27, 143–145
1990, 25, 27
1991, 25
2007, 25–26
development of, 23–26
ethical terminology and, 30
evolution of, 26–29
history of, 21–23
individual protection, 29–30
life pricing, 30
Publication 1, 23
Publication 9, 23
Publication 77, 27
radiological protection systems and, 23–26
on stochastic effects, 143
on worker protection, 132–133
International X-ray and Radium Protection Committee (IXRPC), 17–18, 21
recommendations, 22
IPSN, 320
Ishiba, Shigeru, 193
IXRPC. *See* International X-ray and Radium Protection Committee

J

Japan
bioethics in, 186–187
economic inequality in, 200
electric power sector in, 183–184
nuclear science in, 193
nuclear workers in, 208
poverty in, 200
reactors in, 179–180
Japan Society of Civil Engineers, 178
Jegen, R., 364–365
Johannesburg World Summit on Sustainable Development, 334
Judaism, 57
Justice, 40. *See also* Environmental injustice
Beauchamp on, 62
Childress on, 62
cross-cultural ethics and, 61–62
distributive, 62, 355
procedural, 341, 355
spatial, 191–192
Justification, 54
in clinical radiology, 108–109, 118–121
failures of, 121
of democracy, 338–339
ICRP, 145
worker protection and, 133

K

Kalman, Chris, 7–8
Kant, Immanuel, 41, 59, 77
Karaganda, 245, 247
Karma, 63–64
Kathren, R. L., 151
Kavlock, R. J., 92
Kazakhstan, 241, 246–247
independence of, 250
new economy of, 253–254
radiation risk research in, 248–252
Kennedy, R., 227
Khruchchev, 249
Kirwan, B., 227
Kobe earthquake, 200–201
Kokpektinskii, 249–250
Konechnaia, 244
Kosako, Toshiso, 203–204
Krishna, 59, 65–66
Kurchatov, 244
Kyshtym, 241

L

LAER. *See* Lowest achievable emissions rate
Lancet, 139
Land ethics, 76
Lawn-crossing example, 45
Lawrence, Joseph, 335–336
Lead, 91–93
Leopold, Aldo, 76–77
Leukemia, 22–23, 95, 133–135, 201–202, 356
in Sellafield, 297
worker protection and, 138
Li, N., 247
Life pricing, 30, 152
Life Span Study, 242–243
Light Water Reactors (LWR), 261–262
MOX in, 272
Liland, Astrid, 8
Limit setting, 39–43

Lindell, Bo, 147
on de minimis issue, 147–148
Linear-nonthreshold (LNT), 43, 64–65, 88, 97, 171
Lipophilic toxicants, 93
Liver cancer, 249–250
LNT. *See* Linear-nonthreshold
Local monitoring stations, 163
Lochard, Jacques, 11–12
London Royal Society, 283–284
Lowest achievable emissions rate (LAER), 153
Lung cancer, 44–45, 249–250
Luria, G., 224
Lutz, W. K., 98–99
LWR. *See* Light Water Reactors

M

Mabuchi, K., 242–243
MAFF. *See* Ministry of Agriculture, Fisheries and Food
Mahabharata, 65–66
Malachi, 59
Malone, Jim, 7
Management commitment, 224
Managi, Shunsuke, 80–81
Mayo, H. B., 338–339
Medical confidentiality, 138–139
Medical tourism, 114
Meet-Understand-Modify (MUM), 338
Men, T., 247
Mercury, 92–93
Metallurgical Laboratory, 144
Methylmercury, 95
METI. *See* Ministry of Economy, Trade and Industry
Mill, John Stuart, 338
Minimata, 199–200
Ministry of Agriculture, Fisheries and Food (MAFF), 291, 294, 296–297
Ministry of Economy, Trade and Industry (METI), 180
Ministry of Education, Culture, Sports, Science, and Technology, 180
Minor actinides, 263
Mitigation, 361–363
Mixed Oxide Fuel (MOX), 267, 271–272
in LWRs, 272
Miyasaka, Michio, 8–9
Mohaghegh, Z., 226–227
Moldagaliev, Timur, 10
Molten Salt Reactors (MSRs), 275
Monetary values, 80–81
Monoculture, 184–188
Monosodium glutamate, 95
Moral hazards, 183–184
Moral philosophy, 54
Moral thinking, 33–35. *See also* Ethics
Mosleh, A., 226–227
MOX. *See* Mixed Oxide Fuel
MSRs. *See* Molten Salt Reactors
Muhammad, 62–63, 65–66
Multiculture, 186–188
MUM. *See* Meet-Understand-Modify
Munson, L. H., 151
Murphy, D. M., 226–227
Muslims, 57, 63–64

N

Næss, A., 75
Nagasaki, 25, 193, 209, 242–243
National Academy of Sciences, 99, 264
National Committee on Radiation Protection (NCRT), 144
National Nuclear Laboratory (NNL), 275–276
National Occurring Radioactive Material (NORM), 136
National Radiation Protection Board, 297
National Registry for Radiation Workers (NRRW), 137
consent in, 138
National Research Council, 268
Nazarbaev, Nursultan, 245
NCRT. *See* National Committee on Radiation Protection
Needham, Larry, 90
New build, 361–363
NGOs. *See* Nongovernmental Organizations
NIMBY. *See* Not in my backyard
NNL. *See* National Nuclear Laboratory
Nonaccidental injury, 122
Nongovernmental Organizations (NGOs), 334
Non-maleficence, 109
cross-cultural ethics and, 60–61
Nonmedical exposures, 121–122
Non-occupational exposure limits
Non-Proliferation Treaty, 265
NORM. *See* National Occurring Radioactive Material
Norway, 159–161
Chernobyl and
contamination, 161–162
Cs-137 deposition map, 160f
early lessons from, 162–167
affected people monitoring, 165–166

Norway (*Continued*)
information needs, 162
local monitoring stations, 162
permissible levels settings, 163–164
psychosocial effects, 166–167
foodstuffs in, 160t
long-term challenges, 167–172
affected group involvement, 171
compensation, 167–168
health concerns, 172–174
information needs, 167
social countermeasures, 167–168
technical countermeasures, 167–168
zoning, 167–168
Not in my backyard (NIMBY), 39, 364
No-threshold model, 89
Novaia Zemlia, 242, 244
NRC. *See* Nuclear Regulatory Commission
NRRW. *See* National Registry for Radiation Workers
NSC. *See* Nuclear Safety Commission
Nuclear accidents, 8–10
Nuclear and Industrial Safety Agency, 180
Nuclear Decommissioning Authority, 363
Nuclear Energy Agency, 17–18
Nuclear fuel cycle, 10–11. *See also* Closed fuel cycle, uranium; Reprocessing
proliferation resistance, 274–275
thorium, 273–277
challenges, 275–277
shortcomings of, 275–277
uranium, 261–267
economic viability of, 266–267
plutonium and, 268–269
safety, 262–264
security, 264–265
sustainability and, 265–266
Nuclear power production
pros and cons, 260
safety and, 260–261
security and, 260–261
sustainability and, 260–261
Nuclear power submarines, 133
Nuclear regulations
children and, 204–205
future effects and, 208–210
nuclear workers and, 207–208
poor people and, 201–203
Nuclear Regulatory Commission (NRC), 199, 210
Nuclear Safety Commission (NSC), 199
regulatory rules of, 179
Nuclear science
applied ethics for, 188–194
in Japan, 193
Nuclear waste, 47
high level, 264f, 267
lifetime of, 263
longevity of, 263
thorium and, 275
in UK, 348
Nuremberg trials, 186–187

O

Occupational exposure limits
Occupational radiation health research, 137–139
OECD. *See* Organization for Economic Cooperation and Development
Oh, J. I., 229–230
Olmany, 317, 319–320, 331
diet in, 322t
O'Neill, John, 367–368
Openness, 110, 115
Opinion, public, 334
Optimization, 54
ALARA and, 133–135, 145–146
ALARP and, 133–135
in clinical radiology, 108–109
ICRP, 145
in worker protection, 133
Organization for Economic Cooperation and Development (OECD), 200
Organizational values, 219–223
Organochlorine pesticides, 92–94
Organophosphates, 92–93
Oughton, Deborah, 6, 12
Overutilization, 116

P

PAHs. *See* Polycyclic aromatic hydrocarbons
Paivo, J., 341
Parliament of World Religions, 60–64
Participation. *See* Public participation
Paté-Cornell, M. E., 226–227
PBDEs. *See* Polybrominated diphenyl ethers
PCBs. *See* Polychlorinated biphenyls
Peer review, 115
Perchlorate, 91
Performance shaping factors (PSF), 226–227
Permissible levels, 22
in foodstuffs, 163–164
Personal values, 219–223

Person-rem, 151
Phthalates, 91–93
Physicians for Social Responsibility (PSR), 198–199
Pierce, D. A., 242–243
Pilot Project, 312–313
Placenta, 92
Pluralistic decision-making, 187
Plutonium, 193, 241, 267
  burning up, 269f
  security and, 268–269
  uranium and, 268–269
  weapon-grade, 268–269
Pollard, S. J. T., 226–227
Polybrominated diphenyl ethers (PBDEs), 91, 96
Polychlorinated biphenyls (PCBs), 91–93
Polycyclic aromatic hydrocarbons (PAHs), 94
Poor people
  as FDNPS victims, 200–201
  nuclear regulations and, 201–203
  radiation dose standards and, 201–203
Poverty
  in Japan, 200
  line, 40
Power, 220
Precautionary Principle, 110
  ALARA and, 153
  cross-cultural ethics and, 64–66
  defining, 153
Pregnancy, 23
  clinical radiology and, 123–124
Prepublication notification, 139
Preston, D. L., 242–243
Priestley, Joseph, 35
Prima facie validity, 56
Primary bone cancer, 135f
Prioritarianist approach, 43
Private dissent, 291–298
Proactive Practices, 224
Procedural justice, 341, 355
Production, 225
Proliferation resistance, 274–275
Proportion factors, 152
Prussian blue, 165, 168
Psalms, 62
PSF. *See* Performance shaping factors
PSR. *See* Physicians for Social Responsibility
Psychosocial effects
  early, 166–167
  long-term, 171–172
Public acceptability, 350–351
Public belief, 291–298
Public opinion, 334
Public participation, 11–13, 326, 333–334
  decision-making and, 341
  defining, 335
  in democracy, 335–338
  early involvement in, 341–342
  expert knowledge and, 342–343
  justification of, 338–339
  quality criteria for, 339–343
  representatives, 340
  transparency and, 340–341
Public uptake, 283–285

Q

Quinn, R. E., 221–222, 224–225
Quinn, Warren S., 45
Quran, 57, 59–61

R

RACT. *See* Reasonably available control technology
Radiation hormesis, 139
Radiation monitoring systems, 323
Radiation protection, 3
  consistency in, 136
  cross-cultural ethics and, 62–66
  de minimis issue in, 43–46
  ethics of, 4–6, 34–35
  harms in, 78–79
  ICRP principles on, 145
  ICRP system of, 24
  practice, 6–8
  presumptions of, 88–89
  principles, 88–89
  values in, 78–79
  for workers, 132–137
Radiation risk assessment, 100–101
  Kazakhstan, 248–252
  Semipalatinsk research, 245–248
  zoning and, 249
Radiation science, 139–140
Radioactive waste-management facility (RWMF), 347–350. *See also* Nuclear waste
  bribery and, 358–359, 363–364
    avoidability of, 359–361
  inadequate guidelines, 357–358
  mitigation, 359–361
  negative impacts of, 352–354
  positive impacts of, 354–355
  site selection, 351–352
Radiocesium, 165, 169–170

Radiology. *See* Clinical radiology
Radiotherapy, 139
Radium, 133–135, 135f, 262–263
  dial painting, 133–135
Radon, 262–263
Ramseyer, J. M., 184
RAP. *See* Reference Animals and Plants
Rational choice, 369–370
Rational Goal Model, 225
Rationality, bounded, 224–226
Rawles, Kate, 12
Rawls, John, 48, 63
Reactor Safety Team, 199
Reason, J., 230
Reason blind, 370
Reasonableness, 148–149
Reasonably available control technology (RACT), 153
Reasonably Maximally Exposed Individual (RMEI), 210
Reference Animals and Plants (RAP), 74, 75t, 76
Reference-man models, 205
Referral guidelines, 120–121
Reflective equilibrium, 58
Regan, Tom, 74
Reindeer herding, 164, 169–170, 172
  monitoring, 170f
Representation, 350
Representatives, 340
Reprocessing, 267. *See also* Nuclear fuel cycle
  assumptions concerning, 270–273
  benefits of, 268
  costs of, 270
Resource durability, 265
  closed fuel cycle and, 269–270
  of uranium, 266
Retchissa, 320, 331
Rig Veda, 63
Rights-based ethics, 41
Rio Declaration, 333
  Principle 15 of, 153
Risk
  cancer, 242–243
  chemical assessment, 87–88
    adverse health effects and, 93–98
    substance thresholds, 98–99
  clinical radiology and, 116–118
  ecological, 71–72
  individuals sensitive to, 47–49
  radiation assessment, 100–101
    Kazakhstan, 248–252
    Semipalatinsk research, 245–248
    zoning and, 249
  toxicological, 92, 98
  X-rays, 143
RMEI. *See* Reasonably Maximally Exposed Individual
Rogers, J. M., 92
Rohrbaugh, J., 221–222
Rokkasho, 270
Rollenhagen, Carl, 9–10
Röntgen, Wilhelm, 143
Rossi, Harald H., 147
  on de minimis issue, 147
Royal College of Radiologists, 121
Royal Institute of Technology, 88–89
Royal Society, 271–272
Rudjord, A. L., 160f
Russia, 312
RWMF. *See* Radioactive waste-management facility

S

Safety
  assessing, 227–229
  bias towards, 152
  climate, 223–226
  competence, 218f, 231–232
  FDNPS and, 192–194
  future generational, 263
  IAEA on, 262
  indicators, 232
  nuclear power production and, 260–261
  quality, 218, 218f, 226–227
  uranium nuclear fuel cycle and, 262–264
  values, 224–226
    protecting, 232
Safety culture, 9–10, 215–219, 230–232
  applied ethics and, 186
  assessing, 227–229
  Chernobyl and, 223
  defining, 186, 216
  FDNPS and, 184–188
  IAEA on, 184–185
  ideology, 184–186
  initiatives, 229–230
  as monoculture, 184–188
  organizational values and, 219–223
  personal values and, 219–223
  at TEPCO, 185
SAGE, 323–324
Samuels, R. J., 183–184
Sandman, P., 162
Sardar Sarovar Dam, 367
Schardein, James, 92

SCHAZOP, 227
Schein, E. H., 230–231
Schumpeter, Joseph, 335–338
Schwartz, S. H., 220–222, 229
Scientific knowledge, 288–291
Second International Congress of Radiology, 17–18
Security, 220
  nuclear power production and, 260–261
  uranium nuclear fuel cycle of, 264–265
    plutonium and, 268–269
Self-awareness, 74
Self-direction, 220
Self-referral, 121
Self-torturer, 45
Sellafield, 268, 285–286, 288, 291
  cesium in, 292–294
  leukemia in, 297
  private dissent in, 291–298
  public belief and, 291–298
  sheep farmers, 285–288
Semei, 243
Semipalatinsk, 241–243, 253–254
  cancer in, 247
  nuclear test site, 245–248
  research into radiation effects, 245–248
Sheep farmers
  Sellafield, 285–288
  social identities of, 288–291
Shimizu, Y., 242–243
Shrader-Frechette, Kristin, 9
Singer, Peter, 73–74
Siting regulations, 201–203
Skuterud, Lavrans, 8
Smoking, 46
Soane, E., 226–227
Social countermeasures, 168–170
Social identities, 283–285
  credibility and, 298–302
  scientific knowledge and, 288–291
  of sheep farmers, 288–291
Soviet Centre for Oncology, 247
Soviet Union, 242, 312–313
Spatial justice, 191–192
SPEEDI. *See* System for Prediction of Environmental Emergency Dose Information
Spreitzer, G. M., 224–225
Stability, 221
Stakeholders
  engagement, 316, 323–324, 326, 334
    democracy and, 339
  focus, 131
  identification problem, 337
Stimulation, 220
Stochastic effects, 19, 22, 88
  from FDNPS, 181
  ICRP on, 143
  in worker protection, 132–133
Stolyn District, 313, 317, 319–320
  conference recommendations, 330–331
  ETHOS project and, 318f
Stone, Robert, 44, 144
Substitution principle, 153
Sustainability, 72, 80–81
  nuclear power production and, 260–261
  uranium nuclear fuel cycle and, 265–266
System for Prediction of Environmental Emergency Dose Information (SPEEDI), 182
Systemic ambition, 230–231

T

Taebi, Benham, 10–11
Taishi, Shotoku, 65–66
Talmud, 60–61
Tamoxifen, 95
Taylor, Paul, 74
Technical countermeasures, 168–170
Temporary workers, 208
Tends to zero, 151
TEPCO. *See* Tokyo Electric Power Company
Terebejov, 320, 331
Thalidomide, 95, 99
Thelmann Kolkhoz, 245
A Theory of Justice (Rawls), 48, 63
Thorium, 261
  in Molten Salt Reactors, 275
  nuclear fuel cycle, 273–277
    challenges, 275–277
    shortcomings of, 275–277
  proliferation resistance, 274–275
  uranium and, 273, 278
  viability of, 273–277
  waste management benefits of, 275
Three Mile Island, 3, 193, 199
Three-levelled model, ALARA, 146–148
Threshold model, 89
Thyroid cancer, 95, 251–252
Thyroid hormone, 91
Tissue reactions, 88
Tokyo Electric Power Company (TEPCO), 9, 177–178, 184, 201–202
  report, 178–179
  safety culture at, 185

Tolerance levels, 144
Torah, 57, 60–61, 63–64
Torres Strait Islanders, 65–66
Tort law, 89
Toxicological risk assessment, 92, 98
Trade unions, 340
Tradition, 220–221
Transparency, 110, 115
  democracy and, 340–341
  public participation and, 340–341
Tripling function, 43
Trust, 363–364
Tsunamis, 179–180, 188–189

U

Ugletveit, F., 160f
UK. *See* United Kingdom
Ukraine, 312
Uncertainty, 343
  clinical radiology and, 116–118
Underprivileged, 63
UNDP, 325–326
UNESCO General Conference, 58
United Kingdom (UK)
  cesium in, 292f–293f
  radioactive waste in, 348
United Nations, 17–18, 334
  Conference on Environment and Development, 65
  General Assembly, 58
"Universal Declaration of Human Rights", 58, 60
"Universal Declaration on Bioethics and Human Rights", 58, 60
Universalism, 220, 222
Uranium
  closed fuel cycle, 267–273
    environmental friendliness and, 269–270
    resource durability and, 269–270
    safety and, 268
  dioxide, 261–262
  enriched, 261–262
  environmental friendliness of, 265–266
  highly enriched, 264–265, 273
  nuclear fuel cycle, 261–267
    economic viability of, 266–267
    plutonium and, 268–269
    safety, 262–264
    security, 264–265
    sustainability and, 265–266
  resource durability of, 266
  thorium and, 273, 278
Ust-Kamenogorsk, 244–245
Utilitarian Principle, 110
Utilitarian theories, 33, 54, 65–66
  Bentham on, 35–36
  biocentricism and, 73–74
  on FDNPS, 188–191
    considered values of things, 189–190
    national benefits and, 190–191
  local residents in, 190–191

V

Vaeth, M., 242–243
Valentin, Jack, 4, 19, 27
Valproic acid, 99
Value for preventing a fatality (VPF), 152
Values
  balanced, 224–225
  basis for, 230–231
  competing, 221–228
  ground for, 230–231
  literature about, 219
  monetary, 80–81
  organizational, 219–223
  personal, 219–223
  in psychology, 219
  in radiation protection, 78–79
  safety, 224–226
    protection of, 232
  in sociology, 219
  in utilitarian theories, 189–190
van Hoenhout, P. L. C. H., 229–230
Vedas, 57
Vitamin C deficiency, 245–246
Voluntary evacuation, 182–183
Voluntary relocation zones, 316–317
VPF. *See* Value for preventing a fatality

W

Wallentinus, H. G., 341
Waste. *See* Nuclear waste; Radioactive waste-management facility
Weapon-grade plutonium, 268–269
Weighing, 39–43
Western ethics, 55–56
WHO. *See* World Health Organization
Wikman-Svahn, P., 88–89
Windscale plant, 77. *See also* Sellafield
Wingspread Conference, 65
Wingspread Statement, 153
Wittgenstein, Ludwig, 58–59
Women, 47–48
Woodruff, T. J., 96, 98

Work process, 149–150
Workers
  access, 138–139
  classified status of, 133
  consent, 138–139
  environmental injustice and, 206–207
  as FDNPS victims, 206–207
  information, 138–139
  in Japan, 208
  nuclear regulations and, 207–208
  protection, 131–132
    contingency plans, 136–137
    dose limitation and, 133
    ethical requirements for, 137–139
    hazards and, 132, 136–137
    ICRP and, 132–133
    justification and, 133
    leukemia and, 138
    optimization in, 133
    radiation protection and, 132–137
    radiation science and, 139–140
    statutory requirements, 136–137
    stochastic effects and, 132–133
  recruitment of, 139
  temporary, 208
World Bank, 330
World Health Organization (WHO), 199
  reproductive health study, 251
World Medical Association, 186–187
World Nuclear Association, 55
World War II, 22
Wynne, Bryan, 11

X

X-rays, 21, 149–150
  discovery of, 143
  risks of, 143

Y

Yomiuri Shimbun, 193
Yukawa, Hideki, 193

Z

Zaridze, D. G., 247
Zhana-Semeiskii, 248–249
Zohar, D., 224
Zölzer, Friedo, 5–6, 30
Zoning
  in Norway, 167–168
  radiation risk, 249

FSC
www.fsc.org
MIX
Papier | Fördert gute Waldnutzung
FSC® C083411

Kolibri 360 GmbH
Ferdinand-Jühlke-Straße 7
99095 Erfurt, Deutschland
produktsicherheit@kolibri360.de